KU-075-447

STABILITY, VIBRATION AND CONTROL OF SYSTEMS

Editor-in-chief: Ardéshir Guran
Co-editor: Daniel J. Inman

Professor Ilya Blekhman (left) and Professor Andéshir Guran (right) during the International Symposium on Mechatronics & Complex Dynamical Systems, June 2000, St. Petersburg, Russia.

Dynamics with Friction
Modeling, Analysis and Experiment
Part II

SERIES ON STABILITY, VIBRATION AND CONTROL OF SYSTEMS

Series Editors: Ardéshir Guran & Daniel J. Inman

About the Series

Rapid developments in system dynamics and control, areas related to many other topics in applied mathematics, call for comprehensive presentations of current topics. This series contains textbooks, monographs, treatises, conference proceedings and a collection of thematically organized research or pedagogical articles addressing dynamical systems and control.

The material is ideal for a general scientific and engineering readership, and is also mathematically precise enough to be a useful reference for research specialists in mechanics and control, nonlinear dynamics, and in applied mathematics and physics.

Selected Volumes in Series B

Proceedings of the First International Congress on Dynamics and Control of Systems, Chateau Laurier, Ottawa, Canada, 5–7 August 1999
> Editors: A. Guran, S. Biswas, L. Cacetta, C. Robach, K. Teo, and T. Vincent

Selected Topics in Structronics and Mechatronic Systems
> Editors: A. Belyayev and A. Guran

Selected Volumes in Series A

Vol. 1 Stability Theory of Elastic Rods
> Author: T. Atanackovic

Vol. 2 Stability of Gyroscopic Systems
> Authors: A. Guran, A. Bajaj, Y. Ishida, G. D'Eleuterio, N. Perkins, and C. Pierre

Vol. 3 Vibration Analysis of Plates by the Superposition Method
> Author: Daniel J. Gorman

Vol. 4 Asymptotic Methods in Buckling Theory of Elastic Shells
> Authors: P. E. Tovstik and A. L. Smirinov

Vol. 5 Generalized Point Models in Structural Mechanics
> Author: I. V. Andronov

Vol. 6 Mathematical Problems of the Control Theory
> Author: G. A. Leonov

Vol. 7 Vibrational Mechanics: Theory and Applications to the Problems of Nonlinear Dynamics
> Author: Ilya I. Blekhmam

SERIES ON STABILITY, VIBRATION AND CONTROL OF SYSTEMS

 **Series B** **Volume 7**

Series Editors: **Ardéshir Guran & Daniel J Inman**

Dynamics with Friction
Modeling, Analysis and Experiment
Part II

Editors

Ardéshir Guran
Institute of Structronics, Canada

Friedrich Pfeiffer
Technical University of Munich, Germany

Karl Popp
University of Hannover, Germany

World Scientific
Singapore • New Jersey • London • Hong Kong

Published by

World Scientific Publishing Co. Pte. Ltd.

P O Box 128, Farrer Road, Singapore 912805

USA office: Suite 1B, 1060 Main Street, River Edge, NJ 07661

UK office: 57 Shelton Street, Covent Garden, London WC2H 9HE

British Library Cataloguing-in-Publication Data
A catalogue record for this book is available from the British Library.

DYNAMICS WITH FRICTION: MODELING, ANALYSIS AND EXPERIMENT, PART II

ISBN   981-02-2954-2

Printed in Singapore by Uto-Print

In Memoriam
Heinrich Hertz (1857–1894)
Paul Painleve (1863–1933)
Arnold Sommerfeld (1868–1951)

Preface

The pictures on the front cover of this book depict four examples of mechanical systems with friction: i) dynamic model of normal motion for Hertzian contact, ii) disk with a rotating mass-spring-damper system, iii) planar slider-crank mechanism, iv) dynamic model of a periodic structure. These examples, amongst many other examples of dynamical friction models, are studied in the present volume.

Historically, the exploitation of dynamical friction has had a tremendous effect on human development. In fact, due to the human desire to describe nature, machines, and structures, ideas about friction and dissipation has found their way into scientific thoughts. The science of mechanics is so basic and familiar that its existence is often overlooked. Whenever we push open a door, pick up an object, walk or stand still, our bodies are under the constant influence of various forces. When the laws of the science of mechanics are learned and applied in theory and practice, we achieve an understanding which is impossible without recognition of this subject. Today still we agree with what da Vinci wrote in fifteen century, *mechanics is the noblest and above all others the most useful, seeing that by means of it all animated bodies which have movement perform all their actions.* The science of mechanics deals with motion of material bodies. A material body may represent vehicles, such as cars, airplanes and boats, or astronomical objects, such as stars or planets. For sure such objects will sometimes collide or contact each other (cars more often than stars). One may think of a walking human or animal making frictional contact with the ground, sports such as golf and baseball, where contact produces spin and speed, and mechanical engineering applications, such as the parts of a car engine that must contact each other to transfer force and power. The sub-field of mechanics that deals with contacting bodies is simply referred to as contact mechanics. It is a part of the broader area of solid and structural mechanics and an almost indispensable one since forces are almost always applied by means of frictional contacts.

Contact mechanics has an old tradition: laws of friction, that are central to the subject, were given by Amontons in 1699, and by Coulomb in 1785, and early mathematical studies of friction were conducted by the great mathematician Leonhard Euler. A theory for contact between elastic bodies, that has had a tremendous importance in mechanical engineering, was presented by Heinrich Hertz in 1881. Contact mechanics has seen a revival in recent years, driven by new computer resources and such applications as robotics, human artificial joints, virtual reality, animation, and crashworthiness.

Contact mechanics is the science behind tribology, the interdisciplinary study of friction, wear and lubrication, with major applications such as bearings and brakes, and involving such issues as microscopic surface geometry, chemical conditions, and thermal conditions. Note that while in many tribological applications one seeks to minimize friction to reduce loss of energy, everyday life is at the same time impossible without friction — we would not be able to walk, stand up, or do anything without it. Walking requires adequate friction between the sole of the foot and the floor, so that the foot will not slip forward or backward and the effect of limb extension can be imparted to the trunk. Lack of friction on icy surfaces is compensated for by hobnails on boots or chains on tires. Friction is necessary to operation of a self-propelled vehicle, not only to start it and keep it going but to stop it as well. Crutches and canes are stable due to friction between their tips and the floor; this is often increased by a rubber tip which has a high coefficient of friction with the floor. A wheel-chair can be pushed only because of the friction developed between the pusher's shoes and the floor, and friction must likewise be developed between the wheels and the floor so they will turn and not slide. Many friction devices are used in exercise equipment to grade resistance to movement, as with a shoulder wheel or stationary bicycle. Brakes on wheelchairs and locks on bed casters utilize the principle of friction. Application of cervical or lumbar traction on a bed patient depends on adequate opposing frictional forces developed between the patient's body and the bed. In the operation of machines, sliding friction and damping wastes energy. This energy is transformed into heat which may have a harmful effect on the machine, as with burned-out bearings. To reduce friction, materials having a very smooth or polished surface are used for contacting parts, or a lubricant, such as oil or grease, is placed between the moving parts. Frictional effects are then absorbed between layers of the lubricant rather than by the surfaces in contact. Friction also exists within the human body. Normally ample lubrication is present as tendons slide within synovial sheaths at sites of wear, and the articulating surfaces of joints are bathed in synovial fluid.

Despite this tremendous importance of contact mechanics and frictional phenomena, we still hardly understand it. The present part II of this volume on *Dynamics with Friction* is a continuation of the previous part I, and is designed to help synthesize our current knowledge regarding the role of friction in mechanical and structural systems as well as everyday life. We understand that in the preface of the first part in this book we promised the readers to have a final review chapter with a complete list of references in friction dynamics. However, we soon realized that the knowledge in this field in written form is expanding very rapidly at a considerable rate which makes a comprehensive list almost impossible. The present volume offers the reader only a sampling of exciting research areas in this fast-growing field. In compilation of the present volume, we also noticed, relatively very little is made available in this field to design engineers, in college courses, in handbooks, or in form of design algorithms, because the subject is too complicated. For an expository introduction to the field of dry friction with historical notes we refer the readers to the article by Brian Feeny, Ardéshir Guran, Nicolas Hinrichs, and Karl Popp, published recently in Applied Mechanics Review, volume 51, no. 5 in May 1998, and the list of references at the end of that article.

Every year there are several conferences in this field. Those of longest standing are the conferences of ASME, STLE, IUTAM, and EUROMECH. A separate bi-annual conference, held in U.S., is the Gordon conference in tribology. It is a week-long conference held in June, at which about 30 talks are given. Another separate biannual conference, held in even-numbered years, is the ISIFSM (International Symposium on Impact and Friction of Solids, Structures, and Intelligent Machines: Theory and Applications in Engineering and Science). The proceedings of ISIFSM papers are rigorously reviewed and appeared in volumes published in this series.

Today, research continues vigorously in the description and design of systems with friction models, in quest to understand nature, machines, structures, transportation systems, and other processes. We hope this book will be of use to educators, engineers, rheologists, material scientists, mathematicians, physicists, and practitioners interested in this fascinating field.

Ardéshir Guran	Friedrich Pfeiffer	Karl Popp
Ottawa, Canada	Munich, Germany	Hannover, Germany

Contributors

M. A. Davies
National Institute of Standards
  and Technology
Manufacturing Engineering Laboratory
Gaithersburg, MD 20899
USA

B. F. Feeny
Department of Mechanical Engineering
Michigan State University
East Lansing, MI 48824
USA

Aldo A. Ferri
G. W. Woodruff School of Mechanical
  Engineering
Georgia Institute of Technology
Atlanta, GA 30332-0404
USA

Ardéshir Guran
American Structronics and Avionics
16661 Ventura Blvd
Encino, California 91436
USA

Daniel P. Hess
Department of Mechanical Engineering
University of South Florida
Tampa, Florida 33620
USA

R. V. Kappagantu
Altair Engineering, Inc.
1755 Fairlane Drive
Allen Park, MI 48101
USA

Francesco Mainardi
Department of Physics
University of Bologna
46 Via Irnerio, Bologna 40126
Italy

Dan B. Marghitu
Department of Mechanical Engineering
Auburn University
Auburn, Alabama 36849
USA

J. P. Meijaard
Laboratory for Engineering Mechanics
Delft University of Technology
Mekelweg 2, NL-2628 CD Delft
The Netherlands

F. C. Moon
Department Mechanical and
  Aerospace Engineering
Cornell University
Ithaca, NY 14853
USA

John E. Mottershead
Department of Mechanical Engineering
The University of Liverpool
Livepool, L69 3BX
UK

G. L. Ostiguy
Department of Mechanical Engineering
Ecole Polytechnique
P. O. B. 6079, Succ. "Centre-Ville"
Montreal (Quebec), H3C 3A7
Canada

Contents

Chapter 3: Dynamics of Flexible Links in Kinematic Chains   75
Dan B. Marghitu and Ardéshir Guran

Chapter 4: Solitons, Chaos and Modal Interactions in Periodic Structures   99
M. A. Davies and F. C. Moon

**Chapter 5: Analysis and Modeling of an Experimental Frictionally
Excited Beam**   **125**

R. V. Kappagantu and B. F. Feeny

Chapter 6: Transient Waves in Linear Viscoelastic Media   **155**

Francesco Mainardi

Dynamics with Friction: Modeling, Analysis and Experiment, Part II, pp. 1–27
edited by A. Guran, F. Pfeiffer and K. Popp
Series on Stability, Vibration and Control of Systems, Series B, Vol. 7
© World Scientific Publishing Company

INTERACTION OF VIBRATION AND FRICTION
AT DRY SLIDING CONTACTS

DANIEL P. HESS
Department of Mechanical Engineering
University of South Florida
Tampa, Florida 33620, USA

ABSTRACT

When measuring or modeling friction under vibratory conditions, one should ask
how contact vibrations are influenced by the presence of different types of
friction or one should seek to determine the extent to which vibrations can alter
the mechanisms of friction itself. This paper summarizes results from the
author's work on dry sliding contacts in the presence of vibration. A number of
idealized models of smooth and rough contacts are examined, in which the
assumed sliding conditions, the kinematic constraints, and the mechanism of
friction are well-defined. Instantaneous and average normal and frictional forces
are computed. The results are compared with experiments. It appears that
when contacts are in continuous sliding, quasi-static friction models can be used
to describe friction behavior, even during large, high-frequency fluctuations in
the normal load. However, the dynamics of typical sliding contacts, with their
inherently nonlinear stiffness characteristics, can be quite complex, even when
the sliding system is very simple.

1. Introduction

Surfaces in contact are often subjected to dynamic loads and associated
contact vibrations. The dynamic loading may be generated either external to the
contact region, as in the case of unbalanced moving machinery components, or
within the contact region, as in the case of surface roughness-induced vibration.
Vibrations may be undesirable from the point of view of the stresses that are
induced or noise that is generated and may need to be controlled. Furthermore,
vibrations can affect friction and the outcome of friction measurements.

In this paper, an overview of the author's work on dry friction in the presence
of contact vibrations is given. The reader is referred to other papers[1-7] for
details. Some general observations will be made regarding the interaction of
friction and vibration and the interpretation of friction coefficients under
vibratory conditions.

The models discussed are limited to continuous sliding, although extensions
to loss of contact or sticking could be made. The models accommodate forced

contact vibrations of a rigid rider mass, supported by smooth Hertzian or randomly rough planar compliant contacts undergoing elastic deformation. Initially the rider is constrained to move only along a line normal to the sliding direction. The vibration problem is solved for the normal motions. To allow a well-defined mechanism of friction to be explicitly inserted into the dynamic model, the instantaneous friction force is related to the normal motion through the adhesion theory of friction. Accordingly, the instantaneous friction force is taken to be proportional to the instantaneous real area of contact. While we recognize the limitations of the adhesion theory, it is selected due to its simplicity and its ability to describe many situations of practical interest[8].

A general feature of the results is that as the normal oscillations increase, the average separation of the surfaces increases. This is due to the nonlinear character of the contact stiffness which increases (hardens) as the instantaneous normal load increases from its mean value and decreases (softens) as the load is reduced. This increase in average separation is, under the assumptions stated above, sometimes, but not always, accompanied by a decrease in the average friction force.

A more interesting, yet still simple, model is that of a rough block in planar contact that is allowed to translate and rotate with respect to the countersurface against which it slides. We have developed a modification of the Greenwood-Williamson[9] rough surface model for this purpose. The basic equations are given and general features of the problem are discussed. Some comparisons are made with experiments and with part of the work of Martins et al.[10], in which a similar problem using a phenomenological constitutive contact model is examined.

Before proceeding, we comment on the interpretation of the coefficient of friction under dynamic conditions. If both the load and the friction force at a contact vary with time, the instantaneous friction coefficient, $\mu(t)$, is

$$\mu(t) = \frac{F(t)}{P(t)} \qquad (1)$$

Of particular interest is the interpretation of average friction. One interpretation of average friction is to take the time average of $\mu(t)$, denoted by $\langle \mu(t) \rangle$. Alternatively, one could define an average friction coefficient, μ_{av}, as the average friction force divided by the average normal load, so that

$$\mu_{av} = \frac{\langle F(t) \rangle}{\langle P(t) \rangle} \qquad (2)$$

If the normal load remains constant or the instantaneous friction coefficient does not change with time, the two interpretations are equivalent. Otherwise they are not.

This is readily demonstrated by considering the example of a smooth, massless, circular Hertzian contact to which an oscillating load $P_o\,(1 + \cos\Omega t)$ is applied. This amount of load fluctuation is just enough to give impending contact loss at one extreme of the motion. The friction coefficient is μ_o when the load is at its mean value, P_o. For illustration purposes, the instantaneous friction force is assumed to be proportional to the instantaneous real area of contact. It is easy to show[1] that, in this case, $\frac{\mu_{av}}{\mu_o} = \mathbf{0.92}$ whereas $\frac{\langle\mu\rangle}{\mu_o} = \mathbf{1.84}$. This is illustrated in Fig. 1. The time average of the friction coefficient, $\langle\mu(t)\rangle$, increases while the average friction force decreases. When F, P and μ all vary with time, the coefficient of friction seems to be of limited value. Particular difficulties arise when $P(t) \approx 0$. For defining average friction, the definition of Eq. (2) is preferred.

Sometimes, in friction testing, only the instantaneous friction force is measured. Even this requires a measurement system with sufficient frequency bandwidth to accurately measure the fluctuating forces. The normal load is not monitored. If one incorrectly assumes that the normal load remains constant, when it does not, one obtains an "apparent friction" coefficient which can be quite different from the actual friction. Apparent friction sometimes includes stick or loss of contact which do not represent friction in the usual sense.

2. Normal Vibration and Friction at Hertzian Contacts

As the first and simplest example, the dynamic behavior of a circular Hertzian contact under dynamic excitation is examined. The system is shown in Fig. 2.

The rider has mass, m, and is in contact with a flat surface through a nonlinear stiffness and a viscous damper. The lower flat surface moves from left to right at a constant speed, V. The friction force, F, acts on the rider in the direction of sliding. The rider is constrained to motion normal to the direction of sliding. The model accommodates the primary normal contact resonance. The contact is loaded by its weight, mg, and by an external load, $P = P_o(1 + \alpha \cos\Omega t)$, which includes both a mean and a simple harmonic component. The normal displacement, y, of the mass is measured upward from its static equilibrium position, y_o. The equation of motion during contact, obtained from summing forces on the mass is

$$m\ddot{y} + c\dot{y} - f(\delta) = -P_o(1 + \alpha\cos\Omega t) - mg \qquad for \quad \delta > 0 \qquad (3)$$

where δ is the contact deflection and $f(\delta)$ is the restoring force given by

$$f(\delta) = \frac{4}{3}E'R^{\frac{1}{2}}\delta^{\frac{3}{2}} = K_1(y_o - y)^{\frac{3}{2}} \quad , \quad y_o = \left(\frac{P_o + mg}{K_1}\right)^{\frac{2}{3}} \qquad (4)$$

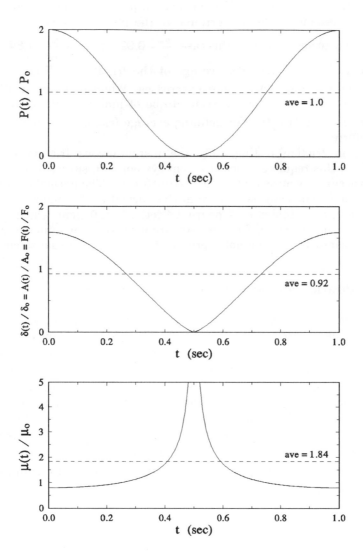

Figure 1. Instantaneous and average load, area, and friction (force and coefficient) for a smooth massless Hertzian contact.

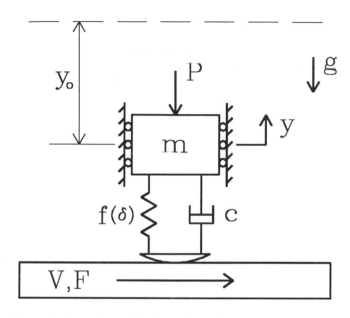

Figure 2. Dynamic model of normal motion for Hertzian contact.

An approximate steady-state solution to this nonlinear system has been obtained[1] using the perturbation technique known as the method of multiple scales.

The contact area, A, is proportional to the contact deflection, $(y_o - y)$. Based on the adhesion theory of friction, the instantaneous friction force is assumed to be proportional to the area of the contact. Therefore,

$$\frac{F}{F_o} - \frac{A}{A_o} = 1 - \frac{y}{y_o} \tag{5}$$

The normal oscillations, $y(t)$, are asymmetrical due to the nonlinear contact stiffness, and give rise to a decrease in average contact deflection, $(y_o - \langle y \rangle)$, (i.e., an increase in separation of the sliding bodies) by an amount $\langle y \rangle$, where $\langle y \rangle$ is the average of $y(t)$. Since Eq. (5) is linear, we can also write

$$\frac{\langle F \rangle}{F_o} = \frac{\langle A \rangle}{A_o} = 1 - \frac{\langle y \rangle}{y_o} \tag{6}$$

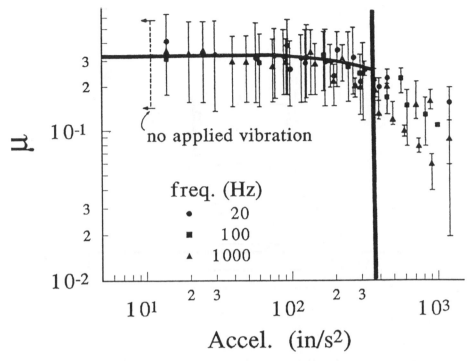

Figure 3. Measurements from Godrey (1967) showing the effect of vibration on friction;
————, present theory.

As oscillations increase, the average contact area, and, by implication, the average friction are reduced. A reduction in average friction force of up to ten percent was shown to occur[1] prior to loss of contact. This is not greatly different from the result obtained without considering inertia forces or damping and illustrated in Fig. 1.

Godfrey[11] conducted experiments to determine the effect of normal vibration on friction. His apparatus consisted of three steel balls fixed to a block that slid along a steel beam and was loaded by the weight of the block. The beam was vibrated by a speaker coil at various frequencies. The normal acceleration of the rider and the friction at the interface were measured. His measurements, under dry contact conditions, are illustrated in Fig. 3. If one assumes that occasional contact loss begins to occur when the normal acceleration reaches an amplitude of one g, one can superimpose the friction reduction predicted by our model as indicated by the heavy line. Reasonably good agreement is obtained. At higher normal accelerations, where there is progressively more intermittent contact loss, a larger reduction in friction occurs.

The dynamic behavior of continuously sliding Hertzian contacts under random roughness-induced base excitation has also been examined[4]. At sufficiently high loads, such contacts can be represented by a smooth Hertzian contact[12]. The role of the surface roughness is only to provide a base excitation as it is swept through the contact region.

By restricting the effective surface roughness input displacement to stationary random processes defined by the spectral density function $S_{y_i y_i}(k) = L\pi^{-1}k^{-4}$ (where L is a constant and k is the surface wavenumber), the Fokker-Planck equation can be used to obtain the exact stationary solution.

Again, one finds a decrease in the mean contact compression under dynamic loading. This also leads to a reduction in the mean contact area and the average friction force, under the assumption that the instantaneous friction force is proportional to the instantaneous area of contact. Based on the analysis the reduction in average friction force when vibration amplitudes approached the limit of contact loss was around nine percent.

A pin-on-disk system with a steel against steel Hertzian contact, excited by surface irregularities, was used to obtain measurements of average friction at various sliding speeds. The normal vibrations increased with sliding speed. The analysis was compared with the experiments by adjusting the parameter, L, so that the analytical model gave initial loss of contact at the same speed (i.e., at 50 cm/s) as observed during the tests. The computed results are shown together with the measurements in Fig. 4.

The measurements show a decrease in friction with increasing sliding speed. Considering that the load criterion of Greenwood and Tripp[12] is not satisfied at all times during the motion, the agreement with the theoretical model is quite good. The measurements illustrate that, only at speeds well above those associated with initial loss of contact, can one obtain large reductions in average friction, of as much as thirty percent.

3. Normal Vibration and Friction at Rough Planar Contacts

The Greenwood and Williamson[9] statistical formulation of the elastic contact of randomly rough surfaces is still the best known and most widely-used model. The normal vibrations of such a contact can be cast in the same form as Eq. (3) with the real contact area and the normal elastic restoring force expressed by

$$A = \pi \eta \tilde{A} \beta \sigma \int_h^\infty (\varepsilon - h)\Phi^*(\varepsilon)d\varepsilon \tag{7a}$$

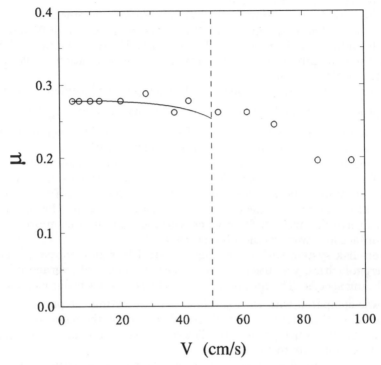

Figure 4. Average coefficient of friction at various sliding speeds: o, measurements; ——— , theoretical.

$$f(\delta) = \frac{4}{3}\eta \, \tilde{A} \, E' \beta^{\frac{1}{2}} \sigma^{\frac{3}{2}} \int\limits_{h}^{\infty} (\epsilon - h)^{\frac{3}{2}} \, \Phi^*(\epsilon) \, d\epsilon \qquad (7b)$$

where

β ≡ asperity radius

σ ≡ standard deviation of asperity height distribution

$h = \dfrac{d}{\sigma} = \dfrac{y + d_o}{\sigma}$ ≡ normalized separation

d_o ≡ static separation

$\epsilon = \dfrac{z}{\sigma}$ ≡ normalized asperity heights

η ≡ surface density of asperities

$\tilde{A}$ ≡ nominal contact area

$\Phi^*(\epsilon)$ ≡ normalized asperity height distribution

The nonlinear vibration problem has been solved[2]. The contact stiffness nonlinearity is stronger than that of the Hertzian model. Again one finds that, on average, the sliding surfaces move apart during sliding. The change in average separation is typically around thirty percent of the vibration amplitude, $|y|$. Although the surfaces on average, move apart, the average friction force obtained by taking the time average of the contact area, remains unchanged in the presence of normal vibrations. This seemingly paradoxical result is not unexpected when one recognizes that the Greenwood-Williamson model leads to a direct proportionality between the normal load and the real contact area at all separations, i.e., a constant instantaneous friction coefficient. While the nonlinear contact vibrations can be complicated, and the instantaneous friction force may change considerably, the friction coefficient is not expected to change. Other rough surface models, may give somewhat different results.

Linearized equations for the normal vibration problem have also been developed[5]. One rather remarkable result of the linearized analysis is that the small amplitude normal natural frequency of a weight-loaded rigid block supported by a Greenwood-Williamson type rough surface is $\omega_{oy} = \sqrt{g/\sigma}$. The natural frequency is independent of the block and countersurface materials. The natural frequency is independent of the block dimensions and, at least on earth, depends only on the standard deviation of the asperity heights, σ. The acceleration spectra of a steel block (a 4.4 cm cube) obtained during sliding against a large steel base at a speed of 3 cm/s are shown in Fig. 5. One finds the normal natural frequency at around 1300 Hz which is in general agreement with the block roughness that was measured($R_a \approx 0.2\,\mu m$). Angular motions, with a resonant frequency of 1070 Hz are also observed and shown in Fig. 5. It is clear that the possibility of angular motions must be included in a model of the problem.

4. Normal and Angular Vibrations at Rough Planar Contacts

A model that allows for both normal and angular motions of a nominally stationary block pressed against a moving countersurface is shown in Fig. 6a in its frictionless equilibrium position and in Fig. 6b in its steady sliding equilibrium position. Some angular displacement, θ_o, and offset, c, of the normal reaction force are necessary to maintain moment equilibrium of the block.

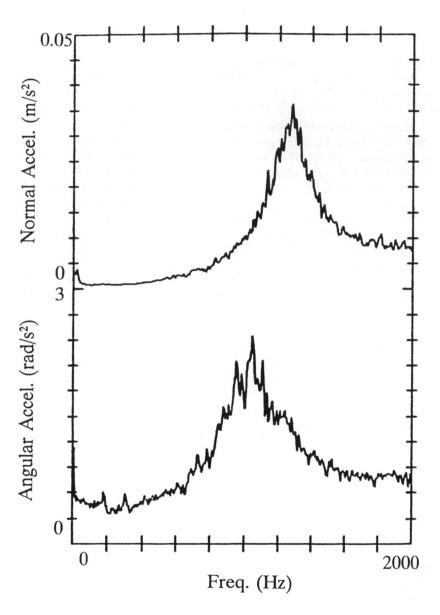

Figure 5. Average rms spectra of measured normal and angular accelerations from sliding block.

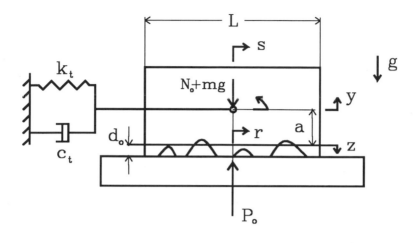

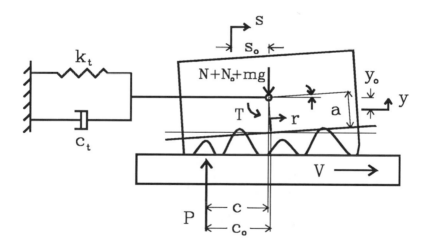

Figure 6. Three degree-of-freedom system model: (a) frictionless equilibrium position, (b) steady sliding equilibrium position with friction.

The Greenwood-Williamson model has been extended[5] to account for angular as well as normal motions. For an exponential distribution of asperity heights, we find that

$$\frac{A(y,\theta)}{A_o} = \frac{P(y,\theta)}{P_o} = \frac{\sigma}{L\theta} e^{-\frac{y}{\sigma}} [e^{\frac{L\theta}{2\sigma}} - e^{-\frac{L\theta}{2\sigma}}] \tag{8}$$

where A_o and P_o denote the real contact area and the normal load at the frictionless equilibrium position. y and θ are measured from that position. Both the contact area and normal load depend on the normal displacement, y, and the angular displacement, θ. However, in the presence of both angular and normal motions, the area of contact remains proportional to the normal load. Therefore, within the assumptions of the analysis, we do not expect the coefficient of friction to change with normal or angular vibrations during continuous sliding, since $A(y,\theta)/A_o = F(y,\theta)/F_o = P(y,\theta)/P_o$.

For other surface topologies, the above result may not hold. For example, it has been shown[3] that for a periodic surface consisting of a regular pattern of hemispherical asperities of equal height, it is the friction force rather than the friction coefficient that remains constant when relative angular motions occur at a given normal load.

The dynamic equations of the three degree-of-freedom sliding system can now be written directly:

$$m\ddot{s} + c_t\dot{s} + k_t s = F(y,\theta) \tag{9a}$$

$$m\ddot{y} + b\dot{y} - P(y,\theta) = -N(t) - N_o - mg \tag{9b}$$

$$J\ddot{\theta} + B\dot{\theta} - c(\theta)P(y,\theta) - (a + d_o)F(y,\theta) = T(t) \tag{9c}$$

Viscous damping terms b and B have been introduced to account for some damping of the motions. Introducing the changes in variables, $q = \dfrac{y}{\sigma}$ and $\phi = \dfrac{L\theta}{2\sigma}$, the equations become[5]

$$\ddot{s} + \frac{c_t}{m}\dot{s} + \frac{k_t}{m}s = \frac{F_o}{m}f_1(q)f_2(\phi) \tag{10a}$$

$$\ddot{q} + \frac{b}{m}\dot{q} - \frac{P_o}{m\sigma}f_1(q)f_2(\phi) = -\frac{1}{m\sigma}(N(t) + P_o) \tag{10b}$$

$$\ddot{\phi} + \frac{B}{J}\dot{\phi} - \frac{L}{2J\sigma}f_1(q)f_3(\phi) = \frac{LT(t)}{2J\sigma} \tag{10c}$$

where

$$f_1(q) = e^{-q} \tag{11a}$$

$$f_2(\phi) = \frac{1}{2\phi}(e^{\phi} - e^{-\phi}) \tag{11b}$$

$$f_3(\phi) = \frac{1}{2\phi}\left(e^\phi - e^{-\phi}\right)\left[P_o\left(\frac{\frac{L}{2\phi}\left[(1-\phi)e^\phi - (1+\phi)e^{-\phi}\right]}{\left(e^\phi - e^{-\phi}\right)} + \frac{2\sigma a\phi}{L}\right) + (a+d_o)F_o\right] \quad (11c)$$

Equilibrium of the rider under steady sliding requires that

$$k_t s_o = F_o\, f_1(q_o)\, f_2(\phi_o) \quad (12a)$$

$$f_1(q_o)\, f_2(\phi_o) = 1 \quad (12b)$$

$$f_3(\phi_o) = 0 \quad (12c)$$

The equilibrium position of the rider under steady sliding can be determined by solving Eq. (12c) for ϕ_o, Eq. (12b) for q_o, and finally Eq. (12a) for s_o.

The equations of motion, Eqs. (10) and (11), clearly reveal nonlinear coupling among the translational and angular motions. Subharmonic, superharmonic, and combination resonances may occur when the system is in forced oscillation. Chaotic motions may also take place[6]. In problems of this type, system stability, i.e., stability of sliding, may be of concern and was studied by Martins et al.[10]. In fact, our equations are similar to those of Martins et al.. For example, Eqs. (10) and (12) of the present paper can be compared to Eqs. (5.8) and (5.6) in their paper[10]. An essential difference between the two approaches is that the normal and angular contact stiffnesses have different forms. Martins et al. used a power law form that has also been used by Back et al.[13] and Kragelskii and Mikhin[14].

5. Stability Analysis

A linear stability analysis of the three degree-of-freedom contact model developed above reveals some interesting aspects of sliding systems. The linearized equations of motion for small perturbations about the steady sliding equilibrium position without forcing terms are[5]

$$\ddot{s}_s + \frac{c_t}{m}\dot{s}_s + \frac{k_t}{m}s_s + \frac{F_o}{m}q_s - \frac{F_o c_2}{m}\phi_s = 0 \quad (13a)$$

$$\ddot{q}_s + \frac{b}{m}\dot{q}_s + \frac{P_o}{m\sigma}q_s - \frac{P_o c_2}{m\sigma}\phi_s = 0 \quad (13b)$$

$$\ddot{\phi}_s + \frac{B}{J}\dot{\phi}_s - \frac{L c_4}{2J\sigma}\phi_s = 0 \quad (13c)$$

where

$$c_2 - \frac{e^{-q_o}}{2\phi_o^2}\left[\phi_o\left(e^{\phi_o} + e^{-\phi_o}\right) - \left(e^{\phi_o} - e^{-\phi_o}\right)\right] \quad (14a)$$

$$c_4 = e^{-q_o} \left\{ \frac{P_o L}{4} \left[\frac{1}{\phi_o} \left(-e^{\phi_o} + e^{-\phi_o} \right) - \frac{2}{\phi_o^3} \left[\left(1 - \phi_o \right) e^{\phi_o} - \left(1 + \phi_o \right) e^{-\phi_o} \right] \right] \right.$$
$$\left. + \frac{\sigma a P_o}{L} \left(e^{\phi_o} + e^{-\phi_o} \right) + \frac{(a + d_o) F_o}{2} \left[\frac{1}{\phi_o} \left(e^{\phi_o} + e^{-\phi_o} \right) + \frac{1}{\phi_o^2} \left(e^{\phi_o} - e^{-\phi_o} \right) \right] \right\} \qquad (14b)$$

These linearized equations, having an asymmetric stiffness matrix, describe a circulatory system. These equations are similar to Eq. (5.15) of Martins et al.[10], which they found to exhibit a high frequency flutter instability at friction values well below that which would result in the block tumbling, when $\mu_o = \frac{L}{2a}$, which is a divergence instability.

This flutter instability does not seem to occur with the model of Eq. (13). The eigenvalues that we have computed with the damping set to zero are always purely imaginary, never exhibiting positive real parts. In Martins et al., the flutter instability occurs when the two eigenvalues associated with the angular and quasi-normal natural frequencies take on the same value. In the present model, the eigenvalue ratio (angular divided by quasi-normal) always remains less than unity, approaching this value only when the block is very long, i.e., when $\frac{2a}{L} < 1$. The differences in the qualitative behavior of the two models seems to be due to the details of the normal and angular restoring forces. These contact stiffnesses are very sensitive to the details of the surface texture. This may largely explain the elusive nature of many high frequency instabilities and squeal phenomena which can occur in sliding systems and can change and appear or disappear as surfaces run-in or wear.

Figure 7 shows the instability regions found by Martins et al. in the absence of damping. In their analysis, and with weight loading, the stability depends only on the height to length ratio of the block, the sliding friction coefficient, and a damping parameter. These results indicate instability over a fairly broad range of aspect ratio (H/L where H=2a). The addition of damping causes the flutter instability boundary to shift to higher friction values for small aspect ratios.

Interestingly, the addition of a linear torsional stiffness, k_ϕ, to our sliding block model can lead to instability. The resulting linearized equations of motion for this case are

$$\ddot{s}_s + \frac{c_t}{m} \dot{s}_s + \frac{k_t}{m} s_s + \frac{F_o}{m} q_s - \frac{F_o c_2}{m} \phi_s = 0 \qquad (15a)$$

$$\ddot{q}_s + \frac{b}{m} \dot{q}_s + \frac{P_o}{m \sigma} q_s - \frac{P_o c_2}{m \sigma} \phi_s = 0 \qquad (15b)$$

$$\ddot{\phi}_s + \frac{B}{J} \dot{\phi}_s + \left(\frac{k_\phi}{J} - \frac{L c_4}{2 J \sigma} \right) \phi_s + \frac{L c_3}{2 J \sigma} q_s = 0 \qquad (15c)$$

where c_2 and c_4 are defined in Eq. (14) and

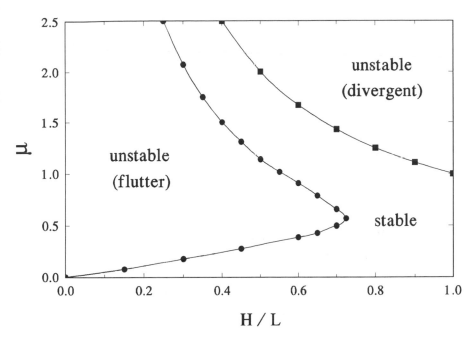

Figure 7. Regions of instability based on Martins et al. (1990).

$$c_3 = e^{-q_o}\left[\frac{P_o L}{4\phi_o^2}\left[(1-\phi_o)e^{\phi_o}-(1+\phi_o)e^{-\phi_o}\right]+\frac{P_o \sigma a}{L}\left(e^{\phi_o}-e^{-\phi_o}\right)+\frac{(a+d_o)F_o}{2\phi_o}\left(e^{\phi_o}-e^{-\phi_o}\right)\right] \quad (16)$$

From the nonlinear governing Eqs. (10), equilibrium of the rider under steady sliding for this case requires that

$$k_t s_o = F_o f_1(q_o) f_2(\phi_o) \quad (17a)$$

$$f_1(q_o) f_2(\phi_o) = 1 \quad (17b)$$

$$-\frac{L}{2\sigma} f_1(q_o) f_3(\phi_o) + k_\phi \phi_o = 0 \quad (17c)$$

From these equations the equilibrium position during sliding can be computed.

The stability of the system defined by Eq. (15) can be assessed by computing the eigenvalues. Fig. 8 shows the stability regions for a particular value of linear torsional stiffness and no viscous damping. The instability region of particular interest is the flutter instability between aspect ratio values of 0.2 and 0.6. As the torsional stiffness is decreased, this region becomes narrower and shifts to

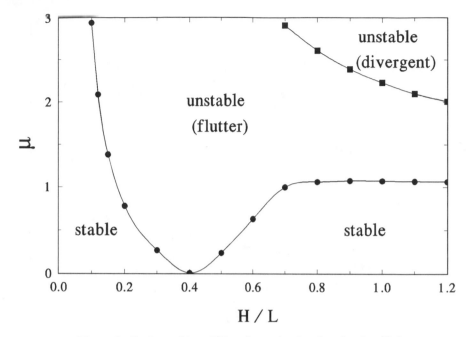

Figure 8. Regions of instability when a torsional spring is added.

lower values of aspect ratio. An alternate means of adding torsional stiffness to the system is to offset the horizontal suspension from the center of mass.

A stability analysis of a contact system, shown in Fig. 9, consisting of a slider with hemispherical legs, in plane, against a stationary flat surface has also been examined[7]. The slider consists of a block of height H=2h and length L with four hemispherical legs of radius R. The center of mass of the slider can be moved vertically. The slider is pulled at its center of mass against a flat surface at a constant velocity, v. The slider is modeled as a rigid body with elastic contacts, and has normal and angular degrees of freedom. The contacts are modeled with a nonlinear Hertzian stiffness in parallel with a nonlinear damping element of the form described by Hunt and Crossley[15].

As before, the stability of the contact system is examined by examining the eigenvalues of the governing equations of motion linearized about the steady sliding equilibrium position. A number of simulations have been performed which predict regions of stable, flutter unstable and divergent unstable behavior.

The position of the slider center of mass is found to strongly influence the instability regions. Figure 10 shows the effect of moving the center of mass of

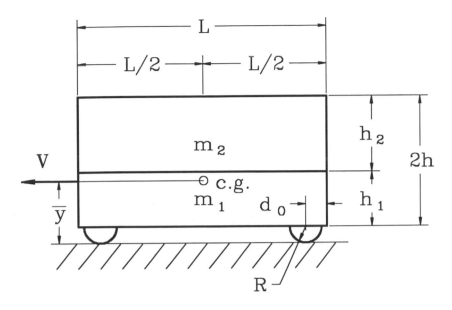

Figure 9 Contact system with hemispherical legs.

the slider vertically from its geometric center. As the center of mass is moved from top to bottom, the divergent unstable region shifts to higher values of friction and aspect ratio, and the flutter instability behavior occurs over a much more significant range of friction coefficient and aspect ratio.

The effect of changing the damping level in the system model was also investigated. It was found that a significant increase in the damping constant was needed before the stability of the system would change notably from that observed without any damping. Figure 11 illustrates how the flutter unstable region decreases as the damping constant c is increased significantly.

The stability of the contact system was also found to depend on the radii of the hemispherical legs. It was generally found that both instability regions increase as the leg radii increases, and the system becomes less stable. Changes in the effective modulus of elasticity did not affect the stability behavior except at extremely low values, the flutter region was found to decrease.

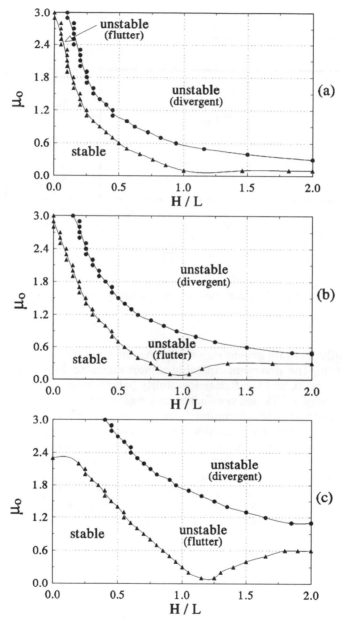

Figure 10   Regions of instability as slider center of mass is moved vertically: (a) $\bar{y} = 0.7H + R$, (b) $\bar{y} = 0.5H + R$, and (c) $\bar{y} = 0.3H + R$ ($\bar{x} = 0.5L$, R = 10 mm, E' = 2 GPa, c = 1.0×10^6 Ns/m$^{5/2}$).

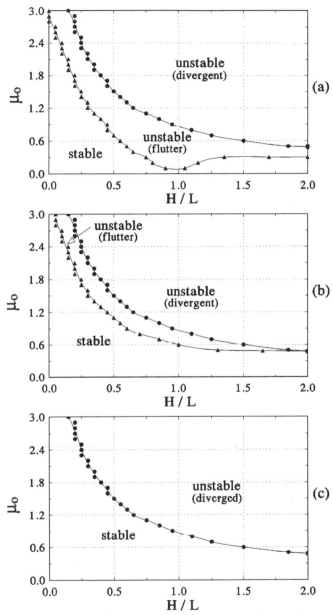

Figure 11   Regions of instability for different damping levels: (a) $c = 0$ Ns/m$^{5/2}$, (b) $c = 5.0 \times 10^{12}$ Ns/m$^{5/2}$, and (c) $c = 1.0 \times 10^{15}$ Ns/m$^{5/2}$ (R = 10 mm, E' = 2 GPa, $\bar{x} = 0.5L$, $\bar{y} = 0.5H + R$).

0. Chaotic Vibration and Friction

Most of the literature dealing with chaotic motion at mechanical interfaces examine sliding systems with motion restricted to the direction of sliding[16-18]. The catalyst for chaos, in these cases, is an abrupt change in friction as the relative sliding speed of the contacting components fluctuates through zero.

Another nonlinearity in mechanical joints results from gaps or play. A recent paper by Moon and Li[19] examined the effect of this type of nonlinearity in cotter pin joints on space structure dynamics. The dynamics were shown to be complicated by joint play, and chaotic vibrations were found to occur.

In general, there are a number of joint nonlinearities that can contribute to the dynamics of mechanical systems. The results of previous studies show that the effect of play dominates for loose joints such as cotter pinned joints, whereas the effect of friction may dominate for contacting components whose relative sliding speed repeatedly fluctuates through zero.

In this section, the joint dynamics resulting solely from nonlinear interface stiffness are examined. The relative sliding speed of the contacting components in our joint model is taken to be constant to avoid the additional complexity of a step change in friction, and gaps are assumed to be minute so that their effect is of secondary importance. In addition, the contacts are assumed to be unlubricated so that the damping normal to the interfaces is small. Examples of joint configurations in which the effect of normal interface stiffness may dominate, include dry slideways moving at constant speed.

The dynamic joint model is shown in Fig. 12. It consists of a rigid block of mass, m, with elastic upper and lower surfaces, constrained between two elastic counter-surfaces which can move at a constant speed, V, as shown, or remain stationary. Two planar contact regions are formed between the block and the counter-surfaces. Both interfaces exhibit contact stiffness and damping. These characteristics are modeled as nonlinear springs in parallel with viscous dampers. The interface stiffness is modeled with either an analytical description, based on a statistical model of interacting rough surfaces, or an empirical power-law description, resulting from actual measurements.

In addition, the block is constrained to motion normal to the interface. Gravity is assumed to act normal to the page, so that the static equilibrium position of the block is at the center between the counter-surfaces. The normal displacement, z, of the block is measured from this position. The block is subjected to a harmonic forcing $N(t) = g\tilde{A}P_o \cos \Omega t$ where g is the gravitational constant in m/s^2, $\tilde{A}$ is the nominal contact area between the block and a counter-surface in cm^2, P_o is the magnitude of the loading in units of kg/cm^2, and Ω is the forcing frequency.

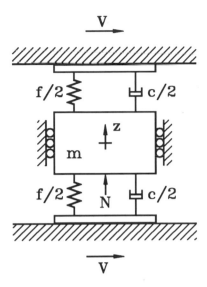

Figure 12   Dynamic joint model.

The equation of motion obtained from summing the forces on the mass is

$$m\ddot{z} + c\dot{z} + f(z) = g\tilde{A}P_o \cos\Omega t \tag{18}$$

where $f(z)$ is the net restoring force of the joint stiffness.   Based on the Greenwood and Williamson[9] rough surface model, the resultant restoring force for a block with two rough surfaces between two counter-surfaces is[6]

$$f(z) = \pi^{1/2}\eta\tilde{A}E^*\beta^{1/2}\sigma^{3/2}\left[e^{-(d_o - |z|)/\sigma} - e^{-(d_o + |z|)/\sigma}\right]sgn(z) \tag{19}$$

The equation of motion of the block in the joint model of Fig. 12 is then

$$m\ddot{z} + c\dot{z} + \pi^{1/2}\eta\tilde{A}E^*\beta^{1/2}\sigma^{3/2}\left[e^{-(d_o - |z|)/\sigma} - e^{-(d_o + |z|)/\sigma}\right]sgn(z) = g\tilde{A}P_o\cos\Omega t \tag{20}$$

In an effort to nondimensionalize, a change in variable $y = z/\sigma$, a normalized separation $h_o = d_o/\sigma$, and a dimensionless damping parameter $\zeta = c/(2m\omega_o)$, are introduced to obtain

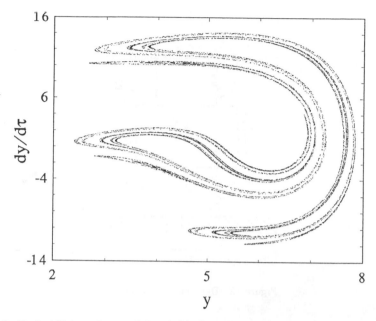

Figure 13.   Typical Poincaré map of chaotic block motion for joint with rough surface stiffness.

$$\ddot{y} + 2\zeta\,\omega_o\dot{y} + \frac{\pi^{1/2}\eta\tilde{A}E^*\beta^{1/2}\sigma^{1/2}}{m}\left[e^{-(h_o-|y|)} - e^{-(h_o+|y|)}\right]sgn(y) = \frac{g\tilde{A}P_o}{m\,\sigma}\cos\Omega t \qquad (21)$$

A quantitative test based on the sign of the Lyapunov exponents of the joint motion is used to identify steady-state chaos. The criterion for chaotic motion is as follows: if the largest Lyapunov exponent is positive ($\lambda_1 > 0$), the motion is chaotic; and if the exponent is zero or negative ($\lambda_1 \le 0$), the motion is periodic; the Lyapunov exponent is also negative for quasi-periodic and strange nonchaotic motions.

Some numerical experiments have been performed for a cubic steel block with run-in ground surfaces between two similar counter-surfaces[6]. Figure 13 shows a typical Poincaré map of chaotic normal motion of the block. The map has a definite structure with fractal or self-similar characteristics which is intrinsic of chaotic motion. If the apparent noise in this data was truly random instead of chaotic, the points in the Poincaré map would be evenly distributed.

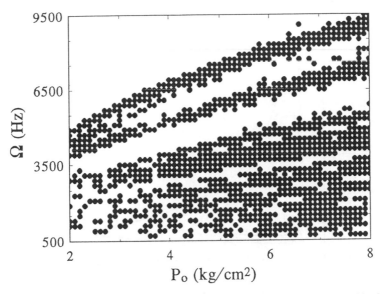

Figure 14. Ω versus P_o parameter diagram for joint with rough surface stiffness (shaded areas signify chaotic behavior).

Figure 14 illustrates a representative parameter-plane diagram of Ω versus P_o. The diagram reveals significant regions of steady-state chaotic motion denoted by the shaded areas. This illustrates that chaotic behavior in mechanical joints can occur over a broad range of frequencies.

The instantaneous friction force at the interfaces is related to the normal motion of the block through the adhesion theory of friction. The real area of contact and interface friction force with respect to their values in the absence of normal motion are[6]

$$\frac{F}{F_o} = \frac{A}{A_o} = \frac{1}{2}\left(e^y + e^{-y}\right) \tag{22}$$

where A_o and F_o are the contact area and friction force at separation, h_o, in the absence of normal vibration. This expression clearly reveals the dependence of friction on normal motion. The importance of this relation, in the present context, is that chaotic normal motion at mechanical joints may result in a chaotic friction force.

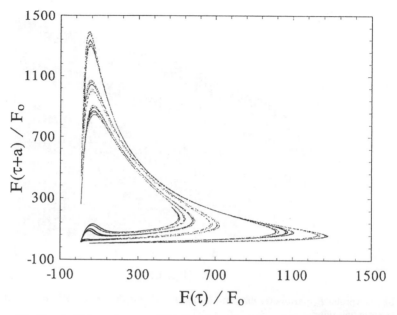

Figure 15. Pseudo-Poincaré maps of friction force for joint with rough surface stiffness.

A pseudo-Poincaré map of friction can be obtained by plotting the friction force, $F(\tau)$, versus itself but delayed by a fixed time constant, a, i.e., $F(\tau)$ versus $F(\tau +a)$. A typical Poincaré map for the joint friction force, when the normal motion of the joint block is chaotic, is shown in Fig. 15. The pseudo-Poincaré map of the friction exhibits a definite fractal structure.

The data from this sample simulation is representative in revealing the significant unsteady fluctuations in friction force resulting from micro-scale normal motions. In addition to influencing the overall dynamics of systems in which such a joint is embedded, the uncertainty in the instantaneous friction force due to chaotic behavior represents major challenges in controlling machines and structures. This analysis reveals that such chaotic fluctuations can be avoided by careful joint design or modification of an existing joint configuration (e.g., increasing pre-load). An excellent reference which reviews progress on the control of machines with deterministic unsteady friction is given by Armstrong-Helouvry[20].

Simulations of joint dynamics have also been performed using an empirical power-law description for the normal stiffness. The restoring force, in Newtons, is given as a function of contact deflection, δ, in meters, as

$$f(\delta) = k\delta^n \qquad (23)$$

where n and k are experimentally determined constants which depend on the surface texture, material, and geometry of the joint. The literature shows that values of the exponent n range from 1.6 to 3.3 for planar contacts of various materials and types of machining[13,14]. Comparable behavior has been found[6] with the empirical power-law description as found with the analytical rough surface model description.

7. Conclusions

It has been shown that, for Hertzian contacts, a maximum average friction reduction of approximately ten percent occurs in the presence of vibration without loss of contact. This is associated with the inherent behavior of the Hertzian contact and is independent of the particular loading and contact parameters. However, for oscillations smaller than those required for loss of contact, the mean separation and friction force are highly dependent on the contact parameters.

The nonlinear contact vibrations of a mass supported by a rough planar contact have been examined. Substantial increases in the mean separation can take place, due to the nonlinear contact compliance. Interestingly, the increase in average separation results in little or no change in average contact area, and consequently in little or no change in average friction. This is a direct result of the Greenwood-Williamson rough surface contact model in which the real area of contact remains nearly proportional to normal load (independent of contact deflection), over a wide range of normal loads. It is concluded that normal oscillations and increases in the mean separation need not be accompanied by significant changes in average friction, as long as continuous contact is maintained between two sliding rough surfaces.

A compliant contact model that accommodates coupled normal and angular motions at rough planar contacts and is suitable for both static and dynamic analyses has been presented. The contact area and resultant normal contact force have been shown to be nonlinearly dependent on both normal and angular displacements, but are found to remain proportional as long as contact is maintained across the entire interface, and may remain so even after some separation has occurred.

It has been shown that high frequency flutter instabilities, which can cause squeal noise and affect apparent friction, are very sensitive to the surface topographies, contact compliances, damping characteristics, and constraints of simple sliding systems such as blocks on flat surfaces. The complex nonlinear dynamic behavior of such seemingly simple sliding systems become even more challenging when such elements are embedded in larger dynamic systems.

A dynamic joint model that accommodates normal motions with either analytical or empirical descriptions of nonlinear interface compliance has also been presented. Numerical analyses of the model using common joint component materials and interface finishes under dynamic loads reveal that chaotic joint vibration can occur over a significant range of contact parameters and operating conditions. Even though such chaotic motions are found to be on the order of microns, the instantaneous friction force is found to fluctuate considerably in the presence of such micro-scale normal motions.

8. References

1.   D. P. Hess and A. Soom, *ASME Journal of Tribology* **113** (1991) 80.
2.   D. P. Hess and A. Soom, *ASME Journal of Tribology* **113** (1991) 87.
3.   D. P. Hess, *Nonlinear Contact Vibrations and Dry Friction at Concentrated and Extended Contacts* (Ph.D. Dissertation, University at Buffalo, 1991).
4.   D. P. Hess, A. Soom and C. H. Kim, *Journal of Sound and Vibration* **153** (1992) 491.
5.   D. P. Hess and A. Soom, *ASME Journal of Tribology* **114** (1992) 567.
6.   D. P. Hess and N. J. Wagh, *ASME Journal of Vibration and Acoustics* **116** (1994) 474.
7.   D. P. Hess and M. Al-Grafi, *Applied Acoustics*, in press.
8.   D. Tabor, *ASME Journal of Lubrication Technology* **103** (1981) 169.
9.   J. A. Greenwood and J. B. P. Williamson, *Proceedings of the Royal Society of London* **A295** (1966) 300.
10. J. A. C. Martins, J. T. Oden and F. M. F. Simoes, *International Journal of Engineering Science* **28** (1990) 29.
11. D. Godfrey, *ASLE Transactions* **10** (1967) 183.
12. J. A. Greenwood and J. H. Tripp, *Journal of Applied Mechanics* **34** (1967) 153.
13. N. Back, M. Burdekin and A. Cowley, *Proceedings of the 13th International Machine Tool Design and Research Conference* (Macmillan, London, 1973) 87.
14. I. V. Kragelskii and N. M. Mikhin, *Handbook of Friction Units of Machines*, (ASME Press, New York, 1988).
15. K. H. Hunt and F. R. E. Crossley, *ASME Journal of Applied Mechanics* **42** (1975) 440.
16. J. Awrejcewicz, J., *Journal of Sound and Vibration* **109** (1986) 178.
17. S. Narayanan and K. Jayaraman, *Nonlinear Dynamics in Engineering Systems - IUTAM Symposium* (Stuttgart, Germany, 1989) 217.

18. K. Popp and P. Stelter, *Nonlinear Dynamics in Engineering Systems - IUTAM Symposium* (Stuttgart, Germany, 1989) 233.
19. F. C. Moon and G. X. Li, *AIAA Journal* **28** (1990) 915.
20. B. Armstrong-Helouvry, *Control of Machines with Friction* (Kluwer Academic Publishers, Boston, 1991).

Dynamics with Friction: Modeling, Analysis and Experiment, Part II, pp. 29–74
edited by A. Guran, F. Pfeiffer and K. Popp
Series on Stability, Vibration and Control of Systems, Series B, Vol. 7
© World Scientific Publishing Company

VIBRATIONS AND FRICTION-INDUCED INSTABILITY IN DISCS

JOHN E. MOTTERSHEAD
Department of Mechanical Engineering,
The University of Liverpool,
Liverpool, L69 3BX, UK

ABSTRACT

The simple shape and common occurrence of discs as machine elements
belies a complexity of behaviour that can be an important factor in the
improvement of modern machine performance. The purpose of this article
is to give a thorough account of the understanding gained from recent
research investigations into the dynamics of discs. Friction-induced vibration
and parametric resonance phenomena are areas of research activity that have
received considerable attention. The approach taken here is to build up to
the discussion of complex topics such as these by firstly establishing the
basic principles of critical speed instability, forward and backward travelling
waves, and destabilisation by a transverse elastic system.

1. Introduction

Discs are essential machine components and can be found in such diverse applications
as automotive disc brakes, turbine discs, circular machine tools and saws, and computer
disc-drives. When the disc is in normal operation as an integral part of a system then
self-excited and parametric vibrations may arise due to interaction with other components.
The vibrations are, in general, undesirable and detrimental to the intended functioning of
the system. Poor cutting accuracy of workpieces, and higher material wastage, by
circular saws is an example of the consequences of large-amplitude disc vibrations, which
further causes excessive wear of components, surface damage and fatigue failure. The
high-frequency squeal noise produced in disc brakes is another example of self-excited
disc vibration. The excitation of computer discs can lead to signal and data losses, and
vibrational resonances may lead to turbine failures.

Quite apart from tackling issues and problems of an engineering nature, there is the
fascinating realm of physics at work in spinning discs. Some of these effects have long
since been demonstrated and understood; others may still be awaiting discovery.
Friction-induced disc vibration, which is, from a theoretical viewpoint, perhaps the most
complex of disc vibration problems, can involve areas such as plate dynamics, follower
forces, and nonlinearities. The interaction of these disciplines leads directly to the
possibility of a much more rich and complex dynamical behaviour. This will ultimately

point towards more effective control methods and will steer designs towards the avoidance of problems currently faced by engineering industry.

In this article we begin by reviewing the established methodologies for the analysis of disc vibrations, critical speeds, and the instabilities caused by the action of a transverse mass-spring-damper system. The treatment of friction as a follower force is explained and its application in the analysis of squeal noise in brakes is discussed. The focus of the article is on the analysis of parametric resonance phenomena in discs under the action of; (i) a distributed elastic system, with friction, that rotates relative to the disc; and (ii) a discrete, rotating frictional load with a negative μ-velocity characteristic. The analysis is illustrated with numerical examples, and a discussion of further research opportunities is included.

2.   Disc Vibrations and Critical Speeds

Lamb and Southwell[1] and Southwell[2] studied the problem of vibrations in a spinning disc. They considered the membrane and flexural effects separately and showed a lower limit on the square of the natural frequency to be given by the sum of the squares of the two frequencies obtained on the extreme suppositions of infinite thinness and infinitely slow rotation.   It appears that the article by Lamb and Southwell[1] was the first to introduce the notion of a forward and a backward travelling wave.

2.1   Flexural Vibrations

McLeod and Bishop[3] gave the equation of motion of a classical annular plate in the form,

$$\rho h \frac{\partial^2 w}{\partial t^2} + D\nabla^4 w = P \ , \quad r\in [a,b], \ \theta\in [0,2\pi], \ t\in [0,\infty) \ , \tag{1}$$

where the biharmonic differential operator when written in the cylindrical coordinates is,

$$\nabla^4 = \left(\frac{\partial^2}{\partial r^2} + \frac{1}{r}\frac{\partial}{\partial r} + \frac{1}{r^2}\frac{\partial^2}{\partial \theta^2} \right)^2 , \tag{2}$$

and the flexural rigidity of the plate is,

$$D = \frac{Eh^3}{12(1-\nu^2)} \ . \tag{3}$$

The inner and outer radii of the plate are denoted by a and b, and the parameters ρ,h,E and ν are the density, thickness, Young's modulus and Poisson's ratio of the plate respectively.

We consider the case of free vibration of the plate, when the external load P is zero. Then by using the method of separation of variables, the transverse vibration of the plate can be written as,

$$w = T(t)\Psi(r,\theta) , \qquad (4)$$

which upon substitution into Eq. (1) yields an ordinary differential equation,

$$\frac{d^2T}{dt^2} + \omega_p^2 T = 0 , \qquad (5)$$

and a partial differential equation,

$$\left(\frac{\partial^2}{\partial r^2} + \frac{1}{r} \frac{\partial}{\partial r} + \frac{1}{r^2} \frac{\partial^2}{\partial \theta^2} \right)^2 \Psi - \kappa^4\Psi = 0 , \qquad (6)$$

where

$$\omega_p^2 = \frac{D\kappa^4}{\rho h} . \qquad (7)$$

The solution of Eq. (6) is seen to be the combined solution of the two differential equations,

$$\left(\frac{\partial^2}{\partial r^2} + \frac{1}{r} \frac{\partial}{\partial r} + \frac{1}{r^2} \frac{\partial^2}{\partial \theta^2} \right) \Psi \pm \kappa^2\Psi = 0 . \qquad (8 \text{ a,b})$$

Upon separating the variable Ψ such that,

$$\Psi(r,\theta) = R(r)\,\Theta(\theta) , \qquad (9)$$

we have the ordinary differential equations,

$$\frac{d^2\Theta}{dt^2} + j^2\Theta = 0 , \qquad (10)$$

and

$$\left(\frac{d}{dr^2} + \frac{1}{r} \frac{d}{dr} - \frac{j^2}{r^2} \pm \kappa^2 \right) R = 0 . \qquad (11 \text{ a,b})$$

Since the variable Θ has to be continuous over the period 2π, j must be an integer i.e., j = 0, 1, 2, ... The Eqs. (11a,b) are recognised as an ordinary Bessel equation and a modified Bessel equation, whose solutions are in the form of Bessel's functions. The solution to the variable R(r) may then be expressed as,

$$R(r) = \alpha\,J_j(\kappa r) + \beta\,Y_j(\kappa r) + \gamma\,I_j(\kappa r) + \delta\,K_j(\kappa r) , \qquad (12)$$

where $J_j(\kappa r)$ and $Y_j(\kappa r)$ are the Bessel functions of the first and second kind of order j, and $I_j(\kappa r)$ and $K_j(\kappa r)$ are the modified Bessel functions of the first and second kind of order j. Certain ratios of the constants $\alpha, \beta, \gamma, \delta$, together with the parameter κ can be found from the boundary conditions of the plate.

An eigenfunction expansion solution from the classical plate equation is found to be,

$$w = \sum_{j=0}^{\infty} \sum_{n=0}^{\infty} R_{jn}(r) \sin(\omega_{p_{jn}} t - \chi_{jn}) \sin(j\theta - \psi_{jn}) , \qquad (13)$$

where $\omega_{p_{jn}}$ represents the natural frequency of the plate when the mode shape consists of j nodal diameters and n nodal circles. The constant χ_{jn} and ψ_{jn} can be determined from the initial conditions. McLeod and Bishop[3] gave the natural frequencies of circular plates with various boundary conditions.

2.2   Vibration of a Spinning Membrane

The transverse vibration of a flexible membrane which is spinning at a constant angular velocity, and therefore under the action of centrifugal forces, can be described by the differential equation (Prescott[4]),

$$\rho \frac{\partial^2 w}{\partial t^2} = \frac{1}{r} \frac{\partial}{\partial r} \left(\sigma_{rr} \, r \, \frac{\partial w}{\partial r} \right) + \frac{\sigma_{\theta\theta}}{r^2} \frac{\partial^2 w}{\partial \theta^2} , \qquad (14)$$

where σ_{rr} and $\sigma_{\theta\theta}$ denote the radial and circumferential stresses respectively. It may be demonstrated (Lamb and Southwell[1]) that the membrane displacement has the solution,

$$w = R(r)\cos j\theta \, \cos\omega_m t \qquad (15)$$

where R(r) is given by,

$$R(r) = C_j \left(\frac{r}{b} \right)^j F\left(\alpha, \beta, \gamma, \frac{r^2}{b^2} \right) \qquad (16)$$

and $F(\bullet)$ is the Gauss or hypergeometric series,

$$F\left(\alpha, \beta, \gamma, \frac{r^2}{b^2} \right) = 1 + \frac{\alpha\beta}{1.\gamma} \left(\frac{r^2}{b^2} \right)^2 + \frac{\alpha(\alpha+1)\beta(\beta+1)}{(1)(2)\gamma(\gamma+1)} \left(\frac{r^2}{b^2} \right)^4 + ..., \qquad (17)$$

with the terms,

$$\alpha = \frac{1}{2}(j-m) , \quad \beta = \frac{1}{2}(j+m+2) , \quad \gamma = j+1 . \qquad (18 \text{ a,b,c})$$

The hypergeometric series is convergent when $r < b$. When $r = b$ the series will diverge unless $(\alpha + \beta) < \gamma$. However, from Eqs. (18a,b,c) it can be seen that $(\alpha + \beta) = \gamma$ and the series is therefore divergent unless it terminates. This can be accomplished by letting $m = j+2n$, where n is a positive integer.

Accordingly, the displacement solution of the spinning membrane is given by

$$
w = C_j \left(\frac{r}{b}\right)^j \left(1 - \frac{(n)(n+j+1)}{1(j+1)} \frac{r^2}{b^2}\right.
$$
$$
\left. + \frac{(n-1)\ (n)\ (n+j+1)\ (n+j+2)}{(1)(2)(j+1)(j+2)} \frac{r^4}{b^4} - \ ... \right) \cos j\theta \ \cos \omega_p t
$$

(19)

where the mode shapes are characterised by n nodal circles and j nodal diameters. The associated natural frequencies are given by,

$$
\frac{\omega^2_{m_{jn}}}{\tilde{\Omega}^2} = (j+2n)\ (j+2n+2)\ A - j^2 B
$$

(20)

where $A = \dfrac{1}{8}(3 + v)$, $B = \dfrac{1}{8}(1 + 3v)$, $\tilde{\Omega}$ is the rotational speed of the disc, and

v is Poisson's ratio. The ratio given in Eq. (20) is independent of the disc diameter and thickness.

2.3   Combined Effects of Centrifugal and Flexural Rigidity

When both the flexural rigidity and centrifugal forces are operative it is possible to obtain an approximate solution by the Rayleigh-Ritz method. The potential energy due to the centrifugal force and the flexural stiffness is given by

$$
V_m = \frac{h}{2} \int \int \left(\sigma_{rr}\left(\frac{\partial w}{\partial r}\right)^2 + \sigma_{\theta\theta}\left(\frac{\partial w}{r\partial\theta}\right)^2\right) rd\theta dr,
$$

(21)

and

$$
V_p = \frac{1}{2} D \int \int \left((\nabla^2 w)^2 + 2(1 - v)\left[\left(\frac{1}{r}\frac{\partial^2 w}{\partial r\partial\theta} - \frac{1}{r^2}\frac{\partial w}{\partial\theta}\right)^2\right.\right.
$$
$$
\left.\left. - \frac{\partial^2 w}{\partial r^2}\left(\frac{1}{r^2}\frac{\partial^2 w}{\partial\theta^2} - \frac{1}{r}\frac{\partial w}{\partial r}\right)\right]\right) rd\theta dr ,
$$

(22)

respectively. The expression for the kinetic energy term for the disc is

$$T = \frac{\rho h}{2} \iint \left(\frac{\partial w}{\partial t}\right)^2 r\,dr\,d\theta \ . \tag{23}$$

By taking the displacement of the disc as,

$$w = \phi(r,\theta) \cos\omega t \ , \tag{24}$$

The Rayleigh-Ritz method gives that,

$$\omega^2 = \frac{V_m(\phi) + V_p(\phi)}{T(\phi)} \ . \tag{25}$$

We note, for the fundamental mode, that when the selected displacement function $\phi(r,\theta)$ is exact, that $\dfrac{V_m(\phi)}{T(\phi)}$ gives an over-estimate of ω_m^2 , and $\dfrac{V_p(\phi)}{T(\phi)}$ gives an

over-estimate of ω_p^2 . Thus the estimate $\omega_m^2 + \omega_p^2$ (from Eqs. (20) and (7)) gives a

lower bound on the fundamental natural frequency of the disc with combined flexural and membrane effects.   This result is given in the article by Lamb and Southwell[1] and attributed to Southwell.

2.4   Travelling Waves and Critical Speeds

The response of a stationary, plane, axisymmetric disc when vibrating in its jth diameter mode can be written, from Eq. (13), in the form,

$$w = R_{jn}(r) \sin j\tilde\theta \cos \omega_{jn}t \ , \tag{26}$$

where the constants χ_{jn} and Ψ_{jn} have been set to zero without loss of generality. By using a trigonometric identity Eq. (26) can be posed in the form,

$$w = \frac{R_{jn}(r)}{2} \sin (j\tilde\theta - \omega_{jn}t) + \frac{R_{jn}(r)}{2} \sin (j\tilde\theta + \omega_{jn}t) \ , \tag{27}$$

which can be interpreted as two travelling waves of shape $\dfrac{R_{jn}(r)}{2} \sin j\tilde\theta,$ and

speed $\pm \dfrac{\omega_{jn}}{j}$. The forward travelling wave is in the direction of advancing $\tilde\theta$, and

the backward wave travels in the opposite direction.

It is observed that for a plane, axisymmetric disc, there will be no preferred orientation of the mode with respect to the disc. We consider the case when the disc is rotated with angular velocity $\tilde{\Omega}$ past a stationary harmonic load of frequency ω, and carries with it the mode shape described in equation (27). In that case the coordinate $\tilde{\theta}$ is fixed to the disc and the space-fixed coordinate is given by

$$\theta = \tilde{\theta} - \tilde{\Omega}t , \tag{28}$$

which appears to move backwards relative to the moving disc. By combining equations (27) and (28) the response can be written in (r,θ) coordinates such that,

$$w = \frac{R_{jn}(r)}{2} \sin (j\theta - (\omega_{jn} - j\tilde{\Omega})t) - \frac{R_{jn}(r)}{2} \sin(j\theta - (\omega_{jn} + j\tilde{\Omega})t) . \tag{29}$$

It is apparent that two resonance conditions are possible when the excitation frequency is given by $\omega_1 = \omega_{jn} - j\tilde{\Omega}$ and $\omega_2 = \omega_{jn} + j\tilde{\Omega}$. An interesting phenomena occurs when the speed of the disc is $\tilde{\Omega} = \dfrac{\omega_{jn}}{j}$. The frequency ω_1 then vanishes to zero and the backward travelling wave remains stationary in space aligned with the non-rotating force. This is the critical speed of the jn^{th} mode which becomes unstable under the action of a static load. It appears that the existence of the forward and backward travelling waves was first identified by Lamb and Southwell[1].

Typical frequency speed diagrams are shown in Figures 1 and 2 the frequency ω and

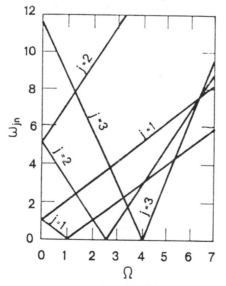

Figure 1: Frequency-speed diagram

speed $\tilde{\Omega}$ are normalised by division by the first critical speed. The lines crossing the diagram trace the values of ω_1 and ω_2 for each of the modes with nodal diameters. When the speed of rotation exceeds the critical speed, then the backward wave travels slower than the disc rotation speed and appears to be travelling forwards. The effect of centrifugal stiffening is shown in Figure 2.

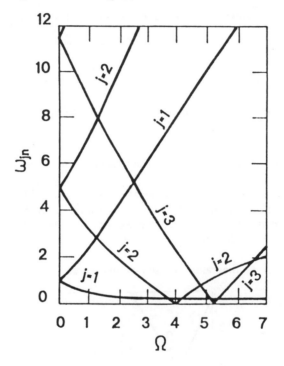

Figure 2: Frequency-speed diagram - centrifugal effects

2.5   Imperfect Discs

Tobias and Arnold[5] studied the vibration of stationary and rotating discs subjected to a transverse concentrated force containing a sinusoidal component. They considered discs which were imperfect in the sense that the nodal diameter modes had preferred orientations and two distinct natural frequencies. It is not unreasonable to assume that imperfections in discs occur frequently in practice and they are attributed to manufacturing errors, material inhomogenities or the presence of holes, slots and grooves. Such imperfections will destroy the axisymmetry of the disc, and their relative positions with respect to each other will determine the preferential configurations of the disc. Tobias and Arnold[5] considered the steady-state forced vibration of an imperfect disc to be given by a linear combination of the responses arising from each of the preferential modes in j nodal diameters. Thus,

$$w(r,\theta,t) = w_1(r,\theta,t) + w_2(r,\theta,t) , \tag{30}$$

where the subscripts 1 and 2 denote the two imperfect preferential modes. In the case of a stationary imperfect disc with a transverse load P cos ωt then the response w_1 can be determined in the form,

$$w_1 = \frac{K_{j1} \, P \, \cos j(\lambda - \alpha_1)}{\sqrt{((\omega_{j1}^2 - \omega^2) + C_{j1}^2 \, \omega^2)}} \, f_j(r) \, \cos j(\theta - \alpha_1) \cos(\omega t - \zeta_1) , \tag{31}$$

and the expression for w_2 can be written by exchanging the subscripts. The constant K_{j1} depends on the dimensions and mass distribution of the disc and the radial position of P. C_{j1} is a damping constant, λ denotes the circumferential position of P, and ζ_1 is a phase angle defined by

$$\zeta_1 = \tan^{-1} \frac{C_{j1} \, \omega}{(\omega_{j1}^2 - \omega^2)} . \tag{32}$$

Depending upon the location of the force it is possible to excite either, or both, of the preferential modes. If the force was situated at $\lambda = \alpha_2 \pm \left(\dfrac{m}{j}\right)\pi$, where m is an integer, then only configuration 1 would be excited, since the force was at a node of configuration 2. Similarly, if the force was situated at $\lambda = \alpha_1 \pm \left(m + \dfrac{1}{2}\right)\dfrac{\pi}{j}$, then only configuration 2 would be excited. When the load is positioned elsewhere both configuration are excited.

In the case of an imperfect disc that rotates past a stationary transverse load P cos ωt, at $\theta = 0$ when t = 0 (where the coordinate θ is attached to the rotating disc) the response w_1 is found to be,

$$w_1 = \frac{K_{j1} \, P \, f_j(r)}{2} \, \cos j\theta \left(\frac{\cos((\omega + j\tilde{\Omega})t + \eta_1)}{\sqrt{((\omega_{j1}^2 - (\omega + j\tilde{\Omega})^2)^2 + C_{j1}^2(\omega + j\tilde{\Omega})^2)}} \right.$$
$$\left. + \frac{\cos((\omega - j\tilde{\Omega})t - \mu_1)}{\sqrt{((\omega_{j1}^2 - (\omega - j\tilde{\Omega})^2)^2 + C_{j1}^2(\omega - j\tilde{\Omega})^2)}} \right), \tag{33}$$

where

$$\eta_1 = \tan^{-1}\left(\frac{C_{j1}(\omega+j\tilde{\Omega})}{\omega_{j1}^2 - (\omega+j\tilde{\Omega})^2}\right), \quad \mu_1 = \tan^{-1}\left(\frac{C_{j1}(\omega-j\tilde{\Omega})}{\omega_{j1}^2 - (\omega-j\tilde{\Omega})^2}\right). \quad (34\text{ a,b})$$

A similar expression is obtained for configuration 2 by exchanging suffix 2 for suffix 1, introducing sines for cosines and reversing the sign of the second term in Eq. (33). The frequency ω_{j1} and damping C_{j1} in Eqs. (33-34) represent the conditions at speed $\tilde{\Omega}$.

The motion of an imperfect rotating disc is complicated and several possibilities can arise. If damping is ignored, then the steady state solution, Eq. (30), can be written as

$$w = \pm (A_1 \cos j\theta \cos j\tilde{\Omega}t + A_2 \sin j\theta \sin j\tilde{\Omega}t) \quad (35)$$

where the plus sign is for $n\tilde{\Omega} < \omega_{j1} < \omega_{j2}$, and the minus sign is for $n\tilde{\Omega} > \omega_{j2} > \omega_{j1}$. Alternatively it is found that,

$$w = -A_1 \cos j\theta \cos j\tilde{\Omega}t + A_2 \sin j\theta \sin j\tilde{\Omega}t \quad (36)$$

when $\omega_{j1} < n\tilde{\Omega} < \omega_{j2}$.

We take, for instance, the case when $n\tilde{\Omega} < \omega_{j1} < \omega_{j2}$ and the amplitude $A_1 > A_2$. The motion of the disc is given by,

$$w = A_2 \cos j(\theta - \tilde{\Omega}t) + (A_1 - A_2)\cos j\theta \cos j\tilde{\Omega}t , \quad (37)$$

where the first term represents a backward travelling wave (relative to the moving disc) of amplitude A_2 and the second component is a fixed vibration term of amplitude (A_1-A_2). Numerical calculations by Tobias and Arnold[5] showed that by increasing the imperfection, the backward wave was reduced at the expense of an increase in the fixed wave component.

Extensive experimental investigations were reported by Tobias and Arnold[5]. The experimental set-up was quite elaborate and consisted of several major components. A high grade steel disc of 12 inches diameter was used in the investigations. It had a thickness of 0.0936" and a tolerance of ±0.0003" which made it a very thin disc in comparison with its other dimensions. Special efforts were made to check its flatness and alignment so that reliable measurements could be taken. Self-aligning ball race bearings were used to support the spindle on which the disc was mounted. The disc was driven by a D.C. motor though a V-shaped belt and the motor was capable of rotation speeds from 400 to 20,000 rpm. One flywheel was attached to the spindle and another to the motor in order to reduce speed variations. A large electromagnet linked to an amplifier and oscillator was used to excite the disc into vibration.

Experiments were conducted with steel cylinders of half an inch diameter and height which were glued to the disc surface in order to create preferred modal configurations. Good correlation for the frequency ratio of the preferential frequencies with theoretical values was found in the stationary disc case. When a static force was applied to the rotating disc (with masses attached), the disc was sufficiently imperfect to cause the two configurations to form two critical speed resonances. While the presence of nonlinearity ruled out a comprehensive explanation using the linear analysis for the resonances that occurred, the experimental results did show that the introduction of a large imperfection would bring about a reduction in the maximum amplitude of the stationary wave. For instance, the reported amplitudes of the critical speeds of the two configurations were only $^1/_9$ and $^2/_{13}$ of the value for the nearly perfect disc.

In a more recent study Strange and MacBain[6] carried out experiments on imperfect discs using holographic interferometry. They reported significant differences in the level of damping present in 'bladed' tuned and mistuned discs in the modes with 3 or 4 nodal diameters. There were unexplained differences in the level of damping in the two preferred configurations of the mode with 2 nodal diameters.

3.   Excitation by a Transverse Mass-Spring-Damper System

A seminal contribution on the dynamics of discs under the action of a transverse mass-spring-damper system was made in the two articles by Iwan and Stahl[7] and Iwan and Moeller[8] . They identified instabilities that were strongly associated with stiffness, damping (at supercritical speeds), modal interactions, and a terminal instability.

The work of Mote and his colleagues on the subject of moving loads on discs, spans three decades. This research work is exhaustive and exceptionally diverse in the treatment of many types of moving load-stationary disc problems. For example, Mote[9,10] examined the free vibration of initially stressed discs induced by rotation and in-plane thermal gradients, which increased the fundamental (or lowest mode) natural frequency. The applicability of Green functions to plates subjected to an arbitrary circumnavigating moving force of harmonic or other excitation form was suggested by Mote[11]. Mote[12] considered the problem of the moving load stability of a circular plate sliding freely on an inner floating central collar, with particular application to the design of circular saws.

Recently the vibration and parametric excitation of stationary asymmetric circular plates containing small imperfections and subjected to moving loads, was discussed by Yu and Mote[13]. The mechanism of instability of such a disc was addressed by Shen and Mote[14] who successively applied classical nonlinear methods like the techniques of slowly varying amplitudes and phase averaging, the method of successive approximation (Valeev[15]) and the Krylov-Bogoliubov-Mitropolsky method to the rotating mass-spring-damper load. The parametric resonance of a stationary circular plate with viscoelastic inclusion subjected to a rotating spring was described by Shen and Mote[16]. A new perturbation method was developed by Shen and Mote[17] based on Valeev's work and demonstrated the parametric resonance of an asymmetric stationary circular plate under the action of a rotating spring. Shen and Mote[18] used the multiple scales approach for an analysis of instability in stationary discs due to a rotating mass.

3.1   Stationary Disc with a Rotating Mass-Spring-Damper System

The equation of motion for a stationary classical annular plate excited by a rotating mass-spring-damper system, illustrated in Figure 3, can be written in the form (Iwan and Stahl[7]),

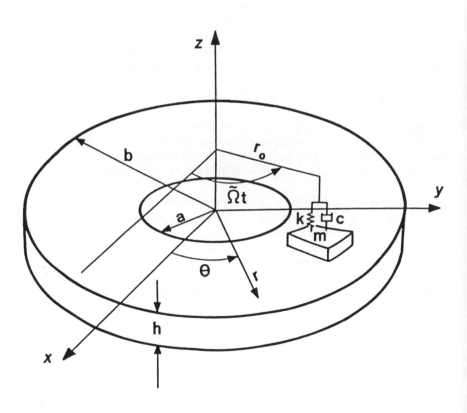

Figure 3: Transverse m-k-c system

$$\rho \frac{\partial^2 w}{\partial t^2} + \frac{Eh^2}{12(1-v^2)} \left(\frac{\partial^2}{\partial r^2} + \frac{1}{r} \frac{\partial}{\partial r} + \frac{1}{r^2} \frac{\partial^2}{\partial \theta^2} \right)^2 w$$

$$= - \frac{1}{hr_o} \delta(\theta - \tilde{\Omega}t)\delta(r-r_o) \left\{ m \left(\frac{\partial}{\partial t} + \tilde{\Omega} \frac{\partial}{\partial \theta} \right)^2 + c \left(\frac{\partial}{\partial t} + \tilde{\Omega} \frac{\partial}{\partial \theta} \right) + k \right\} w \, . \tag{38}$$

The transverse load system consists of mass, m, stiffness, k, and damping, c. The mass slides without separation and traces a circle at radius r_o on the surface of the disc. Thus the deflection of the mass is identically the transverse deflection of the plate at the position defined by the two Dirac delta functions. The transverse motion of the mass is a function of both time and angular position.

For the solution of equation (38), Iwan and Stahl[7] assumed an eigenfunction expansion solution for the disc of a form similar to that for a free disc (c.f. Eq. (13)) and given by,

$$w(r,\theta,t) = \sum_{j=0}^{\infty} \sum_{n=0}^{\infty} T_{jn}(t) \sin(j\theta - \psi_{jn}(t)) R_{jn}(r) \, . \tag{39}$$

The coordinates (r,θ) fixed to the plate were transformed to the system,(r, φ) fixed to the rotating mass-spring-damper arrangement. This transformation allows the equations to be solved more easily, as shall be shown later. The transverse displacement of the plate in the new coordinate system can be expressed in the form of a truncated eigenfunction expansion,

$$w(r,\varphi,t) = \sum_{j=0}^{J} \sum_{n=0}^{N} (A_{jn}(t)\cos j\varphi + B_{jn}(t) \sin j\varphi) R_{jn}(r) \, , \tag{40}$$

where the time-dependent amplitudes are given by,

$$A_{jn}(t) = T_{jn}(t) \sin(j\tilde{\Omega}t - \psi_{jn}(t)) \, , \quad B_{jn}(t) = T_{jn}(t) \cos(j\tilde{\Omega}t - \psi_{jn}(t)) \, . \tag{41 a,b}$$

Transforming Eq. (38) into the rotating frame gives,

$$\rho \left(\frac{\partial^2 w}{\partial t^2} - 2\tilde{\Omega} \frac{\partial^2 w}{\partial t \partial \varphi} + \tilde{\Omega}^2 \frac{\partial^2 w}{\partial \varphi^2} \right) + \frac{Eh^2}{12(1-v^2)}$$

$$\times \left(\frac{\partial^2}{\partial r^2} + \frac{1}{r} \frac{\partial}{\partial r} + \frac{1}{r^2} \frac{\partial^2}{\partial \varphi^2} \right)^2 w = - \frac{1}{hr_o} \delta(r-r_o)\delta(\varphi) \left(m \frac{\partial^2 w}{\partial t^2} + c \frac{dw}{dt} + kw \right) . \tag{42}$$

By multiplying Eq. (42) by $r R_{lp}(r)\sin l\varphi$, integrating over the area of the plate and using the orthogonality property of the eigenfunctions of the free disc, it is found that,

$$\frac{d^2 B_{lp}}{dt^2} + 2l\tilde{\Omega} \frac{dA_{lp}}{dt} + \left(\omega_{lp}^2 - l^2\Omega^2\right)B_{lp} = 0 , \qquad l, p = 0,1,... \qquad (43)$$

Similarly, multiplying Eq. (42) by $r\, R_{lp}(r)\cos l\varphi$, integrating over the area of plate and applying the orthogonality relationships yields,

$$\frac{d^2 A_{lp}}{dt^2} - 2l\tilde{\Omega} \frac{dB_{lp}}{dt} + \left(\omega_{lp}^2 - l^2\tilde{\Omega}^2\right)A_{lp} = -\sum_{j=0}^{J} \sum_{n=0}^{N} R_{jn}(r_o)R_{lp}(r_o)$$

$$\times \left(\frac{m}{\rho h} \frac{d^2 A_{jn}}{dt^2} + \frac{c}{\rho h} \frac{dA_{jn}}{dt} + \frac{k}{\rho h} A_{jn}\right), \qquad l, p = 0,1,.... \qquad (44)$$

Eqs. (43) and (44) can be arranged in state-space form,

$$\dot{x} = D\, x \qquad (45)$$

where

$$x(t) = \left\{ \begin{array}{c} A_{jn}(t) \\ \dot{A}_{jn}(t) \\ B_{jn}(t) \\ \dot{B}_{jn}(t) \end{array} \right\} ; \; j = 0,...,J ; \; n = 0,...,N . \qquad (46)$$

when $x(t) = X\, e^{\lambda t}$ then an eigenvalue problem can be defined as,

$$(D - \lambda I)X = 0 \qquad (47)$$

so that whenever the real part of λ is positive an instability can occur.

It is straightforward to construct a single mode approximation. In this case the non-trivial solution to Eq. (47) is

$$(1 - \gamma_{jn}\bar{m})\lambda^4 + \left[(2 + \gamma_{jn}\bar{m}) + (2 - \gamma_{jn}\bar{m})\left(\frac{j\bar{\Omega}}{\omega}\right)^2 + \frac{\gamma_{jn}\bar{k}}{\omega_{jn}^2}\right]\lambda^2 + \left[1 - \left(\frac{j\bar{\Omega}}{\omega_{jn}}\right)^2\right]$$

$$\times \left[1 + \frac{\gamma_{jn}\bar{k}}{\omega_{jn}} - \left(\frac{j\bar{\Omega}}{\omega_{jn}}\right)^2\right] = 0 \qquad (48)$$

where the non-dimensional parameters are defined as,

$$\bar{m} = \frac{m}{M}, \; \bar{k} = \frac{k}{M\omega_{10}^2}, \; M = \rho h \bar{a}, \quad (49\text{ a,b,c})$$

$$\bar{a} = \pi(b^2 - a^2), \; \bar{\omega}_{jn} = \frac{\omega_{jn}}{\omega_{10}}, \; \bar{\Omega} = \frac{\tilde{\Omega}}{\omega_{10}}, \; \gamma_{jn} = \bar{a}R_{jn}^2(r_o) . \quad (50\text{ a,b,c,d})$$

The dynamics of the disc are found to be unstable in the speed range,

$$\frac{\bar{\omega}_{jn}}{j} \leq \bar{\Omega} \leq \frac{\bar{\omega}_{jn}}{j} \sqrt{\left(1 + \frac{\gamma_{jn}\bar{k}}{\bar{\omega}_{jn}^2}\right)} . \quad (51)$$

It is evident that the instability occurs just above the critical speed of the jn^{th} mode, and

the width of the instability depends upon the stiffness term $\dfrac{\gamma_{jn}\,\bar{k}}{\bar{\omega}_{jn}^2}$. It was referred to

by Iwan and Stahl[7] as the 'stiffness instability'.

The second type of instability, which they termed 'terminal instability' is a function of

both the $\dfrac{\gamma_{jn}\,\bar{k}}{\bar{\omega}_{jn}^2}$ and $\gamma_{jn}\bar{m}$ terms. For this instability, it is necessary to carry out

numerical computations. Some results of case studies can be found in their paper. As is the case with stiffness instabilities, terminal instabilities exist only for rotational speeds above the lowest critical speed.

The numerical results obtained by Iwan and Stahl[7] for higher order approximations showed instabilities that arose from modal interactions. Finally, quite contrary to expectations or intuition, their numerical results revealed that including a viscous damper in the load has the effect of destabilizing the system at supercritical rotational speeds.

The frequency speed diagram reproduced from Iwan and Stahl's paper in Figure 4 shows

quite clearly the various instability regions where $Re(\lambda) > 0$. The first instability (just

above the first critical speed) is the stiffness instability and the terminal instability is shown to the extreme right of the diagram. The other instabilities appear to involve modal interaction.

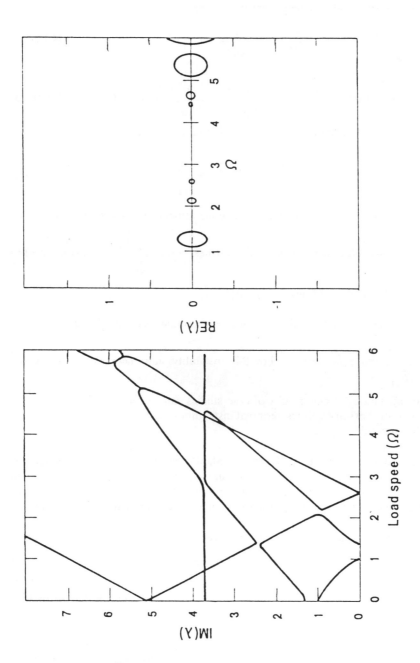

Figure 4:   Frequency-speed diagram showing destabilisation by the transverse m-k-c system

3.2 *Rotating Disc with a Stationary Mass-Spring-Damper System*

Iwan and Moeller[8] defined the differential operator L to be,

$$
L(w) = \frac{\rho h}{2} \left(\frac{\partial^2 w}{\partial t^2} - 2\tilde{\Omega} \frac{\partial^2 w}{\partial t \partial \theta} + \tilde{\Omega}^2 \frac{\partial^2 w}{\partial \theta^2} \right) + \frac{D}{2} \nabla^4 w
$$

$$
- \frac{h}{2r} \frac{\partial}{\partial r} \left(r \, \sigma_{rr} \frac{\partial w}{\partial r} \right) - \frac{h}{2r^2} \, \sigma_{\theta\theta} \frac{\partial^2 w}{\partial \theta^2} ,
$$

(52)

so that the equation of motion may be written in the coordinates (r,θ) attached to the rotating disc as,

$$
L(w) = P(w), \quad w = w(r,\theta,t) , \tag{53 a,b}
$$

where

$$
P(r,\theta,t) = \frac{-H(r,\theta)}{A_L} \left(m \frac{\partial^2 w}{dt^2} + c \frac{\partial w}{dt} + kw \right). \tag{54}
$$

The function $H(r,\theta)$ is defined as

$$
H(r,\theta) = \begin{cases} 1 \ \text{ if } (r,\theta) \in D_L \\[2mm] 0 \ \text{ if } (r,\theta) \notin D_L \end{cases} , \tag{55}
$$

where D_L represents the loaded surface of net area A_L.

An approximate solution to Eq. (53a) is sought, for a vanishingly small load area, in the form,

$$
w(r,\theta,t) \approx \sum_{j=1}^{J} \left(A_j(t)\cos j\theta + B_j(t)\sin j\theta \right) R_j\!\left(\frac{r}{b} \right). \tag{56}
$$

In this case an eigenfunction expansion without nodal circles was evaluated, and the radial function was of the polynomial type,

$$
R_j\!\left(\frac{r}{b} \right) = C_j \left(1 + \beta_j\!\left(\frac{r}{b} \right)^2 \right)\!\left(\frac{r}{b} \right)^j , \tag{57}
$$

where the constant β_j is chosen to minimise the potential energy of the disc and C_j is a normalisation constant.

By the application of Galerkin's method,

$$\int_0^b \int_0^{2\pi} \left(L(w) - P(w)\right)\Psi_{jn} \, r \, dr \, d\theta = 0 \qquad (58)$$

and using the orthogonality property of the eigenfunctions, a set of equations similar in form to Eqs. (43) and (44) are found.

It follows that a similar state-space eigenproblem can be formed as in the preceding section. Indeed the results reached by Iwan and Moeller[8] for numerical simulations are similar to the previous discussion on the instabilities for a stationary plate subject to a rotating mass-spring-dashpot load. They reported that the principal effect of the inclusion of the centrifugal stresses was to stiffen the disc.

Hutton et al.[19] discussed the number and configuration of stationary spring guides for a rotating clamped-free annular disc (while subjected to stationary lateral, concentrated, cutting forces, to improve the design of high-speed circular saws. Specifically they wanted to increase the lowest critical speed by introducing the spring guides. They used a Galerkin approach to obtain approximate solutions to the dynamical equation and found it possible to distribute and locate the springs that can significantly alter the frequency speed characteristics.

The dynamic response of read-write floppy discs subjected to axial sinusoidal movements, as experienced by computers on board ships, aircraft and other moving platforms, was considered by Jiang et al.[20] A floppy disc drive system with a spring stabiliser and including the effects of air film surrounding the disc (modelled as a stiffness, and involving use of the Navier-Stokes equation) was investigated by Chonan et al.[21] in their study of high-speed disc systems suitable for video camera operation.

3.3   Instability Mechanisms

The response of a stationary disc due to the separate effects of a rotating spring, mass or damper was studied by Shen and Mote[14] In the case of a rotating spring Eq.(38) can be reduced to,

$$\ddot{q} + \left[B + \kappa \sum_{j=-\infty}^{\infty} H^{(j)} e^{ij\Omega t}\right] q = 0, \qquad (59)$$

which is of the same form as the equation considered by Valeev[15]. To obtain Eq. (59) it is necessary to transform the differential equation of motion into the modal co-ordinates of the disc. The transformation is explained in Section 5.1. By following Valeev's method, Shen and Mote[14] were able to show that the instability regions occurred at,

$$\bar{\Omega} = (\omega_{jn} + \omega_{lp} + \sigma)/g, \quad g=|j \pm l| \neq 0, \quad j,n,l,p = 0,1,2,..., \qquad \text{(60 a,b)}$$

where σ is a detuning parameter. Whenever $(j,n)=(l,p)$ the instability is a superharmonic parametric resonance, otherwise it is a combination parametric resonance of the summation type. Neither can exist below the first critical speed. The two cases are the stiffness and modal interaction instabilities referred to by Iwan and Stahl[7].

To investigate the effect of a rotating mass Shen and Mote[14] used the method of slowly varying phase and amplitude. For a single mode approximation they used the modal co-ordinates equation,

$$\frac{d^2q}{d\tau^2} + \frac{\omega_{jn}^2}{j^2\tilde{\Omega}^2} q = -\bar{\mu}\left\{(1 + \cos 2\tau)\frac{d^2q}{dt^2} - 2\sin 2\tau \frac{dq}{dt} - (1 + \cos 2\tau)q\right\}, \quad (61)$$

where

$$\bar{\mu} = \frac{m}{\rho h b^2} R_{jn}^2(r_o), \qquad \tau = j\tilde{\Omega}t . \tag{62 a,b}$$

If $\bar{\mu}$ is small and $\dfrac{\omega_{jn}^2}{j^2\tilde{\Omega}^2} \sim O(\bar{u})$, then it follows that $\dfrac{d^2q}{d\tau} \sim O(\bar{\mu})$, and it is

permissible to say that the displacement $q(\tau)$ is slowly varying from the initial

disturbances $q(0)$ and $\dot{q}(0)$.

By taking the temporal average of equation (61) over the range $\dfrac{-\pi}{2} < \tau < \dfrac{\pi}{2}$ it is

found that,

$$\frac{d^2q}{d\tau^2} + \frac{\omega_{jn}^2}{j^2\tilde{\Omega}^2} q = -\bar{\mu}\left\{\frac{d^2q}{d\tau^2} - q\right\} . \tag{63}$$

An instability occurs when $\dfrac{\omega_{jn}^2}{j^2\tilde{\Omega}^2} < \bar{u}$ such that the out-of-plane centrifugal force upon

the mass exceeds the elastic restoring force of the plate (Shen and Mote[14]). This corresponds to the terminal instability of Iwan and Stahl[7]. It is not associated with the modes without nodal diameter modes because they have no out-of-plane component of the centrifugal force.

An interpretation of the instability caused by a rotating viscous damper is provided by a one-mode analysis. From Eq.(27) we find that the backward wave response is given by,

$$ w = \frac{R_{jn}(r)}{2} \sin(j\tilde{\theta} - \omega_{jn} t) , \qquad (64) $$

and the velocity response under the damper is,

$$ \left[\frac{\partial w}{\partial t} \right]_{\theta = \tilde{\Omega} t, r = r_o} = -\omega_{jn} \frac{R_{jn}(r_o)}{2} \cos(j\tilde{\Omega} - \omega_{jn}) t . \qquad (65) $$

The damping force of the dashpot on the disc becomes,

$$ F_D = -c \left[\frac{\partial w}{\partial t} + \tilde{\Omega} \frac{\partial w}{\partial \theta} \right]_{\theta = \tilde{\Omega} t, r = r_o} , $$

or

$$ F_D = -c \frac{R_{jn}(r_o)}{2} \left[(-\omega_{jn} + j\tilde{\Omega}) \cos(j\tilde{\Omega} t - \omega_{jn} t) \right] . \qquad (66) $$

When the critical speed is exceeded   (i.e. $\tilde{\Omega} > \omega_{jn}/j$) then it is observed that the velocity under the damper and the force, F_D, are in phase, so that a negative damping of the backward wave ensues.

4.   Follower Force Friction Models

The non-conservative forces that alter their direction to track the displacements of a system are called follower forces.   Follower forces are a well known source of asymmetry in stiffness matrices and are considered to be responsible for flutter instabilities (Huseyin[22]) in a wide variety of mechanical systems.

4.1   *Follower Force Analysis in Brake Design*

It appears that the first use of a follower force analysis to determine instabilities in disc brakes was due to North[23]. The disc was modelled as a rigid body with a translational and a rotational degree-of-freedom. Springs were introduced to represent the translational and rotational stiffnesses of the disc and the brake pads. Frictional follower forces were considered to act at the contact surfaces between the pads and the disc, and a simple expression was obtained to determine the regions of unstable vibration in the disc.

Millner[24] extended North's model by introducing further degrees of freedom to represent the dynamics of the pad and the calliper. Murakami *et al.*[25] developed a seven degree-of-freedom model with a negative μ velocity characteristic for the friction.   Recently

Brookes *et al.*[26] have developed a sophisticated multi-degree-of-freedom lumped model with frictional follower loads. It is assumed that the disc vibrates in a particular mode having nodal diameters, and that the equivalent mass and moment of inertia of the disc can be obtained for each mode by means of a physical experiment.

4.2 Sensitivity Analysis

Chen and Bogy[27] investigated the sensitivity of the eigenvalues of a state space model to small changes in the various parameters of a spinning disc with a stationary load system. They included a frictional follower force as well as the pitching mass, stiffness and damping terms all of which are additional to the transverse m-k-c system that has been considered previously. Since Chen and Bogy's analysis produced entirely predictable results for the mass, stiffness and damping sensitivities of the eigenvalues, we focus our attention upon the interesting result they obtained for the sensitivity of the eigenvalues with respect to the friction force.

The equation of motion for the system considered by Chen and Bogy[27] belongs to a class of gyroscopic problems described by,

$$M \ddot{u} + G \dot{u} + K u = 0 , \qquad (67)$$

where the stiffness operator contains a term arising from the friction force f_θ,

$$K = ...+ \delta(r - r_o) \, \delta(\theta) \, \frac{f_\theta}{r_o^2 \rho h} \, \frac{\partial}{\partial \theta} . \qquad (68)$$

Eq.(67) can be written in state-space notation as,

$$A \dot{x} = B x , \qquad (69)$$

where

$$A = \begin{bmatrix} M & 0 \\ 0 & K \end{bmatrix}, \quad B = \begin{bmatrix} -G & -K \\ K & 0 \end{bmatrix}, \quad x = \left\{ \begin{matrix} \dot{u} \\ u \end{matrix} \right\} . \qquad (70 \text{ a,b,c})$$

For the freely spinning disc, the state space eigenvalues, λ_{jn} , are purely imaginary and the eigenfunctions are assumed to take the form,

$$u_{jn} = R_{jn}(r) \, e^{\pm ij\theta} , \qquad (71)$$

where $+ij\theta$ is a backward travelling wave, $-ij\theta$ is a forward travelling wave, and

$$x_{jn} = \left\{ \begin{matrix} i \, \omega_{jn} \\ 1 \end{matrix} \right\} u_{jn} . \qquad (72)$$

The sensitivity of an eigenvalue, λ_{jn}, to a small change in a parameter p_k (in the neighbourhood of p_k^0) is given by,

$$
\frac{\partial \lambda_{jn}}{\partial p_k} = \left. \frac{ - < x_{jn}, \left(\lambda_{jn} \dfrac{\partial A}{\partial p_k} - \dfrac{\partial B}{\partial p_k} \right) x_{jn} > }{ < x_{jn}, A\, x_{jn} > } \right|_{p_k = p_k^0} , \qquad (73)
$$

In the case of $p_k = f_\theta^*$ $\left(\text{where } f_\theta^* = \dfrac{f_\theta}{r_o^2 \rho h} \right)$ then by combining Eqs. (70-73) it is found that,

$$
\frac{\partial \lambda_{jn}}{\partial f_\theta^*} = \pm \left. \frac{ j\, \omega_{jn}\, R_{jn}^2(r_o) r_o }{ < x_{jn}, A\, x_{jn} > } \right|_{f_\theta^* = f_\theta^{*0}} , \qquad (74)
$$

where the positive solution applies to the forward wave and the negative solution to the backward wave. It is seen that the presence of an arbitrarily small friction destabilises the forward wave whilst making the backward wave more stable. The magnitude of the derivative is proportional to the number of nodal diameters.

The above result confirmed the findings of earlier research by Ono et al.[28] who also constructed a state-space model and found eigenvalues with positive real parts always to be present whenever a frictional follower load was applied. Ono et al.[28] used a truncated Fourier series to represent the circumferential component of the eigenmode, whilst the radial component was determined from finite elements with cubic shape functions.

4.3   Distributed Frictional Load

Mottershead and Chan[29] analysed the problem of disc with a frictional follower force that was distributed over part of the surface of the disc. They used twenty-noded brick-type finite elements in the construction of the mathematical model and found that the follower force lead to the appearance of an unsymmetric component in the stiffness matrix. The unsymmetric element stiffness matrix (due to the follower load) was found to have the form,

$$a^e = -\int^A N^T f_\theta \, B \, r dr d\theta \qquad (75)$$

where **N** denotes the matrix of shape functions and **B** is the matrix of shape function

derivatives $\dfrac{\partial N}{r \partial \theta}$.

In a numerical simulation the eigenvalues were found to coalesce (thereby indicating a flutter instability) when the friction force was above a threshold value. The higher eigenmodes were found to be more reluctant to flutter than the lower ones, and the modes without nodal diameters were not found to flutter. A physical interpretation of the flutter instability was given in terms of the real and imaginary modes, with equal numbers of nodal diameters, that were held simultaneously in the disc. The nodes of the real eigenmode were in the same position on the disc as the antinodes of the imaginary eigenmode and vice-versa. The flutter instability was shown to result in waves that travelled forwards and backwards around the disc.

The result of Mottershead and Chan[29] is at variance with the findings of Chen and Bogy[27] and Ono *et al.*[28] in respect of the freshold friction force. This may possibly be due to the distributed friction load in Mottershead and Chan's analysis; the friction load acts at a point in the other investigations. Ono *et al.*[28] assumed that the displacement of the disc under the friction load could be projected upon a truncated set eigenvectors of the plate, whereas no such assumption was made by Mottershead and Chan[29].

Mottershead and Chan[29] considered the case when both the disc and the friction force were stationary. This seems to be reasonable in view of Chen and Bogy's result that the friction induced instability did not depend upon rotating speed. In Section 5 we consider the friction-induced parametric resonances in discs, for which it is essential to include rotational effects.

4.4 *Friction with a Negative μ-Velocity Characteristic*

It is well known that a dynamic coefficient of friction that reduces with velocity can lead to a negative damping effect. For the purpose of illustration we consider the single degree-of-freedom mass/spring system. The oscillating mass rubs against a rough moving surface and a negative μ-velocity relationship prevails at the interface. The arrangement is illustrated in Figure 5, and the friction resistance to motion is found to be,

$$f_N = \mu_o N \, (sgn(\dot{x} - v) \times 1 - \sigma(\dot{x} - v)) \, , \qquad (76)$$

where σ denotes the negative slope of μ with the relative velocity $(v - \dot{x})$. The equation of motion of the system is given by,

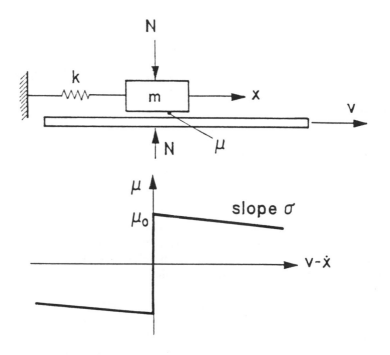

Figure 5: Negative μ-velocity characteristic

$$m\ddot{x} - \mu_\sigma N\sigma\dot{x} - kx = \mu_\sigma N(sgn(v - \dot{x})\times 1 - \sigma v) \ . \tag{77}$$

Since the term $-\mu_0 N\sigma$ is equivalent to negative damping it is clear that the system is capable of performing unstable oscillations without bound. The instability does not depend upon the speed v, normal force N or the static coefficient of friction μ_0. In Section 6 we consider the follower force friction-induced parametric vibrations of a disc with a negative μ-velocity relationship.

5   Friction-Induced Parametric Resonances

Shen[30] analysed the parametric resonances in a stationary disc with a discrete, rotating mass-spring-damper system by the method of multiple time scales (Nayfeh and Mook[31]). Chan et al.[32] carried out a similar analysis but included the effect of a frictional follower force. They found many parametric resonances in the sub-critical speed range.

Ouyang et al.[33] considered a distributed elastic system that rotated around a stationary disc. The equations were generated from a finite element analysis and solved by using a symbolic code for the multiple time scales analysis. Mottershead *et al.*[34] included the effect of friction that was considered to act over the contact surface.

5.1   *Discrete Transverse Load*

The motion of a stationary classical annular disc excited by a rotating mass-spring-damper system is governed by Eq. (38). When a frictional follower load is included, as shown in Figure 6, then the equation may be extended to give,

$$
\rho h \frac{\partial^2 w}{\partial t^2} + D \nabla^4 w = -\frac{1}{r_o} \delta(\theta - \tilde{\Omega}t)\delta(r - r_o) \left[m \left(\frac{\partial}{\partial t} + \tilde{\Omega} \frac{\partial}{\partial \theta} \right)^2 w \right.
$$
$$
\left. + c \left(\frac{\partial}{\partial t} + \tilde{\Omega} \frac{\partial}{\partial \theta} \right) w + kw - f_\theta \frac{\partial w}{r d\theta} \right], \quad t \in [0, \infty), \ r \in [a,b], \ \theta \in [0, 2\pi] ,
$$

(78)

where ∇^4 is the biharmonic differential operator and D represents the flexural rigidity of the disc.

The transverse displacements can be given in terms of a Fourier/Bessel series of eigenfunctions of the disc. Thus,

$$
w = \sum_{n=0}^{\infty} \sum_{j=-\infty}^{\infty} \psi_{jn}(r,\theta) q_{jn}(t) ,
$$

(79)

with the complex eigenfunctions given according to,

$$
\psi_{jn}(r,\theta) = \frac{1}{\sqrt{(\rho h b^2)}} R_{jn}(r) e^{ij\theta} .
$$

(80)

The shape of the $(j,n)^{th}$ mode in the circumferential direction is sinusoidal. The sum on nodal diameters in Eq.79 includes negative numbered diameters because the eigenfunction in Eq. (80) uses the exponential to describe the mode shape in the θ direction. $R_{jn}(r)$ defines the shape of the $(j,n)^{th}$ mode along a radius and is given by the linear combination of Bessel functions satisfying the boundary conditions and the normalising equations,

$$
\int_a^b \int_0^{2\pi} \rho h \, \psi_{jn} \, \bar{\psi}_{LP} \, r \, dr \, d\theta = \delta_{nP} \, \delta_{jL} ,
$$

(81)

$$
\int_a^b \int_0^{2\pi} D\nabla^4 \, \psi_{jn} \, \bar{\psi}_{LP} \, r \, dr \, d\theta = \omega_{jn}^2 \, \delta_{nP} \, \delta_{jL} .
$$

(82)

The overbars denote complex conjugation, and the axisymmetry of the disc requires that,

$$
\omega_{jn} = \omega_{-j,n}, \qquad q_{j,n}(t) = \bar{q}_{-j,n}(t), \qquad R_{jn}(r) = R_{-j,n}(r).
$$

(83 a,b,c)

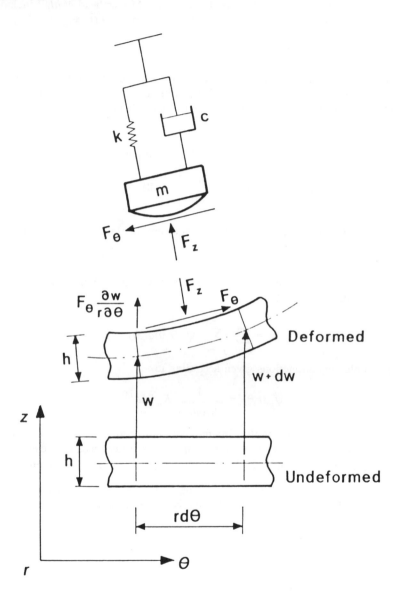

Figure 6: Frictional follower force

By multiplying equation (78) by ψ_{LP}, integrating over the disc area, and using equations (81) and (82), an equation describing the dynamics of the disc system in modal coordinates is obtained. Thus,

$$\frac{d^2 q_{LP}}{d\tau^2} + \beta_{LP}^2 \, q_{LP} = -\sum_{n=0}^{\infty} \sum_{j=-\infty}^{\infty} R_{jn}(r_0) R_{LP}(r_0) e^{i(j-L)\Omega\tau}$$

$$\times \left\{ \frac{m}{\rho h b^2} \left(\frac{\partial}{\partial \tau} + ij\Omega \right)^2 q_{jn} + \frac{c}{\rho h b^2 \omega_{cr}} \left(\frac{\partial}{\partial \tau} + ij\Omega \right) q_{jn} \right. \tag{84}$$

$$\left. + \frac{k}{\rho h b^2 \omega_{cr}^2} q_{jn} - ij \frac{f_\theta}{\rho h b^2 r_0^2 \omega_{cr}^2} q_{jn} \right\} ,$$

$$P = 0,1,2,...;L = 0,\pm1,\pm2,...$$

where the non-dimensional terms,

$$\tau = \omega_{cr} \, t, \quad \beta_{LP} = \frac{\omega_{LP}}{\omega_{cr}}, \quad \Omega = \frac{\tilde{\Omega}}{\omega_{cr}}, \tag{85 a,b,c}$$

have been used, and ω_{cr} is the lowest critical speed given by,

$$\omega_{cr} = \min \left\{ \frac{\omega_{LP}}{L} \; ; P = 0, \, 1, \, 2, \, ...; \, L = 1, \, 2, \, 3,... \right\} . \tag{86}$$

By following the usual multiple scales method and including terms to the first order in the perturbation parameter, ε, the equations that govern the dynamics of the system to $O(\varepsilon)$ can be determined. Chan et al.[32] give the $O(\varepsilon)$ equation as,

$$(D_0^2 + \beta_{LP}^2) \, q_{LP}^{(1)} = -i2\beta_{LP} \, D_1[A_{LP} \, e^{i\beta_{LP}\tau} - B_{LP} \, e^{-i\beta_{LP}\tau}]$$

$$- \sum_{n=0}^{\infty} \sum_{j=-\infty}^{\infty} R_{jn}(r_0) \, R_{LP}(r_0) \, e^{i(j-L)\Omega\tau} \, [D_{jn}^+ \, A_{jn} \, e^{i\beta_{jn}\tau} + D_{jn}^- \, B_{jn} \, e^{-i\beta_{jn}\tau}] , \tag{87}$$

where,

$$D_{jn}^+ = -\gamma \, (C_{jn}^+)^2 + i(\xi C_{jn}^+ - jf) + \kappa , \tag{88}$$

$$D_{jn}^- = -\gamma \, (C_{jn}^-)^2 - i(\xi C_{jn}^- + jf) + \kappa , \tag{89}$$

$$C_{jn}^* = \beta_{jn} + j\Omega \ , \tag{90}$$

and
$$C_{jn}^- = \beta_{jn} - j\Omega \ . \tag{91}$$

In equations (87) - (91)

$$\varepsilon\gamma = \frac{m}{\rho h b^2} \ , \qquad \varepsilon\xi = \frac{c}{\rho h b^2 \omega_{cr}} \ , \tag{92 a,b}$$

$$\varepsilon\kappa = \frac{k}{\rho h b^2 \omega_{cr}^2} , \qquad \varepsilon f = \frac{f_\theta}{\rho h b^2 r_0 \omega_{cr}^2} \ . \tag{93 a,b}$$

By considering the terms on the right-hand-side of Eq. (87) which are resonant to the $q_{LP}^{(1)}$ mode, it can be seen that parametric resonances can arise under the following conditions:

(i)   as **superharmonics** when $2L\Omega = 2\beta_{LP}$, $L > 0$;

(ii)   as **combination resonances** when,

$$(J-L)\Omega = \beta_{LP} + \beta_{JN}, \ J > L \geq 0,$$

$$(J+L)\Omega = \beta_{LP} + \beta_{JN}, \ J > 0 \ , L \geq 0,$$

$$(J-L)\Omega = \beta_{JN} - \beta_{LP}, \ J > L \geq 0,$$

$$(J+L)\Omega = \beta_{JN} - \beta_{LP}, \ J > 0, \ L \geq 0 \ ;$$

(iii)   as resonances which show no functional relationship between the speed of rotation and the natural frequencies of the disc, and hence can occur remotely from the harmonic and combinational effects.

We consider (by way of an example) the combination resonance at $(J - L)\Omega = \beta_{JN} - \beta_{LP} + \varepsilon\sigma$ where σ represents the detuning. The secular terms are eliminated when,

$$- i2\beta_{LP} D_1 [A_{LP} e^{i\beta_{LP}\tau} - B_{LP}e^{-i\beta_{LP}\tau}] - R_{LP}^2(r_0) \left\{ D_{LP}^+ A_{LP} e^{i\beta_{LP}\tau} + D_{LP}^- B_{LP} e^{-i\beta_{LP}\tau} \right\}$$

$$- R_{JN}(r_0) R_{LP}(r_0) e^{i(-\beta_{LP} + \varepsilon\sigma)\tau} \left\{ D_{JN}^- B_{JN} \right\} = 0 .$$

(94)

By separating the coefficients of $e^{i\beta_{LP}\tau}$ and $e^{-i\beta_{LP}\tau}$, and replacing the subscripts LP with JN to represent the other mode in the combination, the conditions under which $A_{LP}(T_1)$ and $B_{LP}(T_1)$ (and $A_{JN}(T_1)$ and $B_{JN}(T_1)$), $T_1 = \varepsilon\tau$, become unstable can be established. The detuning, σ, and hence the transition curve is defined when the imaginary part of the characteristic exponent, λ, is set to zero and,

$$\left(\lambda + \frac{R_{LP}^2(r_0)}{2\beta_{LP}} D_{LP}^- \right) \left(\lambda - \sigma + \frac{R_{JN}^2(r_0)}{2\beta_{JN}} D_{JN}^- \right) - \left(\frac{R_{JN}^2(r_0) R_{LP}^2(r_0)}{4\beta_{JN}\beta_{LP}} D_{JN}^- D_{LP}^- \right) = 0 .$$

(95)

The range of f for which unstable behaviour can occur is obtained from,

$$\xi C_{LP}^+ - Lf < 0 ,$$

(96)

and,

$$\xi C_{JN}^+ - Jf < 0 .$$

(97)

A thorough analysis of the parametric resonances in the disc with a discrete transverse load has been carried out by Chan *et al.*[32]

5.1.1   *Simulated Example*

The transition curves shown in Figure 7 were obtained for a disc having the following non-dimensional ratios,

$$\frac{u}{b} = 0.5, \qquad \frac{r_o}{b} = 0.75,$$

and for the load system,

$$\frac{k}{\rho hb^2\omega_{cr}^2} = 0.7, \qquad \frac{m}{\rho hb^2} = 0.24, \qquad \frac{c}{\rho hb^2\omega_{cr}} = 5 \times 10^{-5}.$$

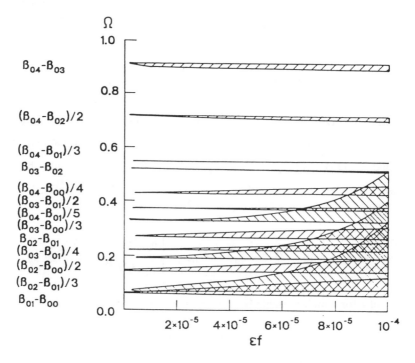

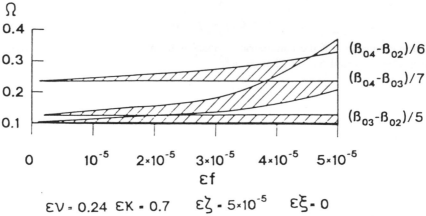

Figure 7: Transition curves for the discrete
transverse load

$R_{LP}(r_0)$ and β_{LP} were obtained by using a symbolic analysis code. A simple formula for the detuning σ in terms of γ, ξ, κ and f is not obtainable from equation (95) (nor from the equations governing stability in the other combination types). The instability regions were determined by computerised symbolic analysis.

Figure 7 is drawn in two parts because the three resonances, $(\beta_{04} - \beta_{02})/6$, $(\beta_{04} - \beta_{03})/7$, and $(\beta_{03} - \beta_{02})/5$ are associated with particularly large instability regions. They are drawn to a different scale on the εf axis of Figure 7. It is clear that the friction is strongly destabilising and that a large number of parametric resonances occur at subcritical speeds.

5.2 Distributed Load System

We now consider the case where the mass-spring-damper system, with friction, is distributed over an annular sector (or 'pad') that rotates around the disc. The system is illustrated in Figure 8.

The pad is assumed to occupy a circumferential interval of $2\varphi_0$ with its centre at $\theta = \varphi$ at time t, and the inner and outer radii of the pad are r_a and r_b. Without loss of generality, φ is taken to be zero when $t = 0$, so that $\varphi = \tilde{\Omega}t$ when $\tilde{\Omega}$ is the radian speed of the pad about the stationary disc. The distributed stiffness, mass, damping and friction are denoted by μ', ρ', ξ' and f' respectively. To describe the influence of the pad, its stiffness, mass, damping and friction may be expressed as a Fourier series in the circumferential direction. Following the analysis by Ouyang et al.[33] we obtain,

$$\mu' = \frac{\varphi_0}{2\pi}\mu' + \frac{\mu'}{2\pi i}\sum_{m=1}^{\infty}\frac{\exp(im\tilde{\Omega}t)}{m}\left\{\exp[im(\varphi_0 - \theta)] - \exp[-im(\varphi_0 + \theta)]\right\} + cc ,$$

$$(98)$$

$$\rho' = \frac{\varphi_0}{2\pi}\rho' + \frac{\rho'}{2\pi i}\sum_{m=1}^{\infty}\frac{\exp(im\tilde{\Omega}t)}{m}\left\{\exp[im(\varphi_0 - \theta)] - \exp[-im(\varphi_0 + \theta)]\right\} + cc ,$$

$$(99)$$

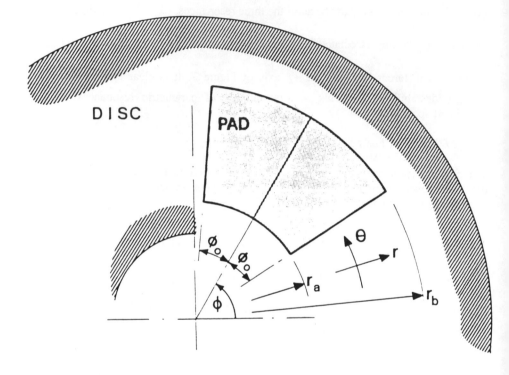

Figure 8: Distributed load arrangement

$$\zeta' = \frac{\varphi_0}{2\pi}\zeta' + \frac{\zeta'}{2\pi i}\sum_{m=1}^{\infty}\frac{\exp(im\tilde{\Omega}t)}{m}\left\{\exp[im(\varphi_0 - \theta)] - \exp[-im(\varphi_0 + \theta)]\right\} + cc \; ,$$

(100)

$$f' = \frac{\varphi_0}{2\pi}f' + \frac{f'}{2\pi i}\sum_{m=1}^{\infty}\frac{\exp(im\tilde{\Omega}t)}{m}\left\{\exp[im(\varphi_0 - \theta)] - \exp[-im(\varphi_0 + \theta)]\right\} + cc \; ,$$

(101)

so that each parameter is given as a time- and space-varying function, and *cc* denotes the complex conjugate.

If the complete area of the pad remains in contact with the disc at all times, then Eqs. (98-101) can be discretised by using finite element shape functions, and the resulting terms can be added to the finite element stiffness, mass and damping matrices of the disc. When the friction is treated as a follower force, then it provides an unsymmetric component to the stiffness matrix given by Eq. (75).

The finite element equations for the complete disc and pad system can then be written as (Mottershead *et al.*[34]),

$$M\ddot{x} + C\dot{x} + Kx = -\frac{\rho'}{2\pi}\left[\varphi_0 I_0 + \sum_{m=1}^{\infty}I_m\exp(im\tilde{\Omega}t)\right]\ddot{x}$$

$$- \frac{\zeta'}{2\pi}\left[\varphi_0 I_0 + \sum_{m=1}^{\infty}I_m\exp(im\tilde{\Omega}t)\right]\dot{x}$$

(102)

$$- \frac{\mu'}{2\pi}\left[\varphi_0 I_0 + \sum_{m=1}^{\infty}I_m\exp(im\tilde{\Omega}t)\right]x$$

$$+ \frac{f'}{2\pi}\left[\varphi_0 J_0 + \sum_{m=1}^{\infty}J_m\exp(im\tilde{\Omega}t)\right]x + cc \; ,$$

where

$$I_0 = \sum_{e=1}^{E}I_{0e}, \qquad I_{0e} = \int_{r_1}^{r_2}\int_{\theta_1}^{\theta_2}N^T(r,\theta)N(r,\theta) \; rdrd\theta \; , \qquad \text{(103 a,b)}$$

$$J_0 = \sum_{e=1}^{E}J_{0e}, \qquad J_{0e} = \int_{r_1}^{r_2}\int_{\theta_1}^{\theta_2}N^T(r,\theta)\frac{\partial N(r,\theta)}{\partial\theta} \; drd\theta \; , \qquad \text{(104 a,b)}$$

$$I_m = \sum_{e=1}^{E} I_{me} \qquad (m = 1,2.....,\infty) , \tag{105}$$

$$I_{me} = \int_{r_1}^{r_2} \int_{\theta_1}^{\theta_2} \frac{\exp[im(\varphi_0-\theta)] - \exp[(-im(\varphi_0+\theta)]}{mi} N^T(r,\theta)N(r,\theta)rdrd\theta, \tag{106}$$

$$J_m = \sum_{e=1}^{E} J_{me} \qquad (m = 1,2.....,\infty) , \tag{107}$$

$$J_{me} = \int_{r_1}^{r_2} \int_{\theta_1}^{\theta_2} \frac{\exp[im(\varphi_0-\theta)] - \exp[(-im(\varphi_0+\theta)]}{mi} N^T(r,\theta)\frac{\partial N(r,\theta)}{\partial\theta}drd\theta, \tag{108}$$

and $M(V \times V)$, $C(V \times V)$, $K(V \times V)$, and $x(V \times 1)$ are the mass matrix, damping matrix, stiffness matrix and nodal displacement vector of the disc.

We consider the matrix, X, of normalised nodal vectors of the disc, which has the property of satisfying the conditions,

$$X^TMX = \rho_d I , \tag{109}$$

and

$$X^TKX = \rho_d \, \text{diag}[\omega_n^2] , \tag{110}$$

where I is the unit matrix.

Assuming that the disc damping is of the proportional type, then Eq. (102) may be written in the modal coordinates of the disc. Thus,

$$\ddot{q}(\tau)+2\xi_n' \, \beta_n \, \dot{q}_n+\beta_n^2 \, q_n(\tau) = - \frac{\varphi_0}{\pi\rho_d} \sum_{j=1}^{V} \left\{ \left[\rho' \ddot{q}_j(\tau) + \frac{\zeta'}{\omega_{cr}} \dot{q}_j(t) + \frac{\mu'}{\omega_{cr}^2} q_j(\tau) \right] Y_{0nj} \right.$$

$$- \frac{f'}{\omega_{cr}^2} Z_{0nj} q_j(\tau) \Big\} - \left\{ \sum_{m=1}^{\infty} \sum_{j=1}^{V} \left\{ \frac{Y_{mnj}}{2\pi\rho_d} \left[\rho' \ddot{q}_j(\tau) + \frac{\zeta'}{\omega_{cr}} \dot{q}_j(t) + \frac{\mu'}{\omega_{cr}^2} q_j(\tau) \right] \right. \tag{111}$$

$$\left. - \frac{f' Z_{mnj}}{2\pi\rho_d\omega_{cr}^2} q_j(\tau) \right\} \exp(im\Omega\tau) + cc \right\} ,$$

where q_n is the participation factor of the nth mode, ξ_n' is the nth modal damping ratio,

$$Y_{m_{nj}} = X_n^T I_m X_j, \qquad Z_{m_{nj}} = X_n^T J_m X_j \qquad (m = 0,1,.....,\infty; \; n,j = 1,2,...,V) \; ,$$

$$(112 \; a,b)$$

and τ, β_n and Ω have been defined in Eq. (85).

As in section 5.1. the method of multiple time scales may be applied. Mottershead *et al.*[34] gave the $O(\varepsilon)$ equation as,

$$(D_0^2 + \beta_n^2)q_n^{(1)} = -2i\beta_n \exp(i\beta_n T_0)(D_1 a_n + \xi_n \beta_n a_n)$$

$$+2\varphi_0 \sum_{j=1}^{V} [(\rho\beta_j^2 - i\zeta\beta_j - \mu)Y_{0_{nj}} + fZ_{0_{nj}}]\exp(i\beta_j T_0)]a_j$$

$$(113)$$

$$+ \sum_{m=1}^{\infty} \sum_{j=1}^{V} \Big\{ [(\rho\beta_j^2 - i\zeta\beta_j - \mu)Y_{m_{nj}} + fZ_{m_{nj}}]\exp[i(\beta_j + m\Omega)T_0]$$

$$+ [(\rho\beta_j^2 - i\zeta\beta_j - \mu)\bar{Y}_{m_{nj}} + f\bar{Z}_{m_{nj}}]\exp[i(\beta_j - m\Omega)T_0]\Big\}a_j + cc,$$

where the over-bar denotes complex conjugation.

It should be noted that equation (113) differs from Eq. (87) with respect to the resulting description of the combination resonances. In equation (87) the sum (or difference) of the number of nodal diameters in the two modes gives the combination index. In equation (113) the finite element discretisation has resulted in the loss of the circle and diameter subscripts on the natural frequencies, and the Fourier decomposition of the mass and stiffness of the pad has led to the appearance of the Fourier index as a multiplier of the rotation speed. Thus for the investigation of combination resonances it is necessary to consider the occasion when $m\Omega$ will be close to the sum or difference of two natural frequencies β_p and β_q . There is an added complication in that the perfect symmetry of the disc results in doublet modes (Mottershead and Chan[29]) whenever a nodal diameter is present. Mottershead *et al.*[34] showed that the combination of a doublet with a singlet mode results in a cubic equation in the characteristic exponent λ, and the combination of two doublet modes results in a quartic equation.

5.2.1. *Simulated Example*

We consider an annular disc of inner and outer radii 67mm and 120mm respectively and thickness 20mm, which is clamped on the inner edge and free at the outer circumference. The disc possesses the following properties: Young's modulus $E = 1.2 \times 10^5 MPa$ Poisson's ratio $v = 0.25$, and density $\rho_d = 140 kgm^{-3}$. Proportional damping is assumed for the disc having the form $\varepsilon\xi_n = \varepsilon\alpha(1000 + 2\times10^{-6}\omega_n^2)/2\omega_n$ so that the disc damping is represented by the single parameter α. For the pad, $r_a = 78mm$, $r_b = 120mm$ and $2\varphi_0 = \dfrac{\pi}{4}$. The dimensions and material properties of the disc and pad are typical of a motor car disc brake.

The finite element model was constructed from 96 QUAD4 elements (MacNeal[35]) and 72 of them were within the annular area $r_b \geq r \geq r_a$ of the pad. The latter have time- and space-varying terms arising from the Fourier series expansion of the mass, stiffness and damping of the pad, and the friction between the pad and disc.

The effect of friction is investigated by allowing εf to vary whilst $\varepsilon\mu = 0.2$, $\varepsilon\rho = 0.05$, $\varepsilon\alpha = 0.15$ and $\varepsilon\zeta = 0.02$. The instability regions are shown in Figure 9 where it can be seen that friction always has a strong destabilizing effect. It seems that the combination resonances involving two doublet modes are more susceptible to the friction-induced instability than those combinations that involve a singlet mode. The effect of varying the other parameters is discussed by Mottershead et al.[34]

It should be pointed out that the equations for the transition curves are cubic or quartic equations in several variable system parameters with complex coefficients. The solution of such equations is obtained by means of a numerical search, and although the imaginary part of at least one λ should be zero (the imaginary parts of all the other roots being negative), one must accept a very small imaginary part (negative or positive) as defining a point on the transition curve.

In the formulation of the QUAD4 element (MacNeal[35]), the strategy is to modify a bi-linear element so that it more closely represents the bi-cubic displacement of a rectangular plate. In order to achieve this it is necessary to compensate for a deficiency in bending energy by increasing the transverse shear energy, which effectively alters the shape functions for displacement and rotations in a way that cannot be determined explicitly. In the construction of the follower force stiffness matrix the bi-linear shape functions for the transverse displacement and the two rotations at each node were used.

As a result of both the numerical search for the instability boundaries, and the finite element approximations, the tips of the transition curves are likely to contain some degree

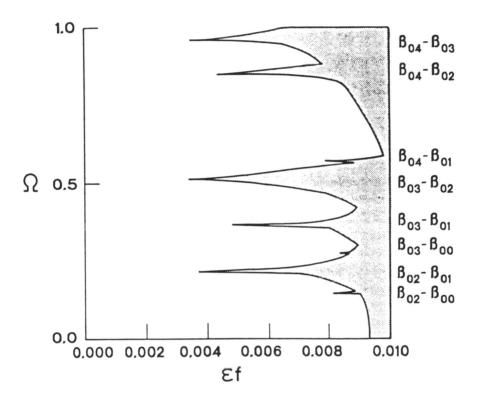

Figure 9: Transition curves for the distributed load

of inaccuracy. In particular the tips of the curves in Figure 9 might extend further to the left than shown because the distance across the instability was possibly too small to be detected in the numerical search. There was also the possibility of inadvertently missing some combination resonances if their instability regions were very narrow.

6.   Parametric Excitation by a Frictional Follower Force with a Negative μ-Velocity Characteristic

Ouyang et al.[36] considered the case of a discrete mass that was driven in a concentric circle over the surface of a stationary disc. The mass was attached to one end of a spring-damper system, the other end of which was rotated at a uniform speed $\tilde{\Omega}$.

When the mass is in a state of oscillation, then its speed of rotation can be expressed as $\tilde{\Omega} + \phi$, where ϕ represents the angular oscillation of the mass with respect to the driven end of the spring-damper. The coefficients of the spring and damper are denoted by k_p and c_p to distinguish them from the transverse spring, k, and damper, c, that were considered in section 5.1, and are also present in the analysis that follows.   The arrangement of the disc and mass-spring-damper system is shown in Figure 10.

Ouyang et al.[36] studied the problem that occurred under persistent oscillation of the mass, when its angular displacement was given by,

$$\varphi(t) = \varphi_o \left[\frac{\alpha\tilde{\Omega}}{2} \left(\exp(i\omega_p t) + \exp(-i\omega_p t)\right) + 1 - \alpha\tilde{\Omega} \right], \tag{114}$$

where $\omega_p = \sqrt{\dfrac{k_p}{m}}$, φ_o is an initial condition on the angular displacement, and α represents the negative slope of the dynamic coefficient of friction with $\tilde{\Omega} + \phi$, $\tilde{\Omega} > >| \phi |$.   This strictly represents the case of zero damping,

$$\left(c_p - \frac{f'\alpha}{r_o} \right) = 0 , \tag{115}$$

but the analysis is arguably relevant to a wider range of problems (Ouyang et al.[36]). The friction force is assumed to vary with the speed of the mass such that,

$$\bar{f} = f'(1 - \alpha(\tilde{\Omega} + \phi)) , \tag{116}$$

where f' represents the static friction.

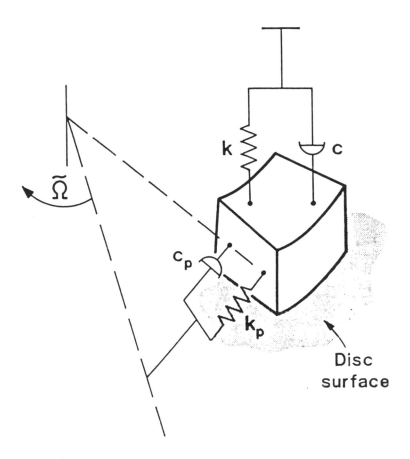

Figure 10: Discrete load arrangement for use with the
negative µ-velocity relation

The transverse motion of the disc is dependent upon the follower force $\bar{f}$, and (from Eq. (116)) upon the angular speed, $\tilde{\Omega} + \phi$, of the rotating mass. The equation of motion, in the modal coordinates of the disc has been written by Ouyang et al.[36] as,

$$
\frac{d^2 q_{LP}}{d\tau^2} + \frac{D' \omega_{cr} \beta_{LP}^2}{D} \frac{dq_{LP}}{d\tau} + \beta_{LP}^2 \, q_{LP} = - \sum_{n=0}^{\infty} \sum_{j=-\infty}^{\infty} R_{jn}(r_o) \, R_{LP}(r_o)
$$

$$
\times \exp[i(j-L)\Omega\tau] \left\{ \frac{m}{\rho h b^2} \left(\frac{\partial^2 q_{jn}}{\partial\tau^2} + j\Omega \left(2i + \alpha\varphi_o\omega_p(\exp(-i\beta_p\tau) - \exp(i\beta_p\tau))\right) \frac{\partial q_{jn}}{\partial\tau} \right. \right.
$$

$$
-j^2\Omega^2 \left(i\alpha\frac{\varphi_o}{2}\omega_p(\exp(i\beta_p\tau) - \exp(i\beta_p\tau)) +1 \right)^2 q_{jn}
$$

$$
\left. -ij\frac{\alpha\varphi_0}{2\omega_{cr}} \omega_p^2\Omega \left(\exp(-i\beta_p\tau) + \exp(i\beta_p\tau)\right) q_{jn} \right)
$$

$$
+ \frac{c}{\rho h b^2\omega_{cr}} \left(\frac{\partial q_{jn}}{\partial\tau} + ij\Omega \left(i\alpha\frac{\varphi_o}{2}\omega_p \left(\exp(i\beta_p\tau) - \exp(-i\beta_p\tau)\right) + 1 \right) q_{jn} \right) + \frac{k}{\rho h b^2\omega_{cr}^2} q_{jn}
$$

$$
\left. - \frac{ijf'}{\rho h b^2 r_o\omega_{cr}^2} \left(1 - \alpha\tilde{\Omega} \left(i\frac{\alpha\varphi_o\omega_p}{2}\left(\exp(i\beta_p\tau) - \exp(-i\beta_p\tau)\right) + 1 \right) q_{rs} \right) \right\},
$$

(117)

where D' is a Kelvin damping coefficient for the disc, and $\beta_p = \dfrac{\omega_p}{\omega_{cr}}$.

By introducing the small parameters,

$$
\varepsilon\gamma = \frac{m}{\rho h b^2}, \; \varepsilon\kappa = \frac{k}{\rho h b^2\omega_{cr}^2} \; , \; \varepsilon\zeta = \frac{c}{\rho h b^2\omega_{cr}} \; ,
$$

(118 a,b,c)

$$
\varepsilon\xi = \frac{D'}{D\omega_{cr}} \; , \; \varepsilon f = \frac{f'}{\rho h b^2 r_o\omega_{cr}^2} \; ,
$$

(118 d,e)

and applying the method of multiple time scales, the $O(\varepsilon)$ equation may be written as,

$$D_o^2 q_{LP}^{(1)} + \beta_{LP}^2 q_{LP}^{(1)} = -D_o(2D_1 + \xi\beta_{LP}^2) (A_{LP}$$

$$\times \exp(i\beta_{LP}T_o) + B_{LP} \exp(-i\beta_{LP}T_o)) \qquad (119)$$

$$+ Y_1(\gamma,\kappa,\zeta,f(1-\alpha\tilde{\Omega})) + Y_2(\gamma,\zeta,f,k_p) ,$$

where $Y_1(\bullet)$ contains terms that associated with the resonances given in section 5.1,

but modified by the μ-velocity relationship. $Y_2(\bullet)$ contains terms that are associated

with new resonances, and can be expressed as,

$$Y_2(\gamma,\zeta,f,k_p) = -\sum_{n=0}^{\infty} \sum_{j=-\infty}^{\infty} j\Omega\alpha\varphi_o\omega_p R_{LP}(r_o)R_{jn}(r_o)$$

$$\times \exp(i(j-L)\Omega T_o) \left(\left(\left(i\gamma j\Omega + \frac{\zeta+f\alpha\omega_{cr}}{2}\right)\right.\right.$$

$$\times \left(\exp(-i\beta_p T_o) - \exp(i\beta_p T_o)\right) - \frac{i\gamma\beta_p}{2}\left(\exp(i\beta_p T_o)\right.$$

$$\left. + \exp(-i\beta_p T_o)\right) + \gamma j\frac{\Omega\alpha\varphi_o\omega_p}{4} \left(\exp(i2\beta_p T_o)\right.$$

$$\left.\left. + \exp(-i2\beta_p T_o)\right)\right) \left(A_{jn}\exp(i\beta_{jn} T_o) + B_{jn}\exp(-i\beta_{jn} T_o)\right)$$

$$+ i\gamma\beta_{jn}\left(\exp(-i\beta_p T_o) - \exp(i\beta_p T_o)\right) \left(A_{jn}\exp(i\beta_{jn} T_o) - B_{jn}\exp(-i\beta_{jn} T_o)\right) \right).$$

$$(120)$$

Ouyang et al.[36] identified twenty types of combination resonance in addition to those considered in section 5.1. We consider by way of an example the case when

$$(j-L)\Omega = \beta_{jn} - \beta_{LP} + \beta_p + \varepsilon\sigma .$$

The secular terms are eliminated from Eq. (119) when,

$$i\beta_{LP}(2D_1 + \xi\beta_{LP}^2)B_{LP} + J\Omega\alpha\varphi_o\omega_p R_{LP}(r_o)R_{JN}(r_o)$$

$$\times \left[i\gamma \left(\frac{\beta_p}{2} + C_{JN}^- \right) - \frac{\zeta + f\alpha\omega_{cr}}{2} \right] B_{JN}\exp(i\sigma T_1) = 0 , \tag{121}$$

and,

$$i\beta_{JN}(2D_1 + \xi\beta_{JN}^2)B_{JN} + L\Omega\alpha\varphi_o\omega_p R_{LP}(r_o)R_{JN}(r_o)$$

$$\times \left[i\gamma \left(\frac{\beta_p}{2} - C_{LP}^- \right) + \frac{\zeta + f\alpha\omega_{cr}}{2} \right] B_{LP}\exp(-i\sigma T_1) = 0 , \tag{122}$$

This leads to a quadratic equation in a characteristic exponent, λ, from which the range of unstable vibrations can be determined for each parametric resonance.

6.1.   Simulated Example

We consider the same disc system that was described in Section 5.1.1. In addition to the parameters given there, we now define the following,

$$\frac{D'}{D\omega_{cr}} = 5\times10^{-13} , \quad \alpha = 3\times10^{-5} , \quad k_p = 5\times10^6 \quad \text{and} \quad \frac{f'}{\rho h b^2 r_o\omega_{cr}^2} = 5\times10^{-5} .$$

Instability regions that only arise when the negative μ-velocity relationship is present are shown in Figures 11 and 12. Once again the friction f is shown to be strongly destabilising. It can be seen from the term $\dfrac{\zeta + f\alpha\omega_{cr}}{2}$ appearing in Eq. (120) that, on these resonances, the friction and viscous damping have a similar effect. This is confirmed in Figure 12 where the viscous damping is shown to be destabilising.

7.   Closure

This article provides a thorough description of research studies on the dynamics of discs, and in particular on the unstable vibrations caused by frictional loading. Whilst the work described here involves quite complex and intricate mathematics, its application relates to the understanding of common machine components such as vehicle brakes, circular saws and computer discs. Significant potential exists for improving the performance of such components by a better appreciation of the dynamics of discs under the action of frictional loads.

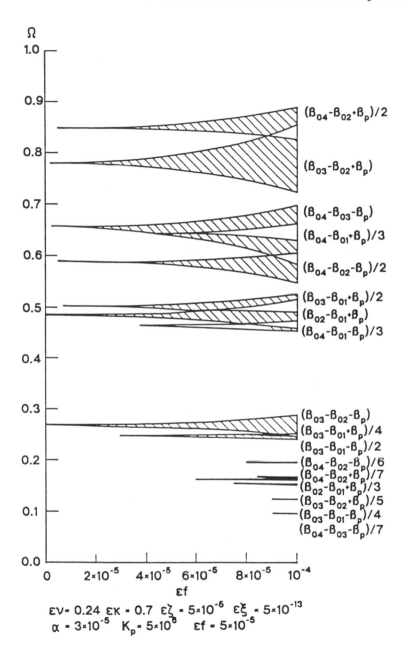

Figure 11: Transition curves showing the effect of friction

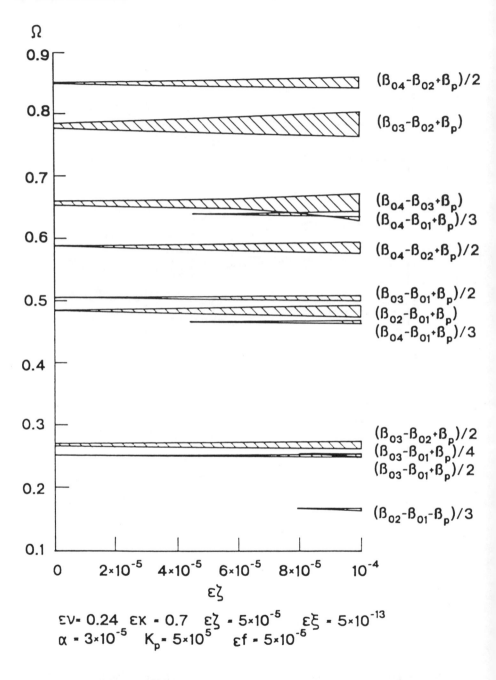

Figure 12: Transition curves showing the effect of damping

Further research remains to be completed on the stick-slip oscillations at low rotational speeds that possibly give rise to low frequency noise problems in vehicle brakes. Other noise problems that have been observed on dynamometer brake tests apparently involve excitation of the disc over a wide range of frequencies and may be attributed to chaos. Further research is needed on methods for suppressing the friction-induced vibrations. Optimisation techniques may be applied in order to design disc systems for the avoidance of unstable regions in the dynamics. The influence of thermal effects needs to be investigated.

Acknowledgement

The author wishes to acknowledge the contribution of M.P. Cartmell, S.N. Chan, M.I. Friswell and H. Ouyang

References

1. H. Lamb and R.V. Southwell, *Proc. Royal Soc. Lon. A.,* 99 (1921) 272-280.
2. R.V. Southwell, *Proc. Royal Soc. Lon. A.,* 101 (1921) 133-153.
3. A.J. McLeod and R.E.D. Bishop, *The forced vibration of circular flat plates*, Mechanical Engineering Science Monograph, No. 1 (1965).
4. J. Prescott, *Applied Elasticity* (Longman, London, 1924).
5. S.A. Tobias and R.N. Arnold, *Proc.I.Mech.E.,* 171 (1957) 669-690.
6. W.A. Stange and J.C. MacBain, *Trans. A.S.M.E., J.Vib.Acoust.,* 105 (1983) 402-407.
7. W.D. Iwan and K.J. Stahl, *Trans. A.S.M.E., J.Appl.Mech.,* 40 (1973) 445-451.
8. W.D. Iwan and T.L. Moeller, *Trans A.S.M.E., J.Appl.Mech.,* 43 (1976) 485-490.
9. C.D. Mote, *Trans. A.S.M.E., J.Engrg.Ind.,* 87B (1965) 258-264.
10. C.D. Mote, *Trans. A.S.M.E., J.Engrg.Ind.,* 89B (1967) 611-618.
11. C.D. Mote, *J.Franklin Institute,* 290 (1970) 329-344.
12. C.D. Mote, *J. Acoust. Soc. Am.* 61 (1977) 439-447.
13. R.L. Yu and C.D. Mote, *J. Sound Vib.,* 119 (1987) 409-427.
14. I.Y. Shen and C.D. Mote, *J. Sound Vib.,* 148 (1991) 307-318.
15. K.G. Valeev, *J.Appl.Math.Mech.* 27 (1963) 1745-1759.
16. I.Y. Shen and C.D. Mote, *J. Sound Vib.,* 149 (1991) 164-169.
17. I.Y. Shen and C.D. Mote, *Int. J. Solids Struct.,* 29 (1992) 1019-1032.
18. I.Y. Shen and C.D. Mote, *J. Sound Vib.,* 152 (1992) 573-576.
19. S.G. Hutton, S. Chonan and B.F. Lehmen, *J. Sound Vib.,* 112 (1987) 527-539.
20. Z.W. Jiang, S.Chonan and H.Abé, *Trans. A.S.M.E., J.Vib.Acoust.,* 112 (1990) 53-58.
21. S. Chonan, Z.W. Jiang and Y.J. Shyu, *Trans. A.S.M.E., J.Vib.Acoust.,* 114 (1992) 326-329.
22. K. Huseyin, *Vibrations and Stability of Multiple Parameter Systems* (Sijthoff and Noordhoff, Alphen aan der Rijn, 1978).
23. M.R. North, *I.Mech.E. paper C38/76,* 169-176.
24. N. Millner, *S.A.E. Paper 780332.*
25. H. Murakami, N. Tsunada and T. Kitamura, *S.A.E. paper 841233.*

26. P.C. Brooks, D.A. Crolla, A.M. Lang and D.R. Schafer, *I.Mech.E. paper C444/004/93*.
27. J.S. Chen and D.B. Bogy, *Trans. A.S.M.E., J. Appl. Mech.* 59 (1992) 5230-5235.
28. K.Ono, J.S. Chen and D.B. Bogy, *Trans. A.S.M.E., J.Vib.Acoust.*, 117 (1995) 161-163.
29. J.E. Mottershead and S.N. Chan, *Trans. A.S.M.E., J.Vib.Acoust.*, 117 (1995) 161-163.
30. I.Y. Shen, *Trans. A.S.M.E., J. Vib. Acoust.*, 115 (1993) 65-69.
31. A.H. Nayfeh and D.T. Mook, *Nonlinear Oscillations* (Wiley Interscience, New York, 1979).
32. S.N.Chan, J.E.Mottershead and M.P.Cartmell, *Proc.I.Mech.E.*, 208C (1994), 417-425.
33. H. Ouyang, J.E. Mottershead, M.I. Friswell and M.P. Cartmell, *Proc. Royal Soc. Lon. A,* submitted.
34. J.E. Mottershead, H. Ouyang, M.I. Friswell and M.P. Cartmell, *Proc. Royal Soc. Lon. A.,* submitted.
35. R.H. MacNeal, *Finite Elements: Their Design and Performance* (Marcel Dekker, New York, 1994).
36. H. Ouyang, J.E. Mottershead, M.I. Friswell and M.P. Cartmell, in preparation.

Dynamics with Friction: Modeling, Analysis and Experiment, Part II, pp. 75–98
edited by A. Guran, F. Pfeiffer and K. Popp
Series on Stability, Vibration and Control of Systems, Series B, Vol. 7
© World Scientific Publishing Company

DYNAMICS OF FLEXIBLE LINKS IN KINEMATIC CHAINS

DAN B. MARGHITU

Department of Mechanical Engineering, Auburn University
Auburn, Alabama 36849, USA

ARDÉSHIR GURAN

American Structronics and Avionics
16661 Ventura Blvd
Encino, California 91436, USA
and
Laboratoire de Méchanique Physique CNRS URA867
Université Bordeaux I Science et Technologie
351, Cours de la Libération
F-33405 Talence FRANCE

ABSTRACT

The main objective of this chapter is to develop analytical models that incorporate the effect of the general motion on the vibration of elastic elements in kinematic chains. Equations for the translational and rotational motions of the links are developed applying Hamilton's principle. Kinetic energy that is required for the application of this principle has been derived by utilizing a generalized velocity field theory for elastic solids. This approach provides means to include the inertia terms directly in the equations of motion. Effects such as centrifugal stiffening and vibrations induced by Coriolis forces are accommodated automatically, rather than with the aid of ad hoc provisions. We use Alfrey and Lee analogy to solve the dynamic problem of viscoelasticity.

In another example we consider the influence of the prismatic joint lubrication film on the planar vibrations of a constant cross-sectional straight flexible link that is attached to the joint. The pressure field exerted through the viscous, Newtonian, lubricant film is obtained from the solution of the Reynolds equation of lubrication. We introduce a scheme to solve the resulting two sets of equations for the vibrations of the link and the motion of the fluid. The pressure field is used to compute the external force exerted by the fluid on the link. The utility of the method is demonstrated by considering a planar mechanism that includes an elastic element with a prismatic joint.

1. Introduction

Structural flexibility is an important factor in the study of dynamics of high performance mechanical systems such as mechanisms, manipulator robots, drilling machines, antennas, etc. In general, flexible mechanical systems have several desirable features relative to stiff mechanisms and manipulators such as lower cost, higher speed, reduced power consumption, improved mobility, and safer operation. Yet, in such systems positioning accuracy depends on effective elimination of oscillations that occur due to flexible effects. Accurate modeling of such effects can greatly facilitate the development of control algorithms that can be used in positioning and trajectory tracking applications. Dynamic analysis of flexible links, however, is complicated by the coupling between the nonlinear rigid body motion and the linear elastic displacements of the link.

The importance of modeling mechanisms systems taking into account the elasticity of the links has long been realized and reviews of the work in this area presented in Erdman et al.[1,2], Turcic and Midha[3], and Nagarajan and Turcic[4]. The nonlinear differential equations of motion derived, consider the rigid body motion and the elastic motion to influence each other and sophisticated numerical algorithms are used solve the problem.

The elastic link of a slider-crank mechanism is treated as a continuous system by Jasinski et al.[5], Chu and Pan.[6], and Badlani and Midha[7]. The effect of internal material damping on the dynamic response of a slider crank mechanism is presented by Badlani and Midha[8]. A linear viscoelastic (Kelvin-Voight) model is assumed for the flexible coupler. Farhang and Midha[9] presented a continuous model for investigating parametric vibration stability in slider crank linkages with flexible couplers. The coupler was assumed to have internal material damping, and small initial curvature.

The dynamics of elastic manipulators with prismatic joints has been investigated by Buffinton and Kane[10,11]. They examine the motion of a single beam as it moves longitudinally over two distinct points. Equations of motion are formulated by treating the beam's supports as kinematical constraints imposed on an unrestrained beam. Gordaninejad et al.[12] examine the motion of a planar robot arm consisting of one revolute and one prismatic joint. All coupling terms between the rigid and the flexible motions of the robot arm are included in the model based on the Timoshenko beam theory. Yuh and Young[13] derived a time varying partial differential equation with boundary conditions for an axially moving beam in rotation.

The one shortcoming of these studies is that they don't examine the load from oil pressure in the revolute and prismatic joints.

In general, a fluid is introduced between two rubbing surfaces in order to separate them and thus reduce their frictional forces. The lubrication or friction reduction of two bodies in near contact is generally accomplished by a viscous fluid moving through a narrow but variable gap between the two bodies. The theory was developed by Reynolds[14].

Benson and Talke[15] investigate the dynamics of a magnetic recording slider of a rigid disk during its transition between sliding and flying. The slider is modeled as a three degree-of-freedom system, capable of lift, pinch, and roll. In addition to considering loads from air pressure, inertia, and the suspension arm, they also consider impulsive load arising from slider/disk collisions.

Our model is different from these models. We study the effect of lubricate film in prismatic joints of mechanism on the vibration of elastic members. We derive the equations of motion by using Hamilton's principle starting from a generalized form of kinetic energy. The equation of motion for a single elastic member attached to a lubricated prismatic joint are presented. The fluid pressure is computed by using a simplified form of Reynolds equation of lubrication. The resulting partial differential equation is transformed to an ordinary integro-differential equation by using the finite Fourier sine transform.

To our best knowledge this class of problems has never been fully discussed nor considered in any previous study in the area of elastodynamics.

Finally, the vibration problem of an elastic link of a slider-crank mechanism is considered. The equations of motion for the first mode of vibration are solved numerically using a Bulirsh-Stoer algorithm with adaptive step-size control. Several plots representing various aspects of the motion are presented.

2. Kinematics and Kinetics of Flexible Bodies in General Motion

The first step of dynamic analysis is the consideration of the relations that govern the kinematics of the linear elastic system. Indeed, effective formulation of equations of motion depends primarily on the ability to formulate proper mathematical expressions for kinematic quantities such as the linear and angular velocities and accelerations. Figure 1 depicts the a linear-elastic body in motion in the three-dimensional space. The fixed coordinate frame is shown at O_1 with the unit vectors $[\mathbf{i}_1, \mathbf{j}_1, \mathbf{k}_1]$. A second moving frame which is attached to the border Σ of the body is shown at O with the unit vectors $[\mathbf{i}, \mathbf{j}, \mathbf{k}]$. For any given time instant, the linear-elastic body occupies the domain $D(t)$. The point C is the center of mass and $\mathbf{r}_C$ is the corresponding position vector. Here we track the motion of a point $P \in D(t)$ on the body. The position vector of this point with respect to the moving frame is expressed in terms of the sum of two vectors $\mathbf{r}$ and $\mathbf{u}$. The vector $\mathbf{r}$ marks the position of the particle subject to the rigid body motion only, whereas, the vector $\mathbf{u}$ represents the displacement as a result of elastic deformations. Furthermore, we attach a local vortex vector $\boldsymbol{\Omega}(r^*, t)$ to the point P. This vector represents the rotational motion of the medium in the neighborhood of the point as a result of rigid body motion and elastic deformations.

2.1. Small Deformations

In the case of small deformations we assume uniform vorticity and drop the functional dependence on position $(\boldsymbol{\Omega}(r^*, t) = \boldsymbol{\omega}(t))$. As a result of this simplification the vortex vector becomes the angular velocity vector that is usually defined for rigid bodies. Now we use Poisson's formula to derive the kinematic relations for velocities and accelerations. Accordingly, the position vector for any particle on the linear-elastic solid $D(t)$ can be written as

$$\mathbf{r}^* = \mathbf{r}_0 + \mathbf{r} + \mathbf{u}, \tag{1}$$

where $\mathbf{r}^*$ is the position vector of P in the $[\mathbf{i}_1, \mathbf{j}_1, \mathbf{k}_1]$ frame, $\mathbf{r}_0$ is the position vector of O in the $[\mathbf{i}_1, \mathbf{j}_1, \mathbf{k}_1]$ frame, and $\mathbf{r}$ is the position vector of P in the frame attached to the linear-elastic body in motion.

Differentiating Eq.(1) with respect to time yields

$$\mathbf{v} = \mathbf{v}_0 + \boldsymbol{\omega} \times \mathbf{r} + \frac{\partial \mathbf{u}}{\partial t} + \boldsymbol{\omega} \times \mathbf{u}, \tag{2}$$

where the terms

$$\mathbf{v}_0 + \boldsymbol{\omega} \times \mathbf{r}$$

corresponds to the rigid body velocity field, and

$$\frac{\partial \mathbf{u}}{\partial t} + \boldsymbol{\omega} \times \mathbf{u}$$

represents the additional component that arises from the deformation of the linear-elastic body.

Differentiating Eq.(2) with respect to time, the acceleration field can be obtained as

$$\mathbf{a} = \mathbf{a}_0 + \boldsymbol{\alpha} \times \mathbf{r} + \boldsymbol{\omega} \times (\boldsymbol{\omega} \times \mathbf{r}) + \frac{\partial^2 \mathbf{u}}{\partial t^2} + 2\boldsymbol{\omega} \times \frac{\partial \mathbf{u}}{\partial t} + \boldsymbol{\alpha} \times \mathbf{u} + \boldsymbol{\omega} \times (\boldsymbol{\omega} \times \mathbf{u}), \tag{3}$$

where

$$\mathbf{a}_0 = \dot{\mathbf{v}}_0 = \ddot{\mathbf{r}}_0$$

$$\boldsymbol{\alpha} = \dot{\boldsymbol{\omega}}$$

The term

$$\mathbf{a}_{rg} = \mathbf{a}_0 + \boldsymbol{\alpha} \times \mathbf{r} + \boldsymbol{\omega} \times (\boldsymbol{\omega} \times \mathbf{r})$$

corresponds to the rigid body acceleration field.

Having obtained the kinematic relations, the linear momentum can be written as

$$\mathbf{L} = \int\int\int_{(D)} \mathbf{v}\rho d\tau$$

$$= M\mathbf{v}_0 + M\boldsymbol{\omega} \times \mathbf{r}_C + \boldsymbol{\omega} \times \int\int\int_{(D)} \mathbf{u}\rho d\tau + \int\int\int_{(D)} \frac{\partial \mathbf{u}}{\partial t}\rho d\tau, \tag{4}$$

where M is the total mass, $\mathbf{r}_C$ is the position vector of the mass center relative to O, ρ is the density, $d\tau$ is the volume element. Note that

$$\int\int\int_{(D)} \frac{\partial \mathbf{u}}{\partial t}\rho d\tau = \frac{\partial}{\partial t} \int\int\int_{(D)} \mathbf{u}\rho d\tau$$

The angular momentum vector at O_1 is given by

$$\mathbf{H}_{O_1} = \int\int\int_{(D)} \mathbf{r}_1 \times \mathbf{v}\rho d\tau = M\mathbf{r}_0 \times \mathbf{v}_0 + M\mathbf{r}_C \times \mathbf{v}_0 +$$

$$\mathbf{r}_0 \times (\boldsymbol{\omega} \times \int\int\int_{(D)} \mathbf{r}\rho d\tau) + J_0(t)\boldsymbol{\omega} + \int\int\int_{(D)} \mathbf{r}_1 \times (\partial \mathbf{u}/\partial t)\rho d\tau +$$

$$\mathbf{r}_0 \times (\boldsymbol{\omega} \times \int\int\int_{(D)} \mathbf{u}\rho d\tau) + \int\int\int_{(D)} \mathbf{r} \times (\boldsymbol{\omega} \times \mathbf{u})\rho d\tau, \tag{5}$$

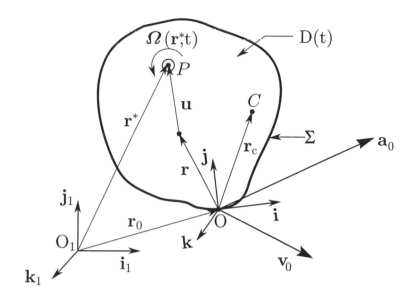

Fig. 1. Spatial deformable solid in motion

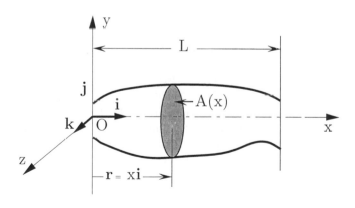

Fig. 2. Rectilinear link

where $\mathbf{J}_0(t)$ is the inertia tensor of the deformable body.

The kinetic energy of the linear-elastic body can be written as

$$T = \frac{1}{2} \int \int \int_{(D)} \mathbf{v}^2 \rho d\tau = T_{rg} + T_{el} + T_s, \tag{6}$$

where

$$T_{rg} = \frac{M\mathbf{v}_0{}^2}{2} + M\mathbf{v}_0 \cdot (\boldsymbol{\omega} \times \mathbf{r}_C) + \frac{1}{2}\boldsymbol{\omega} \cdot (\mathbf{J}_0 \boldsymbol{\omega}) \tag{7}$$

corresponds to a rigid body motion,

$$T_{el} = \frac{1}{2} \int \int \int_{(D)} (\frac{\partial \mathbf{u}}{\partial t})^2 \rho d\tau \tag{8}$$

corresponds to the elastic solid motion without rotation or translation, and

$$T_s = \frac{1}{2} \int \int \int_{(D)} [(\boldsymbol{\omega} \times \mathbf{u})^2 + 2(\mathbf{v}_0 + \boldsymbol{\omega} \times \mathbf{r}) \cdot (\frac{\partial \mathbf{u}}{\partial t} + \boldsymbol{\omega} \times \mathbf{u}) +$$
$$2\frac{\partial \mathbf{u}}{\partial t} \cdot (\boldsymbol{\omega} \times \mathbf{u})] \rho d\tau \tag{9}$$

is the coupling term arising from deformation and the rigid body motion.

2.2. Large Deformations

For the small deformations case we have assumed the points on the linear-elastic solid in rotational and translational motion have a uniform vortex vector. Now, we drop this assumption and consider a vortex vector $\boldsymbol{\Omega}$ in the form

$$\boldsymbol{\Omega}(\mathbf{r}^*, t) = \boldsymbol{\omega}(t) + \boldsymbol{\psi}(\mathbf{r}^*, t), \tag{10}$$

where $\boldsymbol{\omega}(t)$ is the rigid body angular velocity and $\boldsymbol{\psi}(\mathbf{r}^*,t)$ is the local vortex vector and given by

$$\boldsymbol{\psi}(\mathbf{r}^*, t) = \frac{1}{2}\nabla \times \frac{\partial \mathbf{u}(\mathbf{r}^*, t)}{\partial t}, \tag{11}$$

where $\partial \mathbf{u}(\mathbf{r}^*, t)/\partial t$ is the elastic velocity of any particle P. The specific form of the $\boldsymbol{\Omega}(P, t)$ depends on the geometry of the elastic body. For the rectilinear kinematic links that are considered in the present article such as the one depicted in Fig. 2, the linear elastic displacement vector $\mathbf{u}$ can be written as

$$\mathbf{u}(x, t) = u_1(x, t)\mathbf{i} + u_2(x, t)\mathbf{j} + u_3(x, t)\mathbf{k} \text{ for } x \in [0, \mathrm{L}], \tag{12}$$

and the rigid body angular velocity is given by

$$\boldsymbol{\omega}(t) = \omega_1(t)\mathbf{i} + \omega_2(t)\mathbf{j} + \omega_3(t)\mathbf{k}. \tag{13}$$

Then, using Eqs.(10), (12), and (13) the total angular velocity vector becomes

$$\mathbf{\Omega}(x,t) = \omega_1(t)\mathbf{i} + \left[\omega_2 - \frac{1}{2}\frac{\partial^2 u_3(x,t)}{\partial x \partial t}\right]\mathbf{j} + \left[\omega_3 + \frac{1}{2}\frac{\partial^2 u_2(x,t)}{\partial x \partial t}\right]\mathbf{k}. \tag{14}$$

The kinetic energy for the large deformation case is essentially same as the small deformation case. The expression for the energy can be obtained by replacing $\boldsymbol{\omega}(t)$ with $\mathbf{\Omega}$ in Eq.(6).

3. Equations of Motion for Small Deformations in Rectilinear Elastic Links

Figure 2 depicts a rectilinear kinematic link with variable cross-sectional area $A(x)$. We define

$$\mathbf{r} = x\mathbf{i}$$

as the position vector of any particle on the axis of the link,

$$\boldsymbol{\alpha}(t) = \dot{\boldsymbol{\omega}}(t) = \alpha_1(t)\mathbf{i} + \alpha_2(t)\mathbf{j} + \alpha_3(t)\mathbf{k}$$

as the rigid body angular acceleration of the link,

$$\mathbf{v}_O(t) = v_{01}(t)\mathbf{i} + v_{02}(t)\mathbf{j} + v_{03}(t)\mathbf{k}$$

as the translational velocity of the end O of the link,

$$\mathbf{a}_O(t) = a_{01}(t)\mathbf{i} + a_{02}(t)\mathbf{j} + a_{03}(t)\mathbf{k}$$

as the translational acceleration of the end O of the link, and

$$\mathbf{f}(x,t) = f_1(x,t)\mathbf{i} + f_2(x,t)\mathbf{j} + f_3(x,t)\mathbf{k}$$

is the external force per unit length of the link. Also, E is the Young modulus, $I_y(x)$ and $I_z(x)$ are the central moments of inertia of the bar cross-section the axes Oy and Oz respectively.

The equations of motions for the elastic link are developed by applying Hamilton's principle that can be written as

$$\delta \int_{t_0}^{t_1} (V - T - W)dt = 0, \tag{15}$$

where V is the strain energy, T is the kinetic energy, and W is the work of external conservative forces acting on the deformable body (i.e., $\nabla \times \mathbf{f} = 0$).

The strain energy can be written as

$$V = \frac{E}{2}\left[\int_0^L A(x)\left(\frac{\partial u_1}{\partial x}\right)^2 dx + \int_0^L I_z(x)\left(\frac{\partial^2 u_2}{\partial x^2}\right)^2 dx + \int_0^L I_y(x)\left(\frac{\partial^2 u_3}{\partial x^2}\right)^2 dx\right] \tag{16}$$

and the work of external forces

$$W = \mathbf{f} \cdot \mathbf{u}. \tag{17}$$

The kinetic energy T is given by Eqs.(6), (7), (8), and (9). We note that the variational integral of the rigid body kinetic energy is

$$\delta \int_{t_0}^{t_1} T_{rg} dt = 0.$$

Then, using Eqs. (16) and (17) along with the expression for kinetic energy in Eq. (15) and yields the equations of motion of linear-elastic displacements $u_1(x,t)$ $u_2(x,t)$ and $u_3(x,t)$. These equations are given by

$$EA' \partial u_1/\partial x - EA \partial^2 u_1/\partial x^2 + \rho A \partial^2 u_1/\partial t^2 - \rho A u_1(\omega_2^2 + \omega_3^2) +$$
$$\rho A \omega_1 \omega_2 u_2 + \rho A \omega_1 \omega_2 u_3 + \rho A a_{01} -$$
$$\rho A(\omega_3 v_{02} - \omega_2 v_{03}) + \rho A \alpha_2 u_3 + \rho A \alpha_3 u_2 - 2\rho A \omega_3 \partial u_2/\partial t -$$
$$2\rho A \omega_2 \partial u_3/\partial t - \rho A x(\omega_2^2 + \omega_3^2) - f_1(x,t) = 0, \tag{18}$$

$$EI_z'' \partial^2 u_2/\partial x^2 + 2EI_z' \partial^3 u_2/\partial x^3 + EI_z \partial^4 u_2/\partial x^4 + \rho A \partial^2 u_2/\partial t^2 -$$
$$\rho A u_2(\omega_1^2 + \omega_3^2) + \rho A \omega_2 \omega_3 u_3 + \rho A \omega_1 \omega_2 u_1 + \rho A a_{02} -$$
$$\rho A(\omega_1 v_{03} + \omega_3 v_{01}) + \rho A \alpha_3 x - \rho A \alpha_1 u_3 - 2\rho A \omega_1 \partial u_3/\partial t +$$
$$2\rho A \omega_3 \partial u_1/\partial t + \rho A x(\omega_1 \omega_2 + \omega_3^2) - f_2(x,t) = 0, \tag{19}$$

$$EI_y'' \partial^2 u_3/\partial x^2 + 2EI_y' \partial^3 u_3/\partial x^3 + EI_y \partial^4 u_3/\partial x^4 + \rho A \partial^2 u_3/\partial t^2 -$$
$$\rho A u_3(\omega_1^2 + \omega_2^2) + \rho A \omega_1 \omega_3 u_1 + \rho A \omega_2 \omega_3 u_2 + \rho A a_{03} -$$
$$\rho A(\omega_2 v_{01} - \omega_1 v_{02}) + \rho A \alpha_1 u_2 - \rho A \alpha_2 u_1 - 2\rho A \omega_2 \partial u_1/\partial t +$$
$$2\rho A \omega_1 \partial u_2/\partial t + \rho A x(\omega_1 \omega_3 + \omega_2^2) - f_3(x,t) = 0, \tag{20}$$

where primes denote differentiation with respect to x. The effect of rotational inertia of the cross section and the influence of shear forces are neglected in Eqs. (19) and (20). We neglect these terms because of their effect on the primary modes of vibration are negligible. If the rotation inertia of the transverse section is taken into account the term

$$-\rho I_z' \frac{\partial^3 u_2}{\partial x \partial t^2} - \rho I_z \frac{\partial^4 u_2}{\partial^2 x \partial t^2}$$

is added to the Eq.(19) and the term

$$-\rho I_y' \frac{\partial^3 u_3}{\partial x \partial t^2} - \rho I_y \frac{\partial^4 u_3}{\partial^2 x \partial t^2}$$

is added to the Eq. (20). Also, if we consider uniform cross-sectional links, then $A'(x) = I_y'(x) = I''_y(x) = I_z'(x) = I''_z(x) = 0$.

4. Equations of Motion for Large Deformations in Rectilinear Elastic Links

In the following development the general case is considered. This approach leads to a local angular velocity vector which now depends on time as well as position. Thereby the prediction of the present model is expected to be more accurate compared to uniform angular velocity model.

The kinetic energy of a rectilinear link (Fig. 2), for large deformations is

$$
T = \frac{M\mathbf{v_0}^2}{2} + \frac{\rho}{2} \int_0^L \left[(\boldsymbol{\omega} \times \mathbf{r})^2 + (\boldsymbol{\psi} \times \mathbf{r})^2 + 2(\boldsymbol{\omega} \times \mathbf{r}) \cdot (\boldsymbol{\psi} \times \mathbf{r}) \right]
$$

$$
Adx + \frac{\rho}{2} \int_0^L \left[(\boldsymbol{\omega} \times \mathbf{u})^2 + (\boldsymbol{\psi} \times \mathbf{u})^2 + 2(\boldsymbol{\omega} \times \mathbf{u}) \cdot (\boldsymbol{\psi} \times \mathbf{u}) \right] Adx +
$$

$$
\rho \int_0^L \mathbf{v_0} \cdot \boldsymbol{\psi} \times \mathbf{r} Adx + \rho \int_0^L \mathbf{v_0} \cdot \frac{\partial \mathbf{u}}{\partial t} Adx + \rho \int_0^L \mathbf{v_0} \cdot \boldsymbol{\omega} \times \mathbf{u} Adx +
$$

$$
\rho \int_0^L \mathbf{v_0} \cdot \boldsymbol{\psi} \times \mathbf{u} Adx + \rho \int_0^L \boldsymbol{\omega} \cdot \mathbf{r} \times \frac{\partial \mathbf{u}}{\partial t} Adx + \rho \int_0^L \boldsymbol{\psi} \cdot \mathbf{r} \times \frac{\partial \mathbf{u}}{\partial t} Adx +
$$

$$
\rho \int_0^L (\boldsymbol{\omega} \times \mathbf{r}) \cdot (\boldsymbol{\omega} \times \mathbf{u}) Adx + \rho \int_0^L (\boldsymbol{\omega} \times \mathbf{r}) \cdot (\boldsymbol{\psi} \times \mathbf{u}) Adx +
$$

$$
\rho \int_0^L (\boldsymbol{\psi} \times \mathbf{r}) \cdot (\boldsymbol{\omega} \times \mathbf{u}) Adx + \rho \int_0^L (\boldsymbol{\psi} \times \mathbf{r}) \cdot (\boldsymbol{\psi} \times \mathbf{u}) Adx +
$$

$$
\rho \int_0^L \frac{\partial \mathbf{u}}{\partial t} \cdot \boldsymbol{\omega} \times \mathbf{u} Adx + \rho \int_0^L \frac{\partial \mathbf{u}}{\partial t} \cdot \boldsymbol{\psi} \times \mathbf{u} Adx
$$

$$
\rho \int_0^L \mathbf{v_0} \cdot \boldsymbol{\psi} \times \mathbf{r} Adx + \frac{\rho}{2} \int_0^L \left(\frac{\partial \mathbf{u}}{\partial t} \right)^2 Adx + \tag{21}
$$

In the general case, the system equations of motion is coupled and strongly nonlinear, and to find its solutions will be very difficult.

The decoupling, though an approximation, gives a reduced degree of nonlinearity, and thus the obtaining of the solution is simpler. To obtain the system of decoupled equations of motion, the elastic displacements $\mathbf{u} = u_1(x,t)\mathbf{i}$, $\mathbf{u} = u_2(x,t)\mathbf{j}$ and $\mathbf{u} = u_3(x,t)\mathbf{k}$ are substituted successively in the expression of Hamilton's variational principle and the following relation are obtained

$$
c^2 \partial^2 u_1/\partial x^2 - \partial^2 u_1/\partial t^2 + \omega_3^2 u_1 + \omega_3 v_{02} - a_{01} +
$$
$$
\omega_3^2 x + f_1/\rho + c^2 (A'/A) \partial u_1/\partial x = 0 \tag{22}
$$

$$
EI_z'' \partial^2 u_2/\partial x^2 + 2EI_z' \partial^3 u_2/\partial x^3 + EI_z \partial^4 u_2/\partial x^4 - \rho I_z' \partial^3 u_2/\partial x \partial t^2 -
$$
$$
\rho I_z \partial^4 u_2/\partial x^2 \partial t^2 - 2\rho(\omega_3^2 + \omega_1^2) u_2 + 2\rho(\omega_1 v_{03} - \omega_3 v_{01} - x\omega_1\omega_2) -
$$
$$
\rho(xA + 0.5x^2 A') \partial^3 u_2/\partial x \partial t^2 - 0.5x^2 \rho A \partial^4 u_2/\partial x^2 \partial t^2 -
$$
$$
0.5x^2 \rho A u_2 (\partial^2 u_2/\partial x \partial t)^2 - \rho A[(\partial u_2/\partial x)(\partial u_2/\partial t) + u_2 \partial^2 u_2/\partial x \partial t] -
$$
$$
\rho A' u_2 \partial u_2/\partial t - \rho \alpha_3 (2xA + x^2 A') - 2\rho[\omega_3 A u_2 \partial^2 u_2/\partial x \partial t +
$$
$$
\alpha_3 (A u_2 \partial u_2/\partial t + 0.5 A' u_2^2) + \omega_3 [A(\partial u_2/\partial t)(\partial u_2 \partial x) +
$$

$$u_2 \partial^2 u_2/\partial x \partial t + A' u_2 \partial u_2 \partial t]] - \rho a_{02}(xA + xA') + \rho v_{01} A \partial^2 u_2/\partial x \partial t +$$
$$\rho a_{01}(A \partial u_2 \partial x + u_2 A') + \rho v_{01}(\partial^2 u_2/\partial x \partial t + A' \partial u_2/\partial t) +$$
$$\rho(A + xA') \partial^2 u_2/\partial t^2 + 2\rho A(a_{02} - x\alpha_3) + 2\rho A \partial^2 u_2/\partial t^2 - f_2 = 0 \qquad (23)$$

$$EI_y'' \partial^2 u_3/\partial x^2 + 2EI_y' \partial^3 u_3/\partial x^3 + EI_y \partial^4 u_3/\partial x^4 - \rho I_y' \partial^3 u_3/\partial x \partial t^2 -$$
$$\rho I_y \partial^4 u_3/\partial x^2 \partial t^2 - 2\rho(\omega_2^2 + \omega_1^2)u_3 + 2\rho(\omega_2 v_{01} - \omega_1 v_{02} - x\omega_1\omega_3) -$$
$$\rho(xA + 0.5x^2 A') \partial^3 u_3/\partial x \partial t^2 - 0.5x^2 \rho A \partial^4 u_3/\partial x^2 \partial t^2 -$$
$$0.5x^2 \rho A u_3 (\partial^2 u_3/\partial x \partial t)^2 - \rho A[(\partial u_3/\partial x)(\partial u_3/\partial t) + u_3 \partial^2 u_3/\partial x \partial t] -$$
$$\rho A' u_3 \partial u_3/\partial t - \rho \alpha_2(2xA + x^2 A') + \rho[\omega_3 A u_3 \partial^2 u_3/\partial x \partial t +$$
$$2\alpha_2(A u_3 \partial u_3/\partial t + 0.5A' u_3^2) + \omega_2[A(\partial u_3/\partial t)(\partial u_3 \partial x) +$$
$$u_3 \partial^2 u_3/\partial x \partial t + A' u_3 \partial u_3 \partial t]] - \rho a_{03}(xA + xA') + \rho v_{01} A \partial^2 u_3/\partial x \partial t +$$
$$\rho a_{01}(A \partial u_3 \partial x + u_2 A') + \rho v_{01}(\partial^2 u_3/\partial x \partial t + A' \partial u_3/\partial t) +$$
$$\rho(A + xA') \partial^2 u_3/\partial t^2 + 2\rho A(a_{03} - x\alpha_2) + 2\rho A \partial^2 u_2/\partial t^2 - f_3 = 0 \qquad (24)$$

where $c^2 = E/\rho$.

The equations obtained form a decoupled, fourth order system of nonlinear differential equations with partial derivatives. The longitudinal and transverse vibrations of the elastic link can be obtained by solving the systems of equations given by Eq. (22) through (24). For the planar case, however, the response can be obtained from a simpler set of equations given by Eqs. (22) and Eq. (23) without the terms that are associated with Oz axis.

4.1.   Planar Equations of Motion

We consider constant cross-sectional links and let $A'(x) = I_y'(x) = I''_y(x) = I_z'(x) = I''_z(x) = 0$. For a Bernoulli beam in planar motion the equations of motion in $u_1(x,t)$ and $u_2(x,t)$ for small displacements are given by

$$-EA\frac{\partial^2 u_1(x,t)}{\partial x^2} + \rho A \frac{\partial^2 u_1(x,t)}{\partial t^2} -$$
$$2\rho A\omega(t)\frac{\partial u_2(x,t)}{\partial t} - \rho A\omega^2(t)u_1(x,t) -$$
$$\rho A\alpha(t)u_2(x,t) - \rho Ax\omega^2(t) - \rho A\omega(t)v_{02}(t) +$$
$$\rho A a_{01}(t) - f_1(x,t) = 0, \qquad (25)$$

$$EI\frac{\partial^4 u_2(x,t)}{\partial x^4} + \rho A\frac{\partial^2 u_2(x,t)}{\partial t^2} +$$
$$2\rho A\omega(t)\frac{\partial u_1(x,t)}{\partial t} - \rho A\omega^2(t)u_2(x,t) +$$
$$\rho A\alpha(t)u_1(x,t) - \rho Ax\alpha(t) - \rho A\omega(t)v_{01}(t) +$$
$$\rho A a_{02}(t) - f_2(x,t) = 0 \qquad (26)$$

where, $\omega(t) = \omega_3(t)$, $\alpha(t) = \alpha_3(t)$, and $I = I_z$.

5. The Dynamics of Viscoclastic Links

To solve a dynamic problem of viscoelasticity the corresponding solutions of perfectly elastic problems can be used, replacing in the latter the so-called Lamé constants λ and μ by $\bar{\lambda}(s)$ and $\bar{\mu}(s)$ and inverting the Laplace transform. The one side Laplace transform is defined by the formula

$$\bar{f}(x, s) = L[f(x, t)] = \int_0^\infty f(x, t)\, e^{-st} dt. \tag{27}$$

The above elasticviscoelastic analogy was announced by Alfrey[16] and Lee[17].

Using the *elasticvicoelastic* analogy, the mathematical planar model will be described by a system of integro-differential equations

$$\int_0^t E(\tau) \frac{\partial^2 u_1(x, t - \tau)}{\partial x^2} d\tau - \rho \frac{\partial^2 u_1(x, t)}{\partial t^2} +$$
$$2\rho\omega(t) \frac{\partial u_2(x, t)}{\partial t} + \rho\omega^2(t)u_1(x, t) +$$
$$\rho\alpha(t)u_2(x, t) + \rho x \omega^2(t) + \rho\omega(t)v_{02}(t) -$$
$$\rho a_{01}(t) + f_1(x, t)/A = 0, \tag{28}$$

$$\frac{I}{A} \int_0^t E(\tau) \frac{\partial^4 u_2(x, t - \tau)}{\partial x^4} d\tau + \rho \frac{\partial^2 u_2(x, t)}{\partial t^2} +$$
$$2\rho\omega(t) \frac{\partial u_1(x, t)}{\partial t} - \rho\omega^2(t)u_2(x, t) + \rho\alpha(t)u_1(x, t) -$$
$$\rho x \alpha(t) - \rho\omega(t)v_{01}(t) +$$
$$\rho a_{02}(t) - f_2(x, t)/A = 0 \tag{29}$$

where $E(\tau)$ depends on the viscoelastic model (Maxwell model, Kelvin model, etc.). To solve the linear system of Eqs. (28), (29) variant in time we propose an iterative algorithm of successive approximations

We denote the coupling terms depending on time with

$$\Phi_1(x, t) = 2\rho\omega \frac{\partial u_2}{\partial t} + \rho\omega^2 u_1 + \rho\alpha u_2, \tag{30}$$

$$\Phi_2(x, t) = 2\rho\omega \frac{\partial u_1}{\partial t} - \rho\omega^2 u_2 + \rho\alpha u_1. \tag{31}$$

We note with $u_1^{(i)}(x, t)$ and $u_2^{(i)}(x, t)$ the dynamic solution for (i) approximation viz.

$$\int_0^t E(\tau) \frac{\partial^2 u_1^{(i)}(x, t - \tau)}{\partial x^2} d\tau - \rho \frac{\partial^2 u_1^{(i)}}{\partial t^2} + \rho x \omega^2 +$$
$$\rho\omega v_{02} - \rho a_{01} + f_1/A + \Phi_1^{(i-1)} = 0, \tag{32}$$

$$\frac{I}{A}\int_0^t E(\tau)\frac{\partial^4 u_2^{(i)}(x, t - \tau)}{\partial x^4}d\tau + \rho\frac{\partial^2 u_2^{(i)}}{\partial t^2} - \rho x \alpha -$$

$$\rho\omega v_{01} + \rho a_{02} - f_2/A + \Phi_2^{(i-1)} = 0. \tag{33}$$

where $\Phi_1^{(i-1)}$ and $\Phi_2^{(i-1)}$ are obtained from Eqs. (30), (31) with $u_1 \to u_1^{(i-1)}$ and $u_2 \to u_2^{(i-1)}$. For $i = 1$ we have $\Phi_1^{(0)} \equiv 0$ and $\Phi_2^{(0)} \equiv 0$. The systems of Eqs. (32), (33) is linear, invariant in time, and can be solved applying the Laplace transform and iterating $i = 1, 2, 3, \ldots..$

Next we introduce the finite Fourier sine and cosine transforms.

Consider a function $g(x)$ which satisfies Dirichlet's conditions in the interval $[0, L]$ where it has a finite number of maxima, minima, and discontinuities. Infinite discontinuities are not admissible.

The finite Fourier sine transform of the function $g(x)$ which is denoted by $g^{*s}(n)$ is defined by the relation

$$g^{*s}(n) = \int_0^L g(x)sin(\beta_n x)dx,$$

and the finite Fourier cosine transform of the function $g(x)$ which is denoted by $g^{*c}(n)$ is defined by the relation

$$g^{*c}(n) = \int_0^L g(x)cos(\beta_n x)dx$$

where $\beta_n = n\pi/L$, $n = 1, 2, 3 \ldots$

5.1. Application

A double articulated rectilinear link in planar motion is considered. The initial conditions are

$$u_1(x, 0) = \frac{\partial u_1(x, 0)}{\partial t} = u_2(x, 0) = \frac{\partial u_2(x, 0)}{\partial t}, \tag{34}$$

and the boundary conditions are

$$EA\frac{\partial u_1^{(i)}(0, t)}{\partial x} = F_1(t), \quad EA\frac{\partial u_1^{(i)}(L, t)}{\partial x} = F_2(t),$$

$$u_2^{(i)}(0, t) = u_2^{(i)}(L, t) = \frac{\partial^2 u_2^{(i)}(0, t)}{\partial x^2} =$$

$$\frac{\partial^2 u_2^{(i)}(L, t)}{\partial x^2} = 0. \tag{35}$$

where, $F_1(t)$ and $F_2(t)$ are the axial components of the reaction forces at the articulated ends.

We apply to Eqs. (32), (33), (34), (35) the Laplace transform defined by formula (27), and then the finite Fourier sine and cosine transforms. The following displacements are obtained

$$\bar{u}_1^{(i)*c} = \frac{[(-1)^n - 1]\bar{\omega}^2 \dfrac{\rho}{\beta_n^2} + \dfrac{\bar{f}_1^{*c}}{A}}{\rho s^2 + \beta_n^2 \bar{E}} +$$

$$\frac{(-1)^n \dfrac{\bar{F}_2}{A\bar{E}} - \dfrac{\bar{F}_1}{A\bar{E}} + \bar{\Phi}_1^{(i-1)*c}}{\rho s^2 + \beta_n^2 \bar{E}}, \tag{36}$$

$$\bar{u}_2^{(i)*s} = \frac{(-1)^n \bar{\alpha} \dfrac{\rho L}{\beta_n} - [(-1)^n - 1]\dfrac{\rho \bar{a}_{01}}{\beta_n^2}}{\rho s^2 + I\beta_n^4 \bar{E}} -$$

$$\frac{\bar{f}_2^{*s} A - \bar{\Phi}_2^{(i-1)*s}}{\rho s^2 + I\beta_n^4 \bar{E}}. \tag{37}$$

Inverting the finite Fourier sine and cosine transforms in Eqs. (36), (37) we obtain

$$\bar{u}_1^{(i)}(x, s) = \frac{2}{L} \sum_{n=1}^{\infty} \{[(-1)^n - 1]\bar{\omega}^2 \frac{\rho}{\beta_n^2} + \frac{\bar{f}_1^{*c}}{A} +$$

$$(-1)^n \frac{\bar{F}_2}{A\bar{E}} - \frac{\bar{F}_1}{A\bar{E}} + \bar{\Phi}_1^{(i-1)*c}(n, s)\}$$

$$[\rho s^2 + \beta_n^2 \bar{E}]^{-1} \cos(\beta_n x) + \frac{\bar{u}_1^{(i)*c}(0, s)}{2}, \tag{38}$$

$$\bar{u}_2^{(i)}(x, s) = \frac{2}{L} \sum_{n=1}^{\infty} \{(-1)^n \alpha \frac{\rho L}{\beta_n} - [(-1)^n \quad 1]\frac{\rho \bar{a}_{01}}{\beta_n^2} -$$

$$\frac{\bar{f}_2^{*s}}{A} - \bar{\Phi}_2^{(i-1)*s}(n, s)\}[\rho s^2 + I\beta_n^4 \bar{E}]^{-1} \sin(\beta_n x). \tag{39}$$

The inverting of the Laplace transform in Eqs. (38), (39) depends on the specific form of the function $\bar{E}(s)$. The function $\bar{E}(s)$ depends on the viscoelastic model. For ordinary models (Maxwell, Kelvin-Voight, etc.) the following function can be used

$$\bar{E} = \frac{sk_1}{sk_2 + k_3}. \tag{40}$$

For a Maxwell model the above constants are defined as

$$k_1 = 9GK, \quad k_2 = 3K + G, \quad k_3 = 3\tau_0 k$$

$$k = \frac{\mu E}{(1 + \mu)(1 - 2\mu)} + \frac{2}{3}G, \quad \tau_0 = \frac{G}{\eta}, \tag{41}$$

where ν is the Poisson coefficient, G is the shear modulus, and η is the damping constant from Maxwell model.

Inverting the Laplace transform in Eqs. (38), (39) and using Eq. (40) we obtain the final form for the iterative relations:

$$u_1^{(i)}(x,t) = \frac{2}{L} \sum_{n=1}^{\infty} \{ \int_0^t \{ \frac{e^{-\lambda_n D_n (t-\tau)}}{\rho \lambda_n \sqrt{1 - D_n^2}}$$

$$\sin[\lambda_n (t-\tau) \sqrt{1 - D_n^2}] + \frac{k_3 \sqrt{1 - D_n^2}}{\rho \beta \lambda_n} -$$

$$\frac{k_3 e^{-\lambda_n D_n (t-\tau)}}{\rho \beta \lambda_n}$$

$$[\sqrt{1 - D_n^2} \cos[(t-\tau)\lambda_n \sqrt{1 - D_n^2}] -$$

$$D_n \sin[(t-\tau)\lambda_n \sqrt{1 - D_n^2}]\}$$

$$\{[(-1)^n - 1] \frac{\rho \omega^2(\tau)}{\beta_n^2} + \frac{f_1^{*c}(n,\tau)}{A} +$$

$$(-1)^n \frac{F_2(\tau)}{AE} - \frac{F_1(\tau)}{AE} +$$

$$\Phi_1^{(i-1)*c}(n,\tau)\}d\tau\} \cos(\beta_n x) + \frac{u_1^{(i)*c}(0,t)}{2}, \qquad (42)$$

$$u_2^{(i)}(x,t) = \frac{2}{L} \sum_{n=1}^{\infty} \{ \int_0^t \{ \frac{e^{-\mu_n E_n (t-\tau)}}{\rho \mu_n \sqrt{1 - E_n^2}}$$

$$\sin[\mu_n (t-\tau) \sqrt{1 - E_n^2}] + \frac{k_3 \sqrt{1 - E_n^2}}{\rho \beta \mu_n} -$$

$$\frac{k_3 e^{-\mu_n E_n (t-\tau)}}{\rho \beta \mu_n}$$

$$[\sqrt{1 - E_n^2} \cos[(t-\tau)\mu_n \sqrt{1 - E_n^2}] -$$

$$E_n \sin[(t-\tau)\mu_n \sqrt{1 - E_n^2}]\}$$

$$\{(-1)^n \frac{\rho \alpha(\tau)}{\beta_n} + [(-1)^n - 1] \frac{\rho a_{01}(\tau)}{\beta_n^2} -$$

$$\frac{f_2^{*s}(n,\tau)}{A} - \Phi_2^{(i-1)*s}(n,\tau)\}d\tau\} \sin(\beta_n x), \qquad (43)$$

where

$$\lambda_n = \beta_n \sqrt{\frac{k_1}{\rho k_2}}, \quad D_n = \frac{k_3}{2\beta_n} \sqrt{\frac{\rho}{k_1 k_2}},$$

$$\mu_n = \beta_n^2 \sqrt{\frac{Ik_1}{B_n}}, \quad E_n = \frac{k_3}{2\beta_n^2} \sqrt{\frac{\rho}{Ik_1 k_2 k_3}}.$$

5.2. *Computing Algorithm*

Step 1. Determine $\omega(t)$ and $\alpha(t)$ from the rigid body motion of the link.

Step 2. With $\Phi_1^{(0)}(x,t) = \Phi_2^{(0)}(x,t) = 0$ using Eqs. (42), (43), compute the displacements $u_1^{(1)}(x,t)$ and $u_1^{(1)}(x,t)$.

Step 3. With $u_1^{(1)}(x,t)$ and $u_1^{(1)}(x,t)$ determined at Step 2, calculate $\Phi_1^{(1)}(x,t) = \Phi_2^{(1)}(x,t)$.

Step 4. With $\Phi_1^{(1)}(x,t) = \Phi_2^{(1)}(x,t)$ computed at Step 3 determine the displacements $u_1^{(2)}(x,t)$ and $u_1^{(2)}(x,t)$. The iteration process will continue until a desired precision is obtained.

6. The Vibrations of a Flexible Link with a Lubricated Slider Joint

As shown in Fig. 3 we consider a linear-elastic kinematic link on which a slider joint moves along the of axis the member. A lubrication film is located in the lower interface of the link and the translational joint. Our main focus here, is the effect of the pressure at the fluid interface on the fundamental vibration mode of the elastic member in the transverse direction $(u_2(x,t))$. Therefore, we further simplify the equations of motion given by Eqs. (18) through (20) by considering $u_2(x,t)$ only. Then, carrying out the derivations in the previous section for this variable yields

$$EI_z \partial^4 u_2/\partial x^4 + \rho A \partial^2 u_2/\partial t^2 + \omega^2 \rho A u_2 + \rho A a_{02} +$$
$$\omega \rho A v_{01} + \alpha \rho A x - f_2(x,t) = 0 \tag{44}$$

where, $\omega = \omega_3$. The term $f_2(x,t)$ represents the external force per unit length, which includes the fluid pressure and other external effects.

6.1. *Reynolds Equation of Lubrication*

Now to develop the equations that represent the effect of fluid pressure, we define $\xi(X), X \in [0,l]$ as the film thickness in the transverse direction. The length of the slider is l. Using Reynolds equation the differential equation for the pressure $p(X, Z, t)$ is given by

$$\frac{\partial}{\partial X}(\xi^3 \frac{\partial p}{\partial X}) + \frac{\partial}{\partial Z}(\xi^3 \frac{\partial p}{\partial Z}) = 6\mu V_r(t)\frac{\partial \xi}{\partial X}, \tag{45}$$

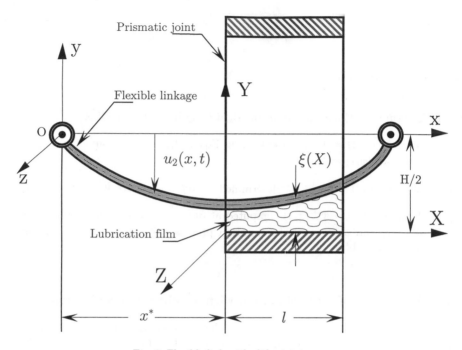

Fig. 3. Flexible link with slider joint

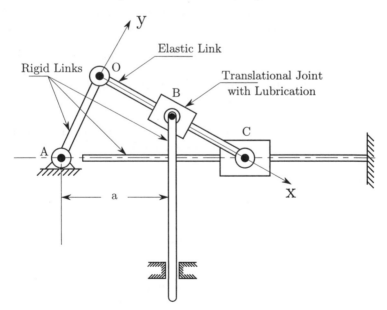

Fig. 4. Planar slider-crank mechanism

where μ is the dynamic viscosity and $V_r(t)$ is the relative velocity between the link and the translational joint. Choosing the thickness of the translational joint as B (i.e. $Z \in [0, B]$) and assuming symmetry between the link and the joint in the cross-sectional plane, as shown in Fig. 3, we obtain the following relations:

$$p(X, B, t) = p(X, 0, t),$$
$$\frac{\partial p}{\partial Z}(X, B, t) = \frac{\partial p}{\partial Z}(X, 0, t).$$

We average the pressure in the transverse direction so the lubricant pressure distribution is reduced to a one dimensional problem introducing a variable $P(X, t)$ given as

$$P(X, t) = \frac{1}{B} \int_0^B p(X, Z, t) \, dZ. \tag{46}$$

Substituting Eq. (46) in (45) yields

$$\xi^3 \frac{\partial^2 P}{\partial X^2} + 3\xi^2 \frac{\partial \xi}{\partial X} \frac{\partial P}{\partial X} = 6\mu V_r(t) \frac{\partial \xi}{\partial X}. \tag{47}$$

Furthermore, using geometry we obtain

$$\xi = \frac{H}{2} - u_2(x^* + X, t)$$

The boundary conditions for $P(X, t)$ are given as

$$X = 0 \Rightarrow P(0, t) = p_{atm}; \quad X = l \Rightarrow P(l, t) = p_{atm},$$

where p_{atm} is the atmospheric pressure.

The general solution for the Eq.(47) is

$$P(X, t) - 6\mu V_r(t) I_1(X, t) + C I_2(X, t) + C_1, \tag{48}$$

where

$$I_1(X, t) = \int dX / [H/2 - u_2(X + x^*)]^2,$$
$$I_2(X, t) = \int dX / [H/2 - u_2(X + x^*)]^3.$$

The constants C and C_1 can be computed from the boundary conditions. The integrals I_1 and I_2 are computed later in the work by using Fourier transform methods.

6.2. Cavitation

Cavitation occurs when the pressure computed from Eq. (48) drops below the ambient pressure. When this occurs, Reynolds equation does not apply inside the cavitation region. The position where cavitation occurs can be found by changing the boundary conditions such that the pressure curve exhibits a smooth transition to the cavitation region. Modifying the boundary conditions gives

$$X = 0 \Rightarrow P(0, t) = p_{atm}; \quad \Rightarrow P(\bar{X}, t) = p_{atm} \text{ and } \left. \frac{\partial P}{\partial X} \right|_{X=\bar{X}} = 0,$$

where $\bar{X}$ is the starting position of the cavitation region. Now the three boundary condition can be used to compute the pressure and the coordinate $\bar{X}$.

In the succeeding section we present an algorithm to solve the system of differential equations given by Eqs. (44) and (47) subject to the corresponding boundary and initial conditions.

6.3. Solution Method for an Elastic Link in a Rigid Mechanism

One of the underlying assumptions in the present study is that the rigid body motion of the elastic member is not affected by the elastic deformations. Using this assumption, the method of solution presented below assumes that the time profiles of rigid body accelerations, velocities, and the dynamic reaction forces at the joints are given. More specifically, the time functions in Eq. (44) v_{01}, a_{02}, ω, α, and f_2 are completely known at the onset of the computations.

The procedure that is followed to solve the equations of motion can be outlined by the following steps:

1. Use the rigid body joint velocities and accelerations, geometry of the mechanism, and the kinematic relations to compute the position $x^*(t)$ of the translational link on the elastic member, force and moment reactions in the slider joint.

2. Transform Eq. (44) by applying finite Fourier sine transform

    (a) Using the boundary conditions

    $$\frac{\partial^2 u_2(0, t)}{\partial x^2} = \frac{\partial^2 u_2(L, t)}{\partial x^2} = u_2(0, t) = u_2(L, t) = 0,$$

    and applying finite Fourier sine transform to Eq. (44) yields the following ordinary differential equation

    $$\frac{d^2 u_2^*(n, t)}{dt^2} + \left[\frac{EI_z}{\rho A} \beta_n^4 + \omega^2(t) \right] u_2^*(n, t) - \frac{f_2^*(u_2^*(n, t), n, t)}{\rho A} =$$
    $$-\frac{1}{\beta_n} \left\{ [a_{02}(t) + \omega(t) v_{01}(t)][1 + (-1)^{(n+1)}] + \alpha(t)(-1)^{(n+1)} L \right\}. \quad (49)$$

    The term $f_2^*(n, t)$ in Eq. (49) can be written as

    $$f_2^*(u_2^*(n, t), n, t) = R^*(n, t) + P^*(u_2^*(n, t), n, t) \quad (50)$$

where $R^*(n,t)$ is the finite Fourier sine transform of the joint reaction forces, which is a known function of time and the transform of the force pre unit length in lubrication film is $P^*(u_2^*(n,t),n,t)$. The pressure response to the nth harmonic of deflection can be approximated by the following expression

$$p^{(n)}(X,t) = 6\mu V_r(t)I_1^{(n)} + CI_2^{(n)} + C_1,$$

where

$$I_1^{(n)} = -\frac{L}{n\pi}(H\gamma_3, -\gamma_4)$$

$$I_2^{(n)} = -\frac{L}{n\pi}\left\{\gamma_3\left[\frac{H^2}{2} + \phi^2(n,t)\right] + \left[\frac{1}{\gamma_2} + \frac{3H}{2\gamma_1}\right]\right\},$$

$$\gamma_1 = \frac{H^2}{4} - \phi(n,t)^2,$$

$$\gamma_2 = -\frac{H}{2} + \phi(n,t)\sin\left(\frac{n\pi}{L}x\right),$$

$$\gamma_3 = \gamma_1^{-3/2}\tan^{-1}\left[\frac{\phi(1,t) - \frac{H}{2}\tan\left(\frac{n\pi}{L}x\right)}{\sqrt{\gamma_1}}\right],$$

$$\gamma_4 = \frac{\phi(n,t)}{2\gamma_1\gamma_2}\cos\left(\frac{n\pi}{L}x\right),$$

$$\phi(n,t) = \frac{2}{L}u_2^*(n,t).$$

Accordingly, the finite Fourier sine transform of the force per unit length can be written as

$$P^*(u_2^*(n,t),n,t) = B\int_{x^*}^{x^*+l}\left[6\mu V_r(t)I_1^{(n)} + CI_2^{(n)} + C_1\right]\sin(\beta_n x)\,dx. \quad (51)$$

(b) Obtain the numerical solution of the nonlinear ordinary differential equation given by Eq. (49) by using a Bulirsh-Stoer algorithm with adaptive step-size control. Application of the differential solver also requires the numerical integration of Eq. (51) for given values of t and $u_2^*(n,t)$. This is realized by using an algorithm that is based on the trapezoidal rule with automatic step adjustment for accuracy.

(c) The final solution for the equation of motion is obtained by inverting finite Fourier sine transform and applying the superposition principle for the harmonics

$$u_2(x,t) = \sum_{n=1}^{n=\infty} \phi(n,t)\sin(\beta_n x).$$

6.4. Application to a Slider Mechanism

In this section we apply the method proposed above to investigate the vibration of the link OC of the slider mechanism presented in Fig. 4. To simplify our presentation, we consider the flexible effects in member OC only.

We consider the motion of the system when the link OA has a constant angular velocity ω_0. Then the kinematic variables of link OC can be written as

$$v_{O1} = -OA \ \omega_0 \sin(\omega_0 t) \cos\theta - OA \ \omega_0 \cos(\omega_0 t) \sin\theta,$$

$$a_{02} = -OA \ \omega_0^2 \sin\theta \cos(\omega_0 t) - OA \ \omega_0^2 \cos\theta \sin(\omega_0 t),$$

$$\omega = -\frac{\lambda\omega_0 \cos(\omega_0 t)}{[1 - \lambda^2 \sin(\omega_0 t)^2]^{1/2}},$$

$$\alpha = -\frac{\lambda^3 \omega_0^2 \cos(\omega_0 t)^2 \sin(\omega_0 t)}{[1 - \lambda^2 \sin(\omega_0 t)^2)]^{3/2}} + \frac{\lambda\omega_0^2 \sin(\omega_0 t)}{[1 - \lambda^2 \sin(\omega_0 t)^2]^{1/2}},$$

where

$$\lambda = OA/L,$$

$$\theta = \sin^{-1}[\lambda \sin(\omega_0 t)],$$

where $OA = 0.07\,m$ and $OC = 1.0\,m$ are the lengths of the members OA and OC respectively, and the distance between joint A and the vertical sliding member is selected as $a = 0.4$ m (see Fig. 4). The elastic link is made of aluminum ($E = 207 \times 10^9 \ N/m^2$, $\rho = 7650 \ kg/m^3$) with equal cross-sectional width and height, i.e. $B = h$. The magnitude of the joint reaction, which include the mass of the vertical slider and prismatic joint is $100N/m$ and is concentrated in the center of the joint.

Throughout the succeeding analysis, only the fundamental mode of transverse oscillations (i.e. $n = 1$) is taken into consideration to simplify our presentation.

Figure 5 depicts the time-space curves of the beam deflection for $h = 0.01m$, $\omega_0 = 10\pi$ rad/s and zero initial conditions. The deflection surface depicted in Fig. 5 demonstrates a recurring pattern which seems to be synchronous with the motion of the crank OA. Actually, the overall motion is nearly periodic with the two frequencies (ω_n and ω) contributing to the oscillation of the beam. The transverse oscillation of the elastic link is influenced by the uniform motion of link OA, by the joint reaction and by the lubrication film pressure.

Figure 6 depicts the time-space curves of the pressure response for the same characteristics of the planar chain. The pressure field exerted through the oil, which is considered as viscous and Newtonian, is obtained from the solution of the Reynolds equation of lubrication. The boundary conditions result in ambient pressure at the edges of the prismatic joint. In order to support the load exerted by the beam, the bearing-film pressure must rise above the ambient. The pressure in the prismatic joint (i.e. plane slider bearing) is a function of the local film thickness, which in our case is dependent on the deflection of the beam. Accordingly the pressure, profile is expected to conform to the deflection profile given in Fig. 5.

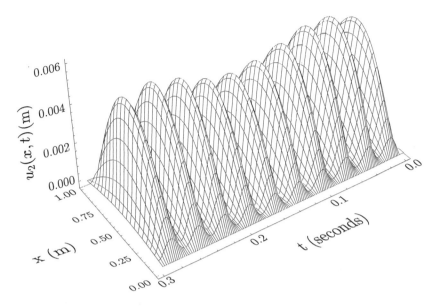

Fig. 5. Deflection of the flexible link in time and space

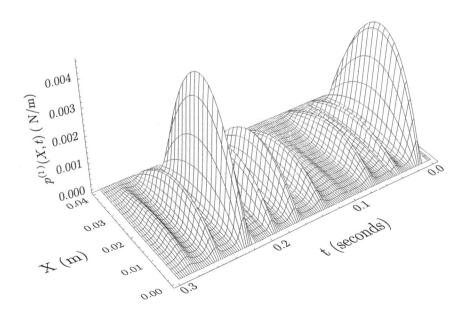

Fig. 6. Pressure response in the lubrication film

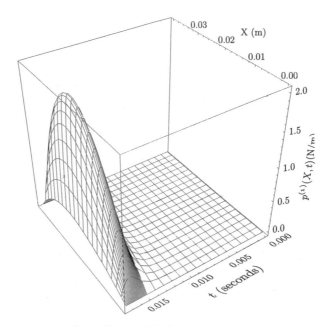

Fig. 7. Breakdown of the lubrication film

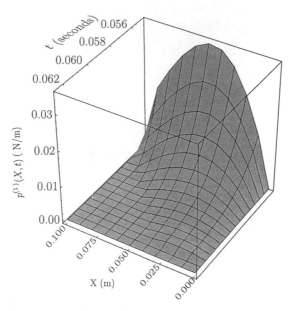

Fig. 8. Cavitation phenomena

Figure 6 shows the variation of the pressure in time and space and its dependence on vibrations of the elastic link in rototranslational motion.

An interesting phenomena occurs during the motion of the present system when the deflection of the beam is as large as the height of the prismatic joint. The lubrication film breaks down, leading to direct contact between the beam and the prismatic joint. The pressure response during the breakdown of the lubrication film at $\omega_0 = \frac{79}{12}\pi$ rad/s is shown in Figure 7. We observe that there is a very sharp increase in the pressure as thickness of the lubrication decreases. Yet, the pressure drops again because of depletion the of fluid between the beam and the joint.

When the pressure drops below the ambient, Reynolds equation does not apply and cavitation occurs. Figure 8 shows the evolution of the pressure and the starting position of the cavitation region for $\omega_0 = \frac{40}{3}\pi$ rad/s and $l = 0.1m$. We follow the methodology that has been presented above, and modify the boundary conditions accordingly when cavitation is detected. Cavitation first occurs at $X = 0.09m$ and since when the Reynolds equation does not apply the pressure will be ambient. The pressure during cavitation is recomputed using the new boundary conditions and the response in time and space can be observed from the figure.

7. References

1. A.G. Erdman and G.N. Sandor, "Kineto-Elastodynamics-A Review of the State of the Arts and Trends," *Mechanisms and Machine Theory* **7** (1972) 19-33.

2. A.G. Erdman, G.N. Sandor, and R.G. Oakberg, "A General Method for Kineto-Elastodynamics Analysis and Synthesis of Mechanisms," *Journal of Engineering for Industry*, **94** (1972) 1193-1205.

3. D.A. Turcic and A. Midha, "Generalized Equations of Motion in the Dynamic Analysis of Elastic Mechanism Systems. Part I: Applications," ASME *Journal of Dynamic Systems, Measurements, and Control*, **106** (1984) 249-254.

4. S. Nagarajan and D.A. Turcic, "Lagrangian Formulation of the Equation of Motion for Elastic Mechanisms with Mutual Dependence Between Rigid Body and Elastic Motion. Part I: Element Level Equations," ASME *Journal of Dynamic Systems, Measurement, and Control*, **112** (1990) 203-213.

5. P.W. Jasinski, H.C. Lee, and G.N. Sandor, "Vibration of Elastic Connecting Rod of a High-Speed Slider Crank Mechanism," ASME *Journal of Engineering for Industry*, **93** (1971) 636-644.

6. S.C. Chu and K.C. Pan, "Dynamic Response of a High-Speed Slider-Crank Mechanism With an Elastic Connecting Rod," ASME *Journal of Engineering for Industry*, **97** (1975) 542-549.

7. M. Badlani and A. Midha, "Member Initial Curvature Effects on the Elastic Slider-Crank Mechanism Response," ASME *Journal of Mechanical Design*, **104** (1982) 159-167.

8. M. Badlani and A. Midha, "Effect of Internal Material Damping on the Dynamics of a Slider-Crank Mechanism," ASME *Journal of Mechanisms, Transmissions, and Automation in Design,* **105** (1983) 452-459.

9. K. Farhang and A. Midha, "Investigation of Parametric Vibration Stability in Slider-Crank Mechanisms with Elastic Coupler," ASME *13-th Biennial Conference on Mechanical Vibration and Noise,* **DE 37** (1991) 167-176.

10. K.W. Buffinton and T.R Kane, "Dynamics of Beams Moving over Supports," *Int. J. Solids Structures,* **21** (1985) 617-643.

11. Buffinton, K.W., "Dynamics of Elastic Manipulators With Prismatic Joints," ASME *Journal of Dynamic Systems, Measurements, and Control,* **114** (1992) 41-49.

12. F. Gordaninejad, A. Azhdari, and N.G. Chalhoub, "Nonlinear Dynamic Modeling of a Revolute-Prismatic Flexible Composite-Material Robot Arm," ASME *Journal of Vibration and Acoustics,* **113** (1991) 461-468.

13. J. Yuh and T. Young, "Dynamic Modeling of an Axially Moving Beam In Rotation: Simulation and Experiment," ASME *Journal of Dynamic Systems, Measurement, and Control,* **113** (1991) 34-40.

14. F.M. White, *Viscous Fluid Flow,* (McGraw-Hill, New York, 1991).

15. R.C. Benson and F.E. Talke, "The Transition Between Sliding and Flying of a Magnetic Recording Slider," IEEE *Transactions on Magnetics,* **MAG-23** (1987) 3441-3443.

16. T. Alfrey, "Non-homogeneous Stresses in Viscoelstic Media," *Quart. Appl. Mech.,* **2** (1944) 113.

17. E.H. Lee, "Stress Analysis in Viscoelastic Bodies," *Quart. Appl. Mech.,* **8** (1955) 183.

18. M.I. Buculei and D.B. Marghitu, "On Kinetoelastodynamic Analysis of Spatial Mechanisms with Rectilinear Elements," RRST *Mécanique appliquée,* **2** (1986) 855-864.

19. W. Nowacki, *Dynamic of Elastic Systems,* (John Wiley, New York, 1963).

Dynamics with Friction: Modeling, Analysis and Experiment, Part II, pp. 99–123
edited by A. Guran, F. Pfeiffer and K. Popp
Series on Stability, Vibration and Control of Systems, Series B, Vol. 7
© World Scientific Publishing Company

SOLITONS, CHAOS AND MODAL INTERACTIONS
IN PERIODIC STRUCTURES

M. A. DAVIES *

National Institute of Standards and Technology, Manufacturing Engineering Laboratory
Gaithersburg, MD 20899, USA

F. C. MOON

Department Mechanical and Aerospace Engineering, Cornell University
Ithaca, NY 14853, USA

ABSTRACT

An experimental structure consisting of nine harmonic oscillators coupled through
buckling sensitive elastica was constructed. The response of the structure to both
periodic and impulse excitation was examined. Due to its relatively large number of
degrees-of-freedom and the strong geometric nonlinearity of the buckling elements,
the structure exhibits highly complex phenomena including modal interactions,
temporally chaotic vibrations, and complex transient motions characterized by the
emergence and subsequent breakdown of soliton-like waves. The experiment was
approximately modeled by a modified Toda lattice, and the dynamic behavior of
the model and the experiment were found to be quite similar. Calculations based
upon Toda's analytical results confirm the soliton-like nature of waves observed
in the structural motions. These findings provide basic insight into the nonlinear
behavior of multiple degree-of-freedom buckling structures. This may be useful
for understanding and predicting the dynamic failure of aircraft, ship and space
structures that are subjected to large dynamic loads.

1. Introduction

Periodically reinforced structures are used in the construction of spacecraft, aircraft and ships to minimize weight while maintaining structural integrity. One of
the common failure modes for such structures is the dynamic buckling of flexible
structural elements. In this chapter, we examine the nonlinear dynamic buckling
of a periodically reinforced structure subjected to both impact and periodic excitation. Experiments and numerical simulations demonstrate a variety of nonlinear
phenomena including modal interaction, low-order temporal chaos and soliton-like
waves that can evolve into complex spatial patterns with chaos-like dynamics. An
understanding of these phenomena is necessary if the dynamics and ultimately the
failure modes of periodic structures are to be predicted. These results may also be
relevant to the interpretation of structural failure tests and large scale numerical

---

*The experimental and numerical work presented here was completed in the Department of Mechan-
ical and Aerospace Engineering, Cornell University, Ithaca NY 14853, USA.

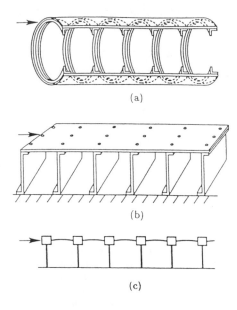

Figure 1: Drawing showing the similarity between practical shell and plate type structures (a) and (b) and our simplified experimental set-up (c).

simulations, perhaps suggesting that an ensemble of tests might be necessary to determine the sensitivity of buckling behavior to small changes in loading and design parameters. Figure 1 illustrates the similarity of our experiment to more complex engineering structures.

Chaotic oscillations have been widely observed in structural systems. These oscillations arise from various nonlinearities; impact and rattling nonlinearities were studied by Moon and Shaw[22], dry friction by Feeny[1], elastoplasticity by Pratap, Mukherjee and Moon[26] and geometric nonlinearities by Cusumano and Moon[5] and Dowell[13]. Erratic oscillations in structural buckling were first documented by Tseng and Dugungi[31] in their study of snap-through oscillations in a pre-buckled arch. Holmes[18] showed that these oscillations were actually chaotic, and that the pre-buckled arch exhibits classic period-doubling bifurcations and fractal-like strange attractors. Shortly after, Moon and Holmes[21] experimentally observed period doubling and chaotic strange attractors in a magnetically buckled beam. In each of these studies, the motions are restricted experimentally and numerically to a low number of degrees-of-freedom (usually a single degree of freedom), while the dynamics of actual engineering systems often involve numerous degrees-of-freedom which may interact in a complex manner.

Much insight into the behavior of higher order nonlinear systems can be found by examining the physics literature on the propagation of energy in atomic lattices. Lattice models typically consist of infinite arrays of periodically spaced masses (atoms) that interact in a nonlinear manner with their neighbors. These systems have been shown to exhibit complex nonlinear wave phenomena including highly localized disturbances known as solitary waves in which the tendency of the wave to disperse is balanced by nonlinear effects.

Modern work on the theory of nonlinear lattices began with the discovery of modal trading phenomena by Fermi, Pasta and Ulam[15] in the early 1950's. This phenomenon of modal interactions and modal trading of energy has subsequently been studied by many authors in both the engineering and physics community[32,16,17,24]. Cusumano has shown that the interaction of modes in structural vibrations can be elegantly demonstrated using the Carhunen-Loéve Decomposition (commonly known as *proper orthogonal decomposition* (POD))[6] a technique that has been used for several years in the study of fluid turbulence[4]. The discovery by Toda of an analytic solution for a lattice with exponential interactions sparked another flurry of research[28,29,30,23,3]. Toda's analytic solutions take the form of both periodic traveling waves and localized soliton waves. A more detailed discussion of waves in periodic lattice structures can be found in Davydov[11].

In addition to having temporally chaotic behavior, there is also strong evidence that nonlinear elastic structures exhibit highly complex, spatially chaotic deformation patterns[10,19,25,27]. However, the work presented here is one of the first to experimentally and numerically examine the dynamics of a higher degree-of-freedom buckling structure in which spatially and temporally complex motions appear to co-exist. Earlier publications by the authors[8,10] concerned with the impact dynamics of this structure have demonstrated strong temporal sensitivity to initial conditions that appear to be related to the emergence and subsequent dispersion of solitary waves. Further work[9] has shown that these solitary waves are similar to Toda solitons, and that their behavior can be approximately analyzed using some of Toda's analytical results. In this chapter, we provide a more complete discussion of the dynamics of this experimental structure examining its forced response and nonlinear modal behavior as well. Proper orthogonal decomposition is the primary technique used to analyze the modal behavior of the experiment.

The second section begins by describing the experimental structure. Section 3 then presents a numerical model for the system that is based upon a set of modified Toda lattice equations. Section 4 describes the nonlinear modal behavior of the structure and discusses the results of the application of POD to the experimental data obtained from the response of the structure to periodic excitation. In section 5 the transient response of the structure to impact excitation is discussed. Wave speed calculations based upon Toda's analytical results show excellent agreement with both experiment and numerical simulation. Section 6 presents our conclusions and suggestions for future work.

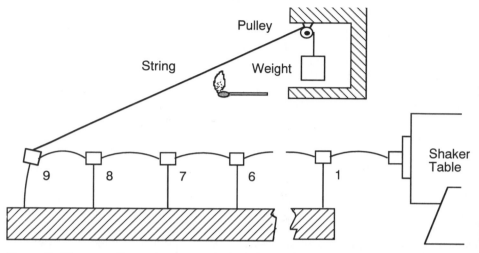

Figure 2: Diagram of experiment consisting of nine elastic oscillators coupled through buckling sensitive elastica.

2. Experiment

The experimental structure consisted of nine elastic oscillators, coupled through buckling sensitive elastica (Figure 2). Each elastic oscillator was constructed by fastening a 157 g aluminum block to the end of an 0.81 mm thick, 25 mm wide, spring steel, cantilevered beam. The length of the oscillators (measured from the fixed end of the beam to the center of mass of the attached block) was approximately 170 mm, and the distance between centers of adjacent blocks was 190 mm. The elastic buckling elements clamped between blocks were made from 0.127 mm thick, steel feeler gage stock. Each was plastically deformed into a shallow arch to enhance its first mode buckling response. One end of the structure was attached through a buckling element to an electromagnetic shaker and the other left free. In this paper, we discuss the response of this structure to both periodic inputs from the electromagnetic shaker and to suddenly released compressive loads applied at the free end.

Bending measurements were taken from a strain gage pair at the base of each cantilevered beam. Strain measurements were correlated to mass displacement for first-mode deformations of the cantilevered beams. Higher-mode vibrations in the beams introduce mechanical signal noise; however, for the results discussed here the ratio of the first mode vibration amplitude to that of the higher modes was generally greater than 5:1. Data was taken by a 16-channel data acquisition system.

Small amplitude vibration tests were used to characterize the linear behavior of the structural components. The first mode vibration frequency of the elastic oscillators was approximately 4.3 Hz, corresponding to an effective spring constant of $k = 117$ N/m. The second mode of the elastic oscillators had a frequency of approximately 50 Hz; as noted above, in our experiments first mode cantilever deformations dominate and therefore the second and higher modes are ignored. For small motions, the behavior of the arched buckling elements was nearly linear with stiffness $\gamma = 362$ N/m. Due to differences in the elements, γ varied by as much as $\pm 6\%$ from element to element. This nonuniformity in the buckling elements is the dominant source of parameter uncertainty in our characterization of the experiment.

Small motion tests were also used to characterize the vibration modes of the full nine-mass experimental set-up. Figure 3 shows the nine, linear, modal resonance frequencies of the structure plotted versus wavenumber. These data points trace out the dispersion relation for the structure. Analytical predictions indicate that, for small motion, the dispersion relation varies periodically with wavenumber, and thus the structure allows only a band of frequencies to propagate without attenuation (see next section, particularly Equation 5). The frequency response of the free-end mass to small amplitude random excitation by the shaker table verifies this prediction, showing a sharp passband ranging from approximately 4.6 Hz to 16.2 Hz (Figure 4). This passband is known as a Brillouin zone (Brillouin, 1946). The observed resonant frequencies and spatial shapes of the structural modes agree well with analytical calculations using linear lattice theory. Structural damping was found to be very low; logarithmic decay rates indicate that all modes are less than 0.1% critically damped.

To characterize the nonlinear behavior of the buckling elements, their static force-displacement behavior was measured. The measurement was done by rigidly clamping one end of the buckling element and attaching the other to an elastic oscillator. Static horizontal loads were applied to the mass, and its displacement was determined from the strain gages. The linear spring behavior of the cantilever beam was subtracted from the result, producing the solid curve in Figure 5. The behavior of the elements was highly nonlinear, rapidly stiffening in tension (positive displacement) and softening rapidly in compression/buckling (negative displacement). Note also that the slope of the static force-displacement curve at the origin nearly matches the value of 362.0 N/m determined in vibration tests.

3. Numerical Model

The experiment was modeled as a one-dimensional lattice of point masses with nearest-neighbor nonlinear interactions (see Figure 6). The accuracy of the model is based upon the following simplifying assumptions.

1. The aluminum blocks behave as point masses (mass m) moving along a hori-

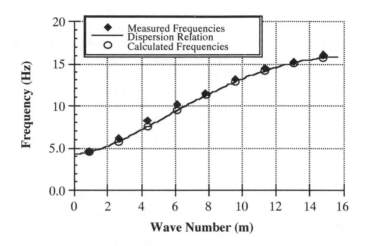

Figure 3: Experimentally measured and analytically predicted dispersion relations for the small motion behavior of the structure and model.

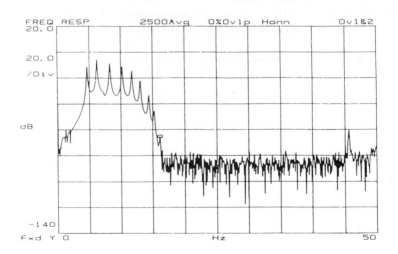

Figure 4: Frequency response of the free-end mass to 0-50 Hz, white noise excitation input from the electromagnetic shaker.

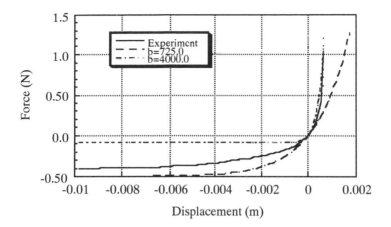

Figure 5: Experimental measurement and mathematical approximation of the force-displacement behavior of the buckling elements. The mathematical approximation is shown for two values of b.

zontal line. Block rotations are ignored.

2. The elastic oscillators (cantilever-mass combinations) deform only in first mode, and can be treated as simple harmonic oscillators with stiffness k and mass m.

3. The buckling elements behave as simple nonlinear springs with the force-displacement behavior shown in Figure 5. Higher buckling modes and moments in the elements are ignored.

4. Experimental components have uniform properties (all blocks have mass m, all cantilevers have stiffness k, etc.).

5. Damping is linear.

The equations of motion for the lattice model shown in Figure 6 are,

$$\ddot{x}_j = -\frac{c}{m}(3\dot{x}_j - \dot{x}_{j+1} - \dot{x}_{j-1}) - \frac{k}{m}x_j + \frac{1}{m}(F(x_{j+1} - x_j) - F(x_j - x_{j-1})), j = 1, \ldots, N \quad (1)$$

with boundary conditions,

$$x_0 = 0 \text{ (fixed-end/zero displacement)}$$

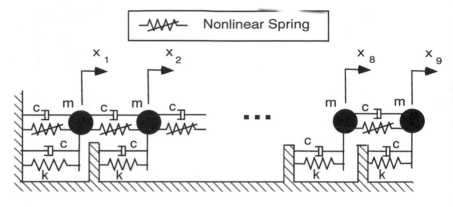

Figure 6: Nonlinear lattice model of experimental structure.

$$(2)$$

$$x_{N+1} = x_N \text{ (free-end/zero force)}.$$

In Equation 1, m is the block mass, k is the cantilevered beam stiffness, $F()$ represents the nonlinear force-displacement relationship for the buckling elements, and the overdot represents differentiation with respect to time, t. The index j ranges from 1 to N, where N is the number of masses, and the free and fixed boundary conditions (Equations 2) are enforced by defining "dummy" masses 0 and $N + 1$. Also note that the $3\dot{x}_j$ damping term in the equations comes about because the dampers between masses and the ones attached between the masses and ground are taken to have identical coefficients.

To complete the model, it was necessary to find a functional expression for the force-displacement behavior of the buckling elements. Although planar elastica theory could have been used to derive this relationship[14] evaluation of the results would require numerical solution of transcendental equations involving elliptic integrals. This approach was too computationally intensive to be practically used in numerical simulation. Instead, we chose to approximate buckling element force, $F(r)$, with an exponential,

$$F(r_j) = \frac{\gamma}{b}(e^{br_j} - 1) \qquad (3)$$

where $r_j = x_{j+1} - x_j$ is the relative displacement between two adjacent masses, γ the element's linearized stiffness $(\frac{dF(r)}{dr}|_{r=0})$, and b determines the strength of the nonlinearity. Figure 5 shows a plot of the exponential force law for $\gamma=362.0$ N/m and $b = 725.0$ m^{-1}. For this approximate value of b, the exponential force law and the experimentally measured behavior are qualitatively similar. Later, we discuss how to refine the value of b, based on experimental measurements, so that our analytical law more closely approximates the tensile behavior of the buckling elements.

The force law in Equation 3 is the same as that used by Toda[28,29,30] to model the interaction forces in an atomic lattice. If k and c are set to zero, Equations 1 and 3 reduce to those of a Toda lattice, which Toda showed to have exact nonlinear "soliton" solutions (see Equation 12 for details). Although the term "soliton" is not precisely defined, Drazin[12] states that it is used to describe any solution of a nonlinear differential equation that (i) represents a wave of permanent form; (ii) is localized, decaying or becoming constant at infinity; and (iii) may interact strongly with other solitons so that after the interaction it retains its form. In a Toda lattice with γ and b both positive, a "soliton" would appear as a localized tensile-wave with a nonlinear speed-amplitude dependence. We show in Section 5 that "soliton-like" disturbances occur in our system.

For the case of small displacement and zero damping $(c=0)$, Equations 1 and 3 may be reduced to the linear conservative system defined by Equation 4.

$$\ddot{x}_j + \frac{k}{m}x_j - \frac{\gamma}{m}(x_{j+1} - 2x_j + x_{j-1}) = 0, j = 1, \ldots, N \tag{4}$$

Substituting a periodic traveling wave solution of the form, $x_j(t) = Ae^{i(\kappa aj - \omega t)}$, into Equations 4 produces the dispersion relation,

$$\omega = \pm\sqrt{\frac{4\gamma}{m}\sin^2(\frac{\kappa a}{2}) + \frac{k}{m}} , \tag{5}$$

where ω is the wave frequency, κ is the wave number and a is the spacing between adjacent masses. The form of Equation 5 indicates that the system allows only a band of frequencies to propagate without attenuation; this band is the Brillouin zone measured in the previous section. Substituting the measured values for γ and k into Equation 5, we predict that the Brillouin zone for this structure should range from 4.3 Hz to 15.9 Hz, in excellent agreement with the experiment. Figure 3 shows that the analytically calculated modal frequencies and dispersion relation agree well with experimental data.

The rate at which energy propagates in the linear system is governed by the slope of the dispersion relation or group velocity, $\frac{d\omega}{d\kappa}$. For the measured parameter values, the maximum possible group velocity for this system is approximately 6.9 m/s, which is the upper limit on the speed of energy propagation for the linearized system. However, as we will see in Section 5, nonlinear wave speeds may greatly exceed this value.

Equations 1, 2 and 3 form a set of N coupled, second-order, nonlinear, differential equations whose solution requires $2N$ initial conditions defining the positions and velocities of the N masses. These equations were integrated using a commercially available 5th-6th order Runge-Kutta-Vernier routine. Based upon previous discussion, the parameters used for the simulation were $m = 157$ g, $\gamma = 362$ N/m, $b = 725$ m^{-1}, $k = 117$ N/m and $c = 0.01$ N·s/m. Accuracy off the numerical integration was verified by checking conservation of energy for the zero-damping $(c = 0)$ case.

4. Forced Vibrations and Modal Interactions

It is now well-known that in many situations, the motions of a multiple degree-of-freedom nonlinear dynamic system can be described in terms of a relatively low number of interacting "nonlinear modes". In these cases, the number of degrees of freedom necessary to describe the motions of the system is reduced, and often the equations of motion for the system can be simplified to a level amenable to low-order nonlinear dynamic analysis. In this section, we discuss motions of our experimental structure and model that are in this category. POD is used to show that the forced vibrations of this structure are in many cases characterized by a few interacting degrees of freedom. In the next section, we discuss another case where nonlinear wave motions resulting from an impact excitation of the structure appear to involve a much larger number of degrees of freedom.

4.1. Numerical Experiment - Modal Trading

In their study of modal trading phenomena, Fermi, Pasta and Ulam[15] showed that a multiple degree-of-freedom nonlinear system may exhibit motions that involve only a limited number of interacting degrees-of-freedom. Following Fermi et al.[15], we conducted a simple numerical experiment with the model defined by Equations 1. The dynamics were simulated beginning with an initial condition that corresponded to one of the "modes" of the linearized Equations 4, and the approximate modal energies were tracked in time. The spatial distribution of energy in the structure was also examined and visualized on a contour plot.

Equations 4 can be solved directly to obtain the modal eigenvectors defined by Equation 6.

$$e_l^j = \sin \frac{(2l-1)j\pi}{2N+1} \qquad (6)$$

where l is the mode number and e_l^j is the jth component of the N-dimensional column vector e_l. Defined in this way, these N-modal eigenvectors $[e_1 \ldots e_N]$ form a set of orthogonal basis vectors for the space of all allowable lattice configurations consistent with the boundary conditions. Thus, any motion of the system can be described by an N-dimensional column vector $\mathbf{x}(t) = [x_1(t) \ldots x_N(t)]^T$, that can be broken down in the following way,

$$\mathbf{x}(t) = \sum_{l=1}^{N} a_l \mathbf{e}_l \qquad (7)$$

where the a_l are time dependent coefficients. Taking the time derivative of $\mathbf{x}(t)$ gives the following expression for the mass velocities.

$$\dot{\mathbf{x}}(t) = \sum_{l=1}^{N} \dot{a}_l \mathbf{e}_l \qquad (8)$$

The time dependent coefficients a_l and $\dot{a}_l$ are given by,

$$a_l = \frac{\mathbf{x} \cdot \mathbf{e}_l}{\mathbf{e}_l \cdot \mathbf{e}_l}$$

$$\dot{a}_l = \frac{\dot{\mathbf{x}} \cdot \mathbf{e}_l}{\mathbf{e}_l \cdot \mathbf{e}_l} \tag{9}$$

where these expressions were determined from the modal orthogonality property, $\mathbf{e}_l \cdot \mathbf{e}_j = \frac{2N+1}{4}\delta_{ij}$, where δ_{ij} is the Kronecker delta function. Using these definitions, the energy of lth mode, denoted E_l, is given by,

$$E_l = \frac{2N+1}{4}\left\{\frac{m}{2}\dot{a}_l^2 + a_l^2\left(\frac{k}{2} + 2\gamma\sin^2\frac{(2l-1)\pi}{2(2N+1)}\right)\right\} \tag{10}$$

If the initial conditions are small (less than about 0.5 mm maximum relative displacement) the nonlinear model will have approximately linear behavior, and the modal energies calculated from Equation 10 will be nearly constant in time. However, for larger initial conditions, the nonlinearities may lead to complex modal interactions.

Figure 7 shows a plot of the time-evolution of the approximate modal energies for the case of a mode 5 initial condition with an amplitude of 1 mm and the damping, c, set to zero. Figure 7 was calculated by projecting the motions onto the basis set defined by Equation 6 and then using Equation 10 to approximate the modal energies. Because of the nonlinearity, the modal energies do not remain constant in time. Rather, mode 5 (shown in blue) trades energy with modes 4 (green) and 6 (violet). The trading is nearly periodic in time (quasiperiodic), with the state of the system returning to very near its initial state approximately every nine seconds. This type of behavior, known as a recurrence phenomenon, has been explained using perturbation methods by Ford[16,17,32]. The black curve in Figure 7 shows the approximate total energy for the system calculated by summing the approximate modal energies. Since the actual total energy remains constant for this case, the variation of the black curve gives the error associated with the approximate calculation of the modal energies. This error will increase as the amplitude of the initial condition is increased.

An examination of the spatial distribution of energy in the structure reveals a quasi-periodic localization of energy accompanying the recurrence phenomenon. This localization is represented pictorially in Figure 8, showing a contour diagram of energy in the structure where each cell consists of a mass and the adjacent linear and nonlinear springs. It is apparent that modal trading manifests itself as the appearance of a spatially localized energy wave in the lattice with a higher local energy value than any occurring during a simple mode 5 oscillation. The speed of this wave is very close to the group velocity associated with mode 5, consistent with a wave packet formed from a combination of modes 4,5 and 6. Thus as the energy spreads in the frequency domain, it localizes in the spatial domain. This is an

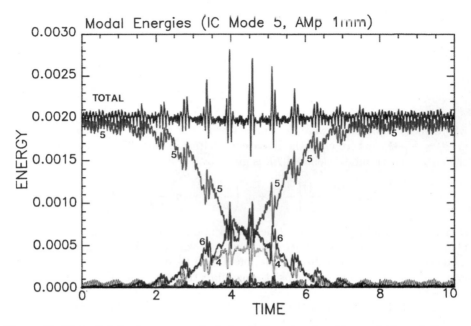

Figure 7: Plot of data from numerical simulation showing the evolution of the approximate modal energies (J) over time (s) for the case where all of the energy was initially placed in mode 5.

important observation since such localization could lead to unexpected damage in nonlinear structures whose behavior has been approximately analyzed using linear techniques.

4.2. Forced Vibrations of the Experimental Structure

Similar types of modal interactions were found to occur in the experimental structure. To demonstrate this, the structure was subjected to a sinusoidal input from the electromagnetic shaker table. For each set of driving conditions 10000 points of data were taken simultaneously for each mass at a sample frequency of 2000 Hz. POD was then applied to the data to determine the degree of spatial coherence in the motion.

Application of POD involves the calculation of the spatial correlation matrix R for the experimental data where R is defined in the following way.

$$R_{ij} = \langle x_i(t)x_j(t) \rangle, \tag{11}$$

Here $\langle \rangle$ represents the time-average over the entire data set and $x_i(t)$ and $x_j(t)$ are the displacements of masses i and j respectively. For discretely sampled experimental data, the integral time-average is approximated by a sum over the set of sampled data points. Defined in this way, R is a Hermitian matrix and thus its eigenvectors

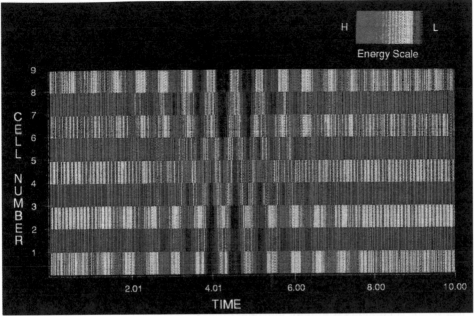

Figure 8: Energy contour plot of the modal trading case. Time axis is in seconds.

form a complete orthogonal set. The eigenvectors of R are called proper orthogonal modes, and the associated eigenvalues represent the mean squared amplitude of each mode. If the modes are ordered according to the size of their associated eigenvalues (from largest to smallest), and the motions are expanded in terms of the eigenvectors, it can be shown that this expansion is optimal in the sense that it captures the most signal power per mode. Because of these properties, POD has become a useful tool for determining the dimensionality of nonlinear dynamic systems.

The application of POD to our experimental structure was first verified by using it to identify the shapes of the linear modes of the structure. For example, for a driving frequency of approximately 12.4 Hz (corresponding to linear mode six) and an amplitude low enough that it did not drive the structure into the nonlinear regime (vibration amplitudes less than about 0.5 mm), the dominant eigenvector of the correlation matrix corresponds quite closely to the analytically calculated shape for linear mode six as shown in Figure 9. For this case, the eigenvalue associated with the dominant mode was eight times larger than that associated with the next most dominant mode.

As the frequency was slowly adjusted to higher values, the character of the motion underwent an abrupt change. Figure 10 shows the two most dominant eigenvectors of the correlation matrix for motions obtained at a driving frequency of 13 Hz and an amplitude approximately double that for the mode 6 case. These modes correlate well with the shapes of linear modes 4 and 6. The eigenvalues associated with these two modes were nearly equal in magnitude and each was

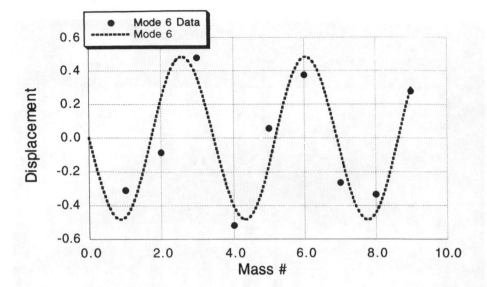

Figure 9: Primary proper orthogonal mode (solid circles) of structural vibrations at 12.4 Hz and the analytically calculated shape of linear mode six (dotted line). The eigenvector is normalized to have unit magnitude.

more than ten times larger than the next largest eigenvalue of the matrix. The power spectrum of the motion of the free-end mass is shown in Figure 11. It shows two dominant peaks at frequencies that correspond to the frequencies of linear modes four and six for this structure. Thus, both the POD and the power spectrum suggest that for these driving parameters, the motions of the structure are apparently characterized by an interaction between modes four and six.

Next, while the frequency was then held constant at 13 Hz, the amplitude of the input was increased by approximately 30%. At this amplitude, the motions of the structure underwent another abrupt change with the vibrations becoming temporally broad band but spatially correlated. Figure 12 shows the power spectrum of mass five. It clearly shows a broad band character. The power spectra of the other masses in the structure showed the same broad band character. The spatial coherence of the motions only became obvious with the application of POD. Figure 13 shows the first proper orthogonal mode for this motion and the close comparison to linear mode one for the structure. The eigenvalue associated with this mode was more than twenty times larger that the next largest eigenvalue, indicating that the first proper orthogonal mode dominates the motion. This spatially coherent, temporally broad band motions may be indicative of low-order chaotic behavior in this system. In addition, these dynamics appear to be an example of a case where energy is downloaded from higher to lower structural modes due to the influence of the nonlinearities[24].

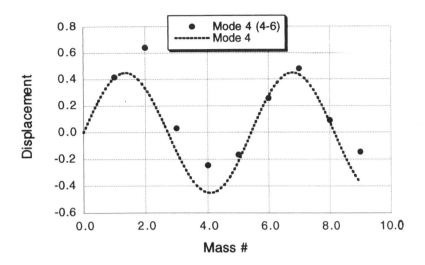

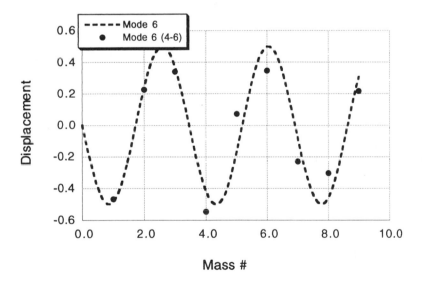

Figure 10: First two proper orthogonal modes of structural vibrations at 13 Hz (solid circles) and the analytically calculated shapes of linear modes four and six (dotted lines) for comparison. The eigenvectors are normalized to have unit magnitude.

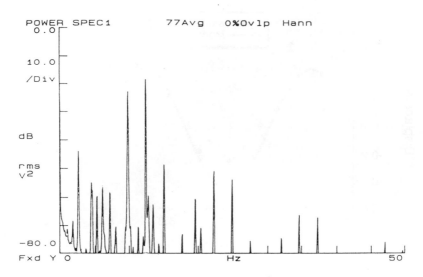

Figure 11: Power spectrum of structural vibrations at a driving frequency of 13 Hz. Two largest peaks at 9.8 Hz and 12.4 Hz correspond to the frequencies of linear modes four and six.

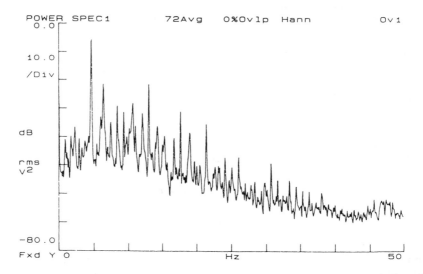

Figure 12: Power spectrum of broad band structural vibrations at a driving frequency of 13 Hz. This data was taken from mass number five.

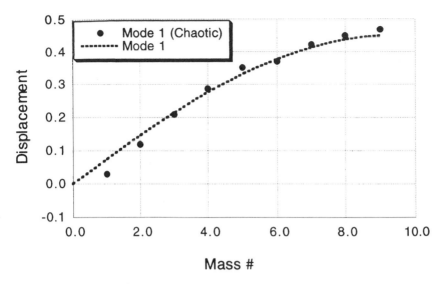

Figure 13: First proper orthogonal mode for broad band motion (solid circles) and the analytically calculated linear first mode (dotted line) of the structure.

5. Impact Response

In this section the impact response of the experiment and model are compared. "Soliton"-like waves are shown to result from the initial impact, and then to break up into spatially and temporally complex patterns. In addition, calculation of wave speeds based upon Toda's analytical results confirm the "soliton"-like nature of the waves and help us to refine our estimates of model parameters, bringing the model behavior closer to that of the experiment.

5.1. Comparison Of Experiment and Model

We began our examination of system dynamics by assessing the predictability of motions induced by sudden loading of the free end. Davies and Moon[8] showed that in both the simulation and experiment, slight differences in the magnitude of the initial load produced large differences in the system state at later times, suggesting the possibility of transient chaos.

To visualize the complex spatial and temporal dynamics that might lead to sensitive behavior in the system, we used contour plots of the compression/tension history of the buckling elements. Figure 14 shows a contour plot of the simulated impact dynamics of the model. Impact excitation was approximated by giving mass nine an initial velocity of -2.0 m/s and setting all other initial conditions to zero. The horizontal axis in the plot is the time and the vertical axis is the buckling element number. The strip of color to the right of each element label shows the relative level of compression (blue) or tension (red) in that element over time. Figure 14 indicates that initially element nine enters an extreme state of buckling or compression. However, because the compressive force of the buckling elements is low, this disturbance remains localized to element nine. After one-half of an oscillation cycle of the ninth mass-linear spring combination (approximately 0.1 s), buckling element nine enters a state of tension. This initiates a high-force tensile disturbance that propagates rapidly through the structure. It reflects from the fixed-end and returns to the free-end, producing a second compression in buckling element nine at 0.2 s. The speed of this tensile-wave is approximately 38 m/s, more than five times the maximum linear group velocity calculated in the previous section!

Figure 15 shows a compression/tension contour of experimental motions. In this case, mass 1 was rigidly fixed in place to ensure the fixed boundary condition, and the masses were re-numbered 1 through 8 starting from the fixed end. The structure was excited by the sudden release of a compressive load applied to the free-end as shown in Figure 2. For this case, the load was approximately 1 N; this load caused a deflection (from equilibrium) of -5 mm for mass 8, -1 mm for mass 7 and negligible amounts of the remaining masses. Release of the compressive load lead to a tensile wave like that seen in the simulation. The speed of this wave was approximately 22 m/s, more than three times the linear group velocity. For both the experiment and the simulation the fact that wave speeds greatly exceed the maximum group velocity is indicative of the strong nonlinearity of the observed motions.

Figures 14 and 15 show evidence of highly complex nonlinear wave dynamics. Particularly in the simulation, the wave breaks apart, splitting, fading and re-emerging to produce complex interaction patterns in space and time. A contour plot of a three-second impact simulation (see Figure 16) clearly illustrates the disordered dynamics which develop. The reason for the break-up of the initially ordered wave patterns is not clear. It may be that the observed tensile-waves are perturbations of Toda "solitons" that break down into disordered patterns because the parameters k and c are nonzero. However, other evidence suggests that wave breakdown is most strongly linked to the boundary conditions. For example, we have found that releasing the fixed-end to produce a symmetric lattice greatly increases the longevity of the ordered wave patterns. Figure 17 shows a three second simulation of the structural dynamics for the case where both ends are free (i.e. in Figure 6 the nonlinear spring and damper connecting mass 1 to the rigid wall is eliminated). It is clear that the longevity of the ordered wave patterns is increased for this case. This data prompts complicated questions about links between soliton-like waves and sensitivity to initial conditions in nonlinear structural vibrations.

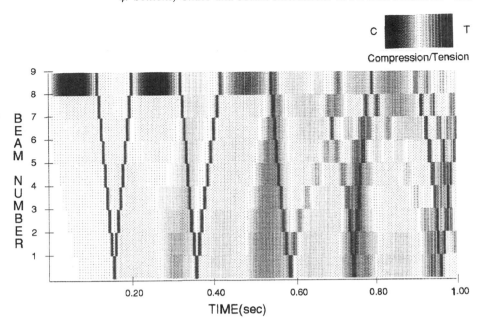

Figure 14: Compression-tension contour of the impact dynamics of the model (one-second simulation).

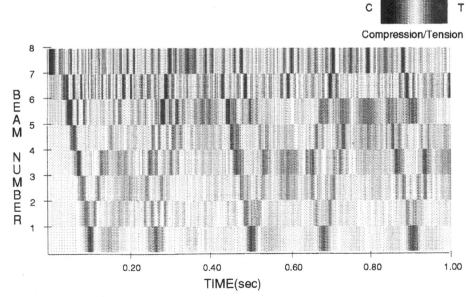

Figure 15: Compression-tension contour of the experimental dynamics induced by a suddenly released compressive load of 1 N.

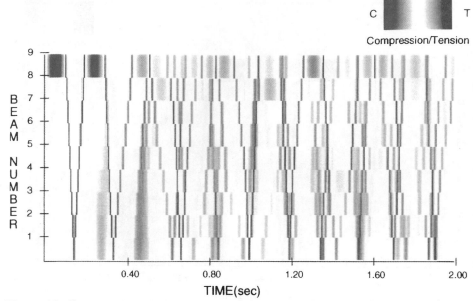

Figure 16: Compression-tension contour of the impact dynamics of the model (three-second simulation).

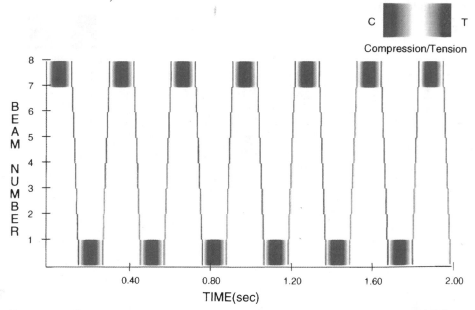

Figure 17: Compression-tension contour of the impact dynamics of the model for the free-free boundary condition case.

5.2. Calculation of Nonlinear Wave Speeds

Further understanding of the observed tensile-waves is gained from a consideration of the forces associated with a wave. From the simulation data, we find that in
the vicinity of a wave, the tensile forces in the nonlinear springs are 300 times higher
than those exerted by the linear springs. In the experiment this ratio is lower, but
the forces in the buckling elements still dominate in the vicinity of the wave. Thus,
the effect of the linear springs on these waves is small, and for short periods of time
they can be treated as Toda solitons.

The functional form of a Toda soliton is given below.

$$r_j = \frac{1}{b} \ln \left(\frac{m\beta^2}{\gamma} \text{sech}^2(\kappa j - \beta t) + 1 \right) \tag{12}$$

Substitution of Equation 12 into Equations 1 and 3 with k and c set to zero, shows
that it solves these equations if the following relation between β and κ is satisfied.

$$\beta = \sqrt{\frac{\gamma}{m}} \sinh \kappa \tag{13}$$

Equation 12 describes a localized wave that travels without changing shape. If $\gamma > 0$
and $b < 0$, $r_j(t) < 0$ for all values of j and t. This corresponds to the compressive
atomic lattice waves studied by Toda. If, as in our case, $\gamma > 0$ and $b > 0$, then
$r_j(t) > 0$ for all j and t and Equation 12 represents a tensile-wave. The speed of
the wave in either case is,

$$c = \frac{\beta a}{\kappa} \tag{14}$$

where a is the distance between masses. The speed of this nonlinear wave is dependent upon its amplitude. In general larger, narrower disturbances propagate faster
than smaller, broader ones.

Equation 14 should provide a reasonable estimate of the speed of the tensile-
waves observed in our system. As an example, we estimate the speed of the first
tensile-wave in Figure 14. From the data, we determined that the amplitude of this
wave is approximately 0.0075 m. The expression for the amplitude of a Toda wave,
denoted $\hat{r}$, is found by setting $\kappa j - \beta t = 0$ in Equation 12.

$$\hat{r} = \frac{1}{b} \ln \left(\frac{m\beta^2}{\gamma} + 1 \right) \tag{15}$$

Substituting values of $\hat{r} = 0.0075$ m, $\gamma = 362.0$ N/m, $m = 0.157$ kg and $b = 725.0$
m^{-1} allows us to determine an approximate value of $\beta = 727.3$ s^{-1} for our wave.
Substituting this value into Equations 14 and 13 gives an approximate velocity
for our tension wave of 40.6 m/s, which is only slightly greater than the observed
velocity of 38 m/s. The difference is probably due to the fact that the linear springs
do exert some small forces on the masses. As the wave passes, these forces work to

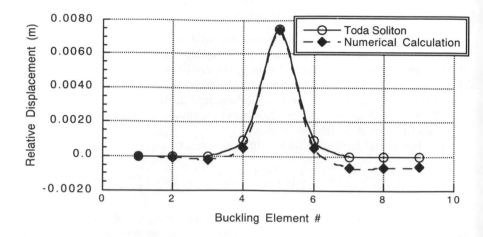

Figure 18: Comparison of the tensile waves observed in the simulation data (t=0.134 s) and a Toda soliton for $b = 725.0$ m^{-1}, $\gamma = 362.0$ N/m, $m = 0.1567$ kg, and $\beta = 727.3$ s^{-1}.

reduce the acceleration of each mass and thus slightly delay the passage of the wave. The similarity of the tensile waves in the model and Toda solitons is illustrated in Figure 18. The propagation and fixed-end reflection of the numerically simulated wave is shown in Figure 19.

Using the experimental data, one can do a similar calculation to determine an improved value for b. From the experimental data shown in Figure 15, approximate values of $c = 22.0$ m/s and $\hat{r} = 9.0 \times 10^{-4}$ m were determined. Using these values in an iterative solution to Equations 13, 14 and 15, we determined that b is approximately 4000 m^{-1} for the experimental system. A plot of the exponential force law for $b = 4000$ m^{-1} is shown in Figure 5. This predicted value of b clearly brings the tensile portion of the analytical curve into excellent agreement with experimental data. The success of these wave speed calculations provide further evidence of the similarity between the waves observed in our system and Toda solitons.

6. Conclusions

In this chapter, several new findings on the dynamics of nonlinear structures are reported. The subject of the study was a periodically reinforced structure with strong buckling nonlinearity. This structure was modeled with a modified Toda lattice.

The numerical model exhibits modal trading phenomena accompanied by a spatial localization of energy in the structure. This phenomenon is important from an

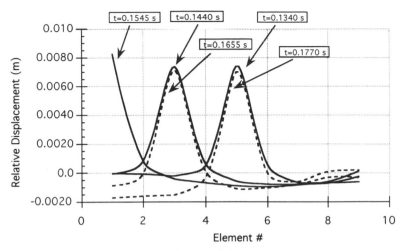

Figure 19: Plot of the propagation of the simulated tensile wave and its reflection from the fixed end. Solid lines indicate the approaching (propagating left) and reflecting waves and dotted lines represent the reflected wave (propagating right). Parameter values are the same as discussed above.

engineering standpoint because it could lead to unexpected localization of energy in structures which would not be predicted by a linear vibration analysis. The experimental system also exhibited modal interactions and this was characterized using the technique of proper orthogonal decomposition. In addition, for driving conditions, structural vibrations were temporally broad-band but still showed tremendous spatial coherence and appeared to have only one dominant proper orthogonal mode. Further, the excitation of this mode seemed to be the result of a downloading of energy from higher frequency modes. This evidence suggests that the motion may involve only a limited number of interacting modes, and that some of these modes may act as conduits that channel energy from higher to lower frequency, a phenomenon that has been observed in other structural systems[24].

The impact response of the structure was found to be characterized by highly localized nonlinear waves. These waves were shown to be similar to Toda solitons through wave-speed calculations based upon Toda's analytical results. The stability of these waves was found to be highly dependent on the boundary conditions of the system, with rapid breakdown of the initial wave occurring for non-symmetric boundary conditions and a much more coherent wave pattern persisting for long times in the case of symmetric boundary conditions. Previous work by the authors[8] provides evidence that the motions of this structure are an example of transient chaos with system states showing sensitivity to initial conditions during the early stages of a transient motion. However, further calculations possibly using Lyaponov exponents or entropy measures are necessary to determine the degree to which these motions are actually chaotic in time as well as the level of spatial disorder in the structural dynamics. The results of this paper are relevant to a broad array of nonlinear structures and prompt further inquiry into the role of nonlinear waves in generating spatio-temporal complexity.

7. Acknowledgments

This work was supported by the Air Force Office of Scientific Research, University Research Initiative, Grant No. AFOSR-90-0211, and by The Fannie and John Hertz Foundation.

8. References

1. B. F. Feeny, *Physics Letters A* **141**(8,9) (1989) 397.
2. L. Brillouin, *Wave Propagation in Periodic Structures* (McGraw-Hill Book Company Inc., London, New York, 1946).
3. G. J. Brinkman and T. P. Valkering, *Journal of Applied Mathematics and Physics (ZAMP)*, **41** (1990) 61-78.
4. G. Berkooz and J. Elezgaray and P. Holmes and J. Lumley and A. Poje, *Applied Scientific Research (Netherlands)*, **53**(3-4) (1995) 321-338.
5. J. Cusumano and F. C. Moon, in *Proceedings of the IUTAM Symposium, Stuttgart Germany*, ed. Schielen, W.O. (Springer-Verlag, 1989).
6. , J. P. Cusumano and B.-Y. Bai, it Chaos, Solitons and Fractals, **3**(5) (1993) 515-535.
7. M. A. Davies and F. C. Moon, *Chaos* **3**(1) (1993) 93.
8. M. A. Davies and F. C. Moon, *Chaos, Solitons and Fractals* **3**(1) (1994) 275.
9. M. A. Davies and F. C. Moon, to appear *Journal of Applied Mechanics* (1995).
10. M. A. Davies, *Spatio-Temporal Chaos and Solitons in Nonlinear Periodic Structures* (Ph.d. Dissertation, Cornell University, Ithaca NY, 1993).
11. A. S. Davydov, *Solitons in Molecular Systems* (Reidel, Dordrecht, 1985).
12. P. G. Drazin, *Solitons* (Cambridge University Press, 1984).
13. E. H. Dowell, *Journal of Sound and Vibration* **85**(3) (1982) 333.
14. R. Frisch-Fay, *Flexible Bars* (Butterworth and Co. Limited, London, 1962).
15. E. Fermi and J. Pasta and S. Ulam, in *Collected Papers of Enrico Fermi, Volume 2*, E. Fermi, (The University of Chicago Press, Chicago, 1965).
16. J. Ford, *Journal of Mathematical Physics*, **2**(3) (1961) 387
17. J. Ford and J. Waters, *Journal of Mathematical Physics*, **4**(10) (1963) 1293-1306.
18. P. J. Holmes, *Philosophical Transactions of the Royal Society of London* **292** (1979) 419.
19. A. Meilke and P. J. Holmes, *Archive for Rational Mechanics and Analysis* **101** (1988) 319.
20. F. C. Moon, *Chaotic and Fractal Dynamics* (John Wiley & Sons, 1992).

21. F. C. Moon and P. J. Holmes, *Journal of Sound and Vibration* **18** (1983) 465-477.
22. F. C. Moon and S. W. Shaw, *J. Nonlinear Mech.* **18** (1983) 465-477.
23. V. Muto and A. C. Scottland and P. L. Christiansen, *Physics Letters A*, **136**(1) (1989) 33-36.
24. A. H. Nayfeh and B. Balachandran, *Applied Mechanics Review* **42**(11) (1989) S175-S201.
25. M. S. Naschie, *Journal of the Physical Society of Japan*, **58** (1989) 4310-4321.
26. R. Pratap and S. Mukherjee and F. C. Moon, *Journal of Sound and Vibration*, bf 172(3) (1994) 321-337.
27. J. M. T. Thompson and L. N. Virgin, *Physics Letters A*, (1988) **126**(8,9) 491-496.
28. M. Toda, *Journal of the Physical Society of Japan*, (1966) **22**(2) 431-436.
29. M. Toda, *Journal of the Physical Society of Japan*, (1967) **23**(3) 501-506.
30. M. Toda, *Theory of Nonlinear Lattices* (Springer-Verlag, Berlin-Heidelberg,1989).
31. W. Y. Tseng and J. Dugungi, *ASME Journal of Applied Mechanics* (June 1971) 467-476.
32. J. Waters and J. Ford, *Journal of Mathematical Physics*, **7**(2) (1966) 399.

Dynamics with Friction: Modeling, Analysis and Experiment, Part II, pp. 125–153
edited by A. Guran, F. Pfeiffer and K. Popp
Series on Stability, Vibration and Control of Systems, Series B, Vol. 7
© World Scientific Publishing Company

Analysis and Modeling of an Experimental Frictionally Excited Beam

R. V. KAPPAGANTU
Altair Engineering, Inc.
1755 Fairlane Drive
Allen Park, MI 48101 U.S.A.

and

B. F. FEENY
Department of Mechanical Engineering
Michigan State University
East Lansing, MI 48824 U.S.A.

Abstract.
   The friction characteristics and dynamics of an experimental frictionally excited beam are
investigated. Beam displacements are approximated from strain gage signals. The friction is
shown to adhere to a model with an idealized massless contact compliance. The system dynamics
are rich, including a variety of periodic, quasi-periodic and chaotic responses. Proper orthogonal
decomposition is applied to chaotic data to obtain information about the spatial coherence of
the beam dynamics. Responses for different parameter values result in a different set of proper
orthogonal modes (POMs). The number of POMs that account for 99.99 % of the signal power is
compared to the corresponding number of linear normal modes, and it is verified that the POMs
are more efficient in capturing the dynamics.
   A set of POMs is then used for order reduction of the beam equations. The distributed model
is based on an Euler-Bernoulli beam with frictional excitation. The friction is modeled with
Coulomb's law and contact compliance, and the contact surface undergoes an imposed oscillation.
The POMs are extrapolated into proper orthogonal modal functions (POMFs) by using the linear
normal modes of the model as basis functions. The POMFs are used as the basis for projection
the partial differential equation of motion to a low-order set of ordinary differential equations.
Simulated responses based on the POMFs and linear normal modes are compared to that of a
"truth set" simulation, which is based on ten linear normal modes. Including contact compliance
versus a rigid contact is also investigated regarding the effect on the response.

1   Introduction

This paper involves a dynamical study of a frictionally excited experimental beam
and its proper orthogonal decomposition. Proper orthogonal decomposition (POD)
is a method of extracting a spatial characterization of dynamical distributed media.
POD, which will be described later as applied in our system, yields the optimal
energy distribution in a set of measured time histories. The resulting proper orthogonal modes (POMs) had been applied to turbulence as a tool for characterizing
spatial coherence [1, 2]. In some studies, the modes have been treated as "empirical modes" and used for projecting distributed models to low-dimensional models
[2]. This idea has carried over to the reduced order modeling of structural systems
[3, 4, 5, 6].

   Other applications of POD to structural dynamics have focused on determin
ing spatial coherence and estimating the dimension of the active state space [7],
and projecting empirical modes [3]. More recent studies on structural dynamics

include references [8, 4, 9, 10, 11, 5, 6]. Among these studies, it has been observed that the POMs tend to lie close to the linear normal modes [8]. Indeed, it has been shown that for symmetric linear vibration systems with uniform mass distributions, POMs of multi-modal responses converge to the linear normal modes [12]. Furthermore, the dominant POM of a synchronous single-mode nonlinear modal motion represents a best fit of the nonlinear normal mode [12]. Results of Ma *et al.* [6] suggest that this property may carry over to multi-modal responses.

Our purpose in this study is to examine the use of POD in the case of frictional excitation. It is hoped that POD can be applied to industrial problems such as chatter and squeal to draw out active degrees of freedom and assist in determining mechanisms and effective models of such phenomena. Since friction is often considered a nonsmooth force, it is associated with high-frequencies, which could have significant effects in the excitation of distributed-parameter systems. Impacting systems also involve similar issues, and are being studied by POD by Azeez and Vakakis [5].

We consider a cantilever beam subjected to frictional excitation, in which the frictional contact is provided by a solid mass of steel connected to a shaker. We characterize the friction characteristics and the dynamical response. A chaotic response is exploited in the POD study to obtain the empirical system modes. We use the response characteristics and the POMs to assist in the modeling and verification. The effect of details of our friction characteristics on the dynamic response are also investigated.

This chapter is organized as follows. The next section provides a description of the experimental set up. We then discuss the characterization of the friction. Section 4 provides a study of the dynamical response. This is followed by a spatial characterization by means of POD. Section 6 involves the mathematical modeling of the beam, and the modal reduction. We then discuss the numerical simulation strategy and results. The chapter is capped off with supplementary discussions and conclusions.

2   Experimental Setup

The schematic of experimental setup is shown in Figure 1. The cantilever wass built by taking a $0.4000 \times 0.0128 \times 0.00086$ m^3 mild steel beam (E=128 $\times$ 10^9 N/m^2, ρ=7488 kg/m^3) and fixing it at one end. At the free end of the cantilever, we attached a small beam ($0.064 \times 0.0134 \times 0.00056$ m^2, E = 126 $\times$ 10^9 N/m^2, $\rho = 6777$ kg/m^3) to act as a leaf spring. It is referred to as a loading beam. The mass of the fixture (not shown in the figure) is 0.01226 kg. The presence of fixture made the effective length of loading beam 0.076 m. We define two coordinate systems XYZ and xyz as shown in Figure 1. The bending axes of the two beams are perpendicular to each other, *i.e* the length of loading beam is along the X axis, the thickness along the Z axis and the width along the Y axis. The loading beam had low transverse (Z) rigidity and high lateral (Y) rigidity. This

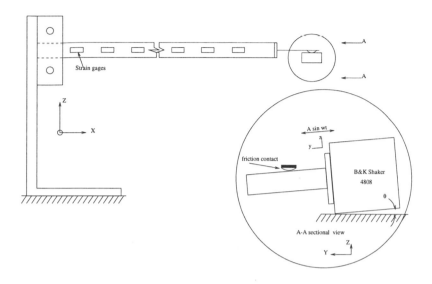

Fig. 1.   Schematic diagram of the experimental setup

configuration allowed for the transmission of any force on the tip of loading beam in the Y direction to the cantilever without significant change. A small hemispherical rubber nub was glued to the free end of the loading beam. This nub made contact with an oscillating surface. The loading beam was used to provide a spring-loaded normal load. The normal load was monitored by attaching strain gages to this loading beam (the amount of strain is directly proportional to the moment which in turn is proportional to the normal force).

As the oscillating surface moved to and fro, the friction force between the rubber nub and the surface was transmitted to the beam. An angle between the plane of beam deflection and the oscillating surface would allow a variation in normal load. This affected the force transmitted to the main beam. To make this adjustable, we incorporated a tilt θ in the shaker axis by lifting the rear end of the shaker. The tilt changes the effective coefficient of friction from μ to $\frac{\sin\theta + \mu\cos\theta}{\cos\theta - \mu\sin\theta}$ where θ is the tilt angle. The presented results, however, involve a negligible angle.

Using an impulse response test we noted that the first six modes of the cantilever setup have frequencies of approximately 3.75, 18.4, 53.75, 102.5, 172.5 and 265 Hz. In the next two sections we give the details of measuring friction and displacements at various locations.

3   Friction Measurement

A schematic of the apparatus used to get the friction characteristic is presented in Figure 2. The apparatus consisted of the friction surface used in the main experiment, and two slender beams (similar to the one used in the main experiment) with rubber nubs attached at the end and a fixture to connect these beams to the shaker (B&K model 4808). The fixture was designed to minimize the moments due to normal loads. Strain gages on the beams were used to sense the normal load. As shown in the figure, the sliding contact surface was then fixed to the steel bracket through a load cell (PCB 208B). Thus the axial component of any force on this mass could be measured using the load cell. The two rubber nubs attached to the beams slid over this block, exerting normal (to the surface) and frictional (along the surface) forces onto the block. The fixture was mounted on to the shaker through a roller bearing so that the normal forces balanced each other and did not produce a resultant moment which would contaminate the force transducer measurement. For determining the friction coefficient we need the normal load acting between the surfaces. This was measured using the strain gages attached to the beams at appropriate locations. The strains measured were proportional to the moments in the beams at the gage locations which in turn were directly proportional to the normal load. Thus one can obtain both the friction and normal loads.

As the rubber nubs rode on the surfaces, the normal load variation was found to be 0.8%. Thus, the axis of oscillation was considered to be parallel to the rubbing surfaces.

The load cell signal incorporated the sum of the friction forces from the two sliding surfaces. The sum of the two normal loads (taking the sign into account) is then taken to be the effective normal load between the sliding surfaces. During the friction measurement this mean normal load was found to be 1.86 N.

For measuring the input displacement and velocity we made use of an accelerometer (PCB 309A). At 10 Hz excitation frequency, the phase difference between measurements of accelerometer PCB 309A and force transducer PCB 208B was found to be 50.4 degrees (0.88 radians) with the accelerometer signal leading. This phase has been adjusted in our measurements.

We first obtained the acceleration using the accelerometer. As we are measuring the acceleration of the input whose frequency is known, we high-pass filtered the acceleration signal to take care of any drifts due to poor ground looping. We took a Fourier transform of this signal and divided this by the frequency component once to obtain velocity Fourier transform and twice to obtain the displacement Fourier transform. We took the inverse Fourier transform of the two resulting signals to get velocity and displacement. We cross checked this information using a laser transducer B&K 6808 with power amplifier B&K 2815. The laser transducer has a low-frequency limit of 0.3 Hz in the displacement measurement mode. The phase difference between the accelerometer and the force transducer was determined and

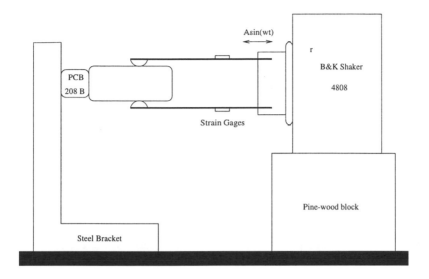

Fig. 2.   Configuration of friction measurement setup

accounted for in the measurements.

After running the shaker at 10 Hz for about an hour, we recorded the data for 32 seconds at a sampling rate of 1000 Hz. We used Laboratory WorkBench on a Masscomp 5550 for this data acquisition. The four channels of data we collected are the acceleration, both the normal loads and the friction force.

A plot of the relative displacement (of the shaker) versus the friction force is shown in Figure 3 and relative velocity versus friction force is shown in Figure 4.

The kinetic friction force limit is close to 1.03 N. Dividing this by the mean normal load of 1.86 N gives us the kinetic friction coefficient of 0.55. The static to kinetic transition, if it exists, is generally at high frequencies and gets filtered.

This motivated a separate experiment consisting of a mass of 250 grams under which we attached three rubber nubs (similar to the one used in the experiment) on an inclined plane made of a similar surface as the one used in the experiment. We increased the tilt of the inclined plane until the mass began to slide. The average tilt at which the mass began to slide was 37.2 degrees (standard deviation across seven different readings was 2.7% of the mean) which amounted to an average coefficient of static friction $\mu_S = 0.75$ (standard deviation of μ_S is 3.7% of the mean). Thus we have $\mu_S \approx 1.38\mu_K$. Whether this is representative of the friction during 10-Hz oscillations is not confirmed.

Close examination of the friction-displacement reveals a steep slope in the re-

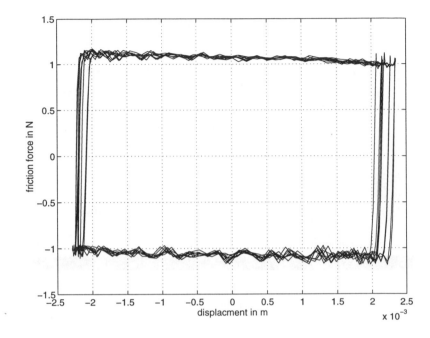

Fig. 3.   Friction-displacement relation

gion where sliding changes its direction. This has been described as "spring-like" sticking behavior in the literature [13]. Experimental evidence of this kind of behavior is reported by Dahl [14] and Courtney-Pratt and Eisner [15]. Both the reports indicate a junction compliance which induces small displacements prior to the actual slip. Efforts for modeling these elastic contact problems have been put forth by many researchers [16, 17, 13, 18]. We use the term "microsticking" as during this sticking event the elastic displacement is really small if the contact stiffness is high.

The hysteresis in the friction-velocity plot in Figure 4 suggests that the friction characteristic may be close to a Coulomb model with contact compliance. A schematic of the model is shown in Figure 5. A linear contact stiffness is used, which can approximate the stiffness of asperity-born contacts [19, 20], and also model structural compliances, such as those often used in turbine-blade studies [21, 22, 23, 24, 25]. The mean slope of the nearly vertical lines of direction reversal in the displacement-friction plots gives the contact stiffness K_C, which is -20.2

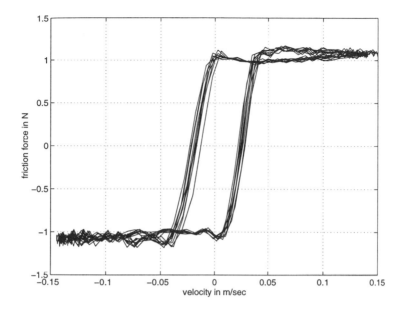

Fig. 4.   Friction-velocity relation

kN/m (this is the average slope with a standard deviation of 10%). This slope can be used to predict the velocity with which the rubber nub slips, which is discussed in the following paragraphs.

This model effectively describes the observed phenomenon in our experiment. $Y_T(t)$ describes the displacement of the shaker. $Y_C(t)$ represents the displacement of the contact point (the point is on the rubber nub). The dynamics of the block and the mass of the nub are negligible. However due to contact compliance there is a relative motion between the contact point (on the rubber nub) and the shaker attachment. During pure sliding motion, the relative motion $\dot{Y}_T(t) - \dot{Y}_C(t)$ is negligible since the contact is nearly massless and its dynamics are damped out immediately. When $\dot{Y}_C(t)$ reaches zero, we have sticking event. However, $\dot{Y}_T(t)$ does not remain at zero. When stuck, $\dot{Y}_T(t)$ evolves differently from $\dot{Y}_C(t)$ which is zero, until the static friction force can no longer sustain the stiffness force exerted by the compliant-contact joint, *i.e* $|K_C(Y_T(t) - Y_C(t))| > |F_s|$. When the contact elements slip we again have $Y_T(t) = Y_C(t)$. F_s is the static friction force bound. Thus the slope in the friction-displacement plot is the effective contact stiffness K_C.

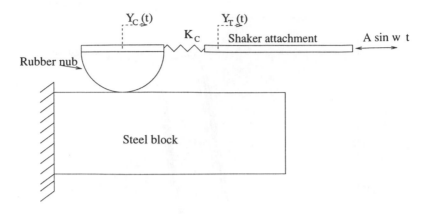

Fig. 5.   A schematic diagram showing a massless compliant-contact model

Now we know from the friction-displacement plot the change in $Y_T(t)$ during the microsticking event. We also know that micro-sticking begins when the relative velocity is zero, in our case $\dot{Y}_T = \dot{Y}_C(t) = 0$ and ends when stiffness forces overcome the friction forces. Suppose t_1 and t_2 are the times when microsticking begins and ends respectively, and F_k is the kinetic friction force just before the contact elements stick. Then the relative displacement $Y_T(t_2) - Y_T(t_1)$ is effectively $(F_s - F_k)/K_C$. Let us assume harmonic excitation of the shaker attachment, *i.e.* $Y_T(t) = A_0 + A\sin\omega t$. Then $\dot{Y}_T(t_1) = \dot{Y}_C(t_1) = 0$ as prior to $t = t_1$ we have $\dot{Y}_T(t) = \dot{Y}_C(t)$. This implies $Y_T(t_1) = A_0 + A$. Also $Y_T(t_2) = A_0 + A\sin\omega t_2$. This in turn equals $Y_T(t_1) + (F_s - F_k)/K_C = A_0 + A + (F_s - F_k)/K_C$. Thus we have an estimate of t_2. From this we calculate $\dot{Y}_T(t_2)$. Going through simple mathematics we get $\dot{Y}_T(t_2) = -\omega A\sqrt{1 - (1 - (F_s - F_k)/K_C A)^2}$.

From the friction-displacement plot we have $A = 0.0023$ m, $K_C = -20$ KN/m, $F_s = 1.03$ N and $F_k = -1.0$ N. We know the value of $\omega = 20\pi$ rad/s. Substituting all these parameters in the equation for $\dot{Y}_T(t_2)$ we arrive at a slipping velocity equal to 0.044 m/s which is consistent with the the friction-velocity plot, in which a microstick-slip transition occurs at approximately $V_s = \pm 0.045$ m/s. Thus the compliant-contact stiffness and the slip velocities are consistent within the compliant contact model. This confirmation supports prior work on the usage of this model [16].

4 Displacement Measurement

We based our displacement measurements on strain gage sensing. We chose strain gages because they have frequency response characteristics that allow the detection of both low-frequency and high-frequency events, and they are inexpensive. We used six half-bridge strain-gage circuits to obtain the strains at six uniformly distributed locations. We attached the strain gages on either side of the beam to form a half bridge. The gages were spaced at 0.05 m, with the first gage located at 0.05 m from the fixed end. Based on numerical simulations we found that this distribution approximately optimizes the conversion from strain to displacement. We numbered the gages from 1 to 6, starting from the gage at 0.05 m and ending at 0.30 m. All the strain gages used are of the type Micro-Measurements Precision Strain-Gages model CEA-06-250UN-350. These have a resistance of $350.0 \pm 0.3\%$, a gage factor $2.1 \pm 0.5\%$ and a transverse sensitivity of $(+0.1 \pm 0.2)\%$ all at $24°$ C. For minimizing the effect of wiring, we used Micro-Measurements single conductor wire type 134-AWP, twisted the three wires coming out of the each half bridge and ran the wires along the beam to the fixed end. Four strain-gage conditioners were of the type 2120 and two of the type 2210, both are from the Micro-Measurements Group Inc. The excitation voltage used was 10 V for both the models. A gain of 2000 was used on model 2120 and gain of 1000 was used on model 2210. The difference in gains have been accounted later in the post-processing programs.

For data acquistion we used Laboratory Workbench of Concurrent Computer Corporation on a MASSCOMP 5550. We acquired six strain signals on the main beam, one strain signal of the loading beam and one input acceleration signal using an accelerometer PCB 309A. We used a sampling frequency of 1000 Hz for the acquisition.

4.1 *From Strain to Displacement*

The strain $\epsilon(x)$ in a beam subjected to pure bending is related to the transverse displacement $y(x)$ by

$$\epsilon_{xx} = c\frac{\partial^2 y}{\partial x^2} \tag{1}$$

where c is half the width of the beam. We approximate the displacement of the beam as a function of x as

$$y(x,t) = \sum_{i=1}^{n} \phi_i(x)u_i(t), \tag{2}$$

and the discrete set of displacements as

$$\mathbf{y} = \mathbf{\Phi}(x)\mathbf{u}(t),$$

where $\phi_i(x)$ form a basis satisfying the geometric boundary conditions, $u_i(t)$ are the generalized coordinates, $\mathbf{y}$ is a vector of n values of $y_j = y(x_j, t)$ evaluated at particular values of x_j, and where $\boldsymbol{\Phi}$ and $\mathbf{u}$ are the appropriate $n \times n$ matrix and $n \times 1$ vector. Thus

$$\epsilon_{xx}(x, t) = c \sum_{i=1}^{n} \psi_i(x) u_i(t) \qquad (3)$$

or

$$\epsilon = c\boldsymbol{\Psi}(x)\mathbf{u}(t)$$

where $\psi_i(x) = \partial^2 \phi_i(x)/\partial x^2$ and ϵ is a vector of n values of $\epsilon_{xx}(x_k, t)$ evaluated at the strain gage locations x_k, and $\boldsymbol{\Psi}$ is defined appropriately. By measuring the strain ϵ at n different points on the beam we get n such equations which can be solved for $u_i(t)$ at each t. These in turn can be used to determine $y(x, t)$ using equation (2). For the basis $\{\phi_i\}$ we used the linear natural modes of a cantilevered beam and we used $n = 6$.

In our calibration, we applied a high normal load at the contact so that the contact remained in the stick mode at that operating frequency. We operated at 8.45 Hz frequency during this stage. We measured the displacement of the beam at a location of 0.35 m from the fixed end using the laser transducer and simultaneously obtained the strain signals, and applied equations (2) and (3) to obtain the displacement at the same location. The laser transducer sensitivity is 1 mm/V. We took Fourier transforms of the laser displacement and the computed displacement and took the ratio of the amplitudes at the operating frequency. This gave us the calibration constant, which, based on equation (1), should equal half the thickness of the beam. At 8.45 Hz we found this constant to be 99.9% of the thickness. At an operating frequency of 12.0 Hz (with normal load high enough to result in a pure stick motion), we found the constant to be 99.84% of the thickness. We used this calibration constant to verify the displacements at six different locations. Across six different locations, the standard deviation of these constants was found to be 0.71% of the mean of 98.57%, with the worst match occuring close to the fixed end where the displacements are very small (where we suspect the laser measurement to be not so accurate).

The filtered direct and indirect measurements were compared [26] in the time and frequency domains. At an excitation frequency of 12 Hz, we found the ratio between the peaks in the frequency domain to be 99.98%. At 41.1 Hz operating frequency, the direct and indirect measurements did not match as well especially in the high frequency regions (above 120 Hz). The ratio between the peak amplitudes of power spectra at this operating frequency was 81.11%.

The normal load variation along the slider surface has been measured using a quasi-static approach. The variation was fairly linear, between 0.375 N and 0.330 N in a displacement range from -0.02 m to 0.01 m.

5 Dynamical Responses

In order to explore the dynamics we carried out frequency sweeps. For commercially available shakers, as the input frequency to the shaker is increased, the amplitude decreases. For the shaker (B&K 4808) that we used, we curve-fit the following equation to relate the frequency f in Hz to the amplitude a in meters: $a = \exp(-4.4514 - 0.9037\log(f))$.

During the sweeps, the two signals that we made use of were the acceleration of the slider (input) and the strain at location 1 on the beam. A LabVIEW program on a Macintosh was used to run the sweep. We sampled both the signals simultaneously (to keep the phase constant) at a sampling rate of 48 samples/cycle and omitted the first 500 cycles to account for transients. We collected the next 100 cycles of data and processed the data to obtain 100 "quarter phase strain" signals, meaning the strain value when the input acceleration was at the peak and the excitation velocity was passing through zero. We carried out this operation at 450 slider frequencies uniformly spaced between 5 and 50 Hz. Figure 6 shows the resulting diagrams in the forward and reverse sweeps. These diagrams plot an observable consisting of a single strain-gage signal, which does not correspond directly to a physical displacement.

In the sweeps of Figures 6, at each frequency we have 100 points. A single tightly packed cluster of points (ideally the cluster should be of zero dimension) at any given frequency indicates the response is periodic with period one. Multiple clusters indicate higher periods. A smear of points would indicate either a quasi-periodic or chaotic orbit. Sweeping up and down reveals hysteresis in the dynamical responses.

We conducted finer frequency sweeps near the frequencies of 12.0 Hz and 40.0 Hz, where nonperiodic responses were observed. These are presented in Figure 7. The rich dynamics are more pronounced in the latter zone which was accompanied with slight audible chatter. First we suspected normal vibrations to be the cause of this chatter and made an attempt to visualize the loss of contact using a stroboscope. The stroboscopic probe did not reveal a visible loss of contact. We suspect the jump from the static friction to kinetic friction to be the cause of this chatter.

The deviation in the plots during finer sweeps might be attributed to the wear on which we did not have any control.

In the frequency sweeps, we note that in the vicinity of 12.0 Hz frequency, the system shows a deviation from a periodic orbit and there is a change in sign of the quarter-phase strain. Observation of the tip deflection showed that in this frequency range the system went from period-one motion to nonperiodic motion back to period-one motion. Similarly there are interesting dynamics in the vicinity of 40 Hz. In this zone as the frequency is varied from 35 Hz to 45 Hz, the system goes from period one to nonperiodic, period two, nonperiodic and period one in that order. We also observed extremely low-frequency motions among these

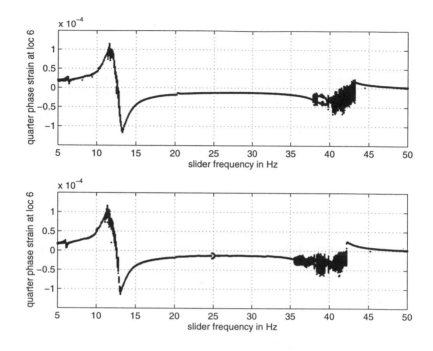

Fig. 6. Frequency sweeps (forward and reverse): the beam response is denoted by the strain at the location 1 and measured at quarter phase of the input acceleration. The sweep parameter is the input frequency.

excitation frequencies, and are exploring the details of these responses in on-going work.

We present the phase-portraits using the displacement and "velocity" of the contact point of the beam, at a few selected frequencies (11, 12 and 41.1 Hz). We measured the tip displacement using the method described earlier. We used a simple forward-difference scheme to differentiate these displacements to determine the velocity of the beam at the contact point. We have not accounted for any phase deviation that may result due to this process. We present the resulting phase portraits and the corresponding power spectra in Figure 8.

The dynamics at 11 Hz were taken to be periodic and were presented in this study as a reference for the noise level. (The noise level in the strain gage of the unforced system was approximately -110 dB with isolated spikes rising as high as -80 dB.) The dynamics at 12.0 and 41.1 Hz are non-periodic. The power spectra

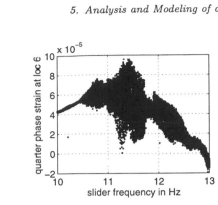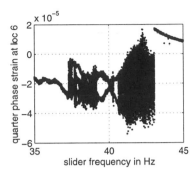

Fig. 7. Finer frequency sweeps around 12 and 41.1 Hz : the beam response is denoted by the strain at the location 1 and measured at quarter phase of the input acceleration. The bifurcation parameter is the input frequency.

are broad band and well above the noise floor, as are the power spectra from the strain gage at location 1, which are not shown.

From the phase portraits and broad band nature of the FFT of the contact point dynamics and the strain signal at 12.0 Hz and 41.1 Hz, we suspect the dynamics at 12.0 and 41.1 Hz to be chaotic.

We first carried out the phase space reconstruction studies on the time series consisting of strain at location 1 on the beam for the 12-Hz case. From the mutual information [27] we realized a time delay, which was then used in determining the embedding dimension by means of the method of false nearest neighbors [28]. We decided that a neighbor is false when the proportional change in the distance (with the additional dimension) is greater than 10. The embedding dimension was found to be five. Thus five is the number of delay coordinates necessary to describe the data. There is not a clear relationship between delay coordinates and dimension of the state space. But it suggests the number of active modes to be in the ballpark of half the number which is close to three in this case. With the embedding dimension of five and the above stated time delay we computed the largest Lyapunov exponent to be 3.73 (using Wolf's algorithm [29]), thus indicating the chaotic nature of signal.

A similar study of the dynamics at 41.1 Hz resulted in a shorter time delay, an embedding dimension of eight and a positive Lyapunov exponent of 1.91, indicating the chaotic nature of the dynamics at 41.1 Hz.

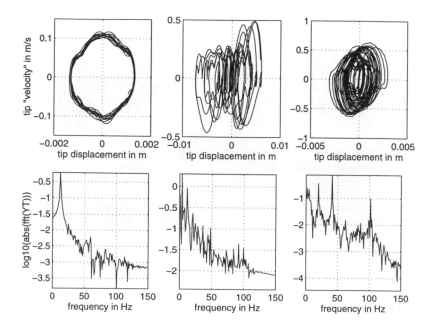

Fig. 8.   Dynamics of the contact point of the beam. The top row indicates the phase-portraits of the contact point at frequencies 11, 12 and 41.1 Hz. The bottom row indicates corresponding power-spectra.

6   Proper Orthogonal Modes

In this section we describe the spatial content of the beam dynamics in terms of proper orthogonal modes and proper orthogonal values. We present results for the case of 12.0 Hz excitation, where the dynamics are chaotic. The selection has been made as we intend to use the dominant modes in these regimes in building a reduced-order model for the beam.

We collected the six strain-signal time series at the operating frequencies 12.0 Hz. Using the method described in the previous sections we have converted the six strain signals into displacement signals at locations 1 − 6. After subtracting the mean displacement from each of these displacement time series, we arranged them to form a matrix $\mathbf{X}$ of six columns. Each column represents a displacement history at a particular beam location. We computed the covariance matrix $\mathbf{R} = \mathbf{X}^T\mathbf{X}$. This matrix is a positive definite matrix and the singular values and vectors of

TABLE I
Contribution of LNMs to POMFs at 12.0 Hz.

POM #	LNM 1	LNM 2	LNM 3	LNM 4	LNM 5	LNM 6
1	0.7113	0.2474	0.0296	0.0089	0.0021	0.0007
2	0.2494	0.7265	0.0108	0.0096	0.0035	0.0002
3	0.1978	0.2343	0.5538	0.0053	0.0051	0.0038

this 6×6 matrix yield the proper orthogonal values (POVs) and proper orthogonal modes. Using the proper values we identified the dominant modes in the system. We then curve-fit the discrete modes to get continuous modal functions. Though the modes are orthonormal, by doing a curve-fit we lose the orthogonality, as the definition of orthonormality is changed as we move from discrete vectors to continuous functions. In order to make the continuous functions orthonormal with reference to the distributed mass matrix, we use the Gram-Schmidt reorthonormalization technique, to yield what we call proper orthogonal modal functions (POMFs). We have used the linear natural modes (LNMs) of the cantilever for the curve fit. The motive is to identify the content of the natural modes in the dominant proper orthogonal modes.

In Figure 9 we present the POMFs obtained by Gram-Schmidt orthonormalization of the continuous versions of the POMs derived from the dynamics at 12.0 Hz.

Three of the original POMs were found to be dominant with their proper values accounting for 0.9913, 0.0083 and 0.0003 of the total energy, and thus satisfy the 99.99% criterion. This is consistent with the result from the FNN analysis which led to an estimate of about three active nodes. The computation of FNN involved some tolerances and these may have some relation to the criterion of 99.99% "energy". Cusumano and Bai [7] and Cusumano *et al.* [30] presented more detailed examples of such comparisons. We should note that the POD resembles singular systems analysis [31], and there are examples when singular systems analysis produce unreliable dimension estimates [32].

The effect of orthonormalization is to remove components from successive functions. As such it would be wrong to assign the same energy content to the POMFs, though the order of dominance is still maintained in this system. We have seen this to be the case when the orthonormalization procedure was replaced by straightforward POMs computed from additional, redundant "pseudo sensors" that were built into the system by evaluating the bases functions at additional points on the beam [26].

In Table I we give the percentage contribution of the natural modes in the POMFs. From this table we realize that the POMFs are not purely the LNMs. The six natural modes sufficiently describe the POMFs as indicated in Table I. In fact, the first five LNMs account for 99.99% energy of the first three POMFs. This

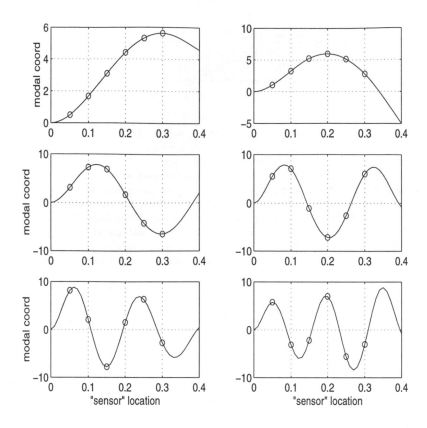

Fig. 9.   POMFs (obtained by orthonormalization of the curves fit on POMs) from chaotic dynamics at a driving frequency of 12.0 Hz.

indicates that we need the first five natural modes to describe the same dynamics (at least from the 99.99% energy perspective) as that of the three POMF model. This observation has also been made in an independent work by Azeez and Vakakis for an impacting rotor [5].

It is certainly not surprising that the POMFs are more efficient than the LNMs in describing the energy distribution in the signals. The POMFs are derived from the POMs, which are optimal in this sense [2]. In cases in which the difference between the number of significant POMFs and LNMs is clear, we would suspect that the POMFs would be more effective in model reduction, for example by

Galerkin's method. This is the focus in subsequent sections.

This result was more pronounced when the procedure was carried out at 41.1 Hz [26]. To this regard, we should be aware that different parameter regimes, in our case 12-Hz and 41.1 Hz excitation frequencies, produce different sets of dominant POMFs, and that this should be kept in mind when using the reduced-order model to make predictions outside the parameter regime.

Finally, the POMFs presented here are based on POMs on displacements evaluated at the strain-gage sites. These displacement locations were chosen arbitrarily, and these are not likely to be the optimal ones. The POMFs involve some extrapolation when describing the mode shapes toward the beam tip.

7 Mathematical Model

The system under consideration [26] consists of a cantilever with an attachment for convenient variation of normal load. Because of the attachment, we have the friction force F acting not on the cantilever, but at the end of an extension. This force can be transferred to the cantilever tip, by an equivalent force and moment combination (the force is F and the moment is Fl, where l is the length of the extension). To take the inertial effects of the extension and the fixture we lump the masses together at the cantilever tip. Hence we need the beam equation where both external forces and moments are present. Assuming that there is no shear, we use the extended Hamilton's principle [33] to get the standard PDE for a Euler-Bernoulli beam and write our system equation as

$$\frac{\partial^2}{\partial x^2}[EI(x)\frac{\partial^2 y(x,t)}{\partial x^2}] + \rho(x)\frac{\partial^2 y(x,t)}{\partial t^2} = p(x,y(x,t),\dot{y}(x,t)) - \frac{\partial q(x,y(x,t),\dot{y}(x,t))}{\partial x}$$
$$(4)$$

where $y(x,t)$ is the deflection at the point x at time t. $E(x), I(x)$ and $\rho(x)$ are the elastic modulus, the area moment of inertia and the linear density respectively of the system at location x. Linear viscous damping is neglected. Quantities $p(x,y,\dot{y})$ and $q(x,y,\dot{y})$ are the external forces and moments applied at location x on the beam at time t. Note that p and q are implicitly time varying functions of relative velocity, normal load and other parameters in case of frictional excitation. The parameter values were estimated in the experimental study [26].

With this assumption of "an Euler-Bernoulli cantilever beam with a lumped mass at the free end" for the system, we numerically estimated frequencies of 3.13, 21.97, 64.72, 130.64, 220.55 and 334.12 Hz. These frequencies are higher than the experimental frequencies (3.75, 18.4, 53.75, 102.5, 172.5 and 265 Hz), with the exception of the fundamental frequency. Discrepencies may be rooted in the beam-model assumptions, the neglected rotational inertia of the loading-beam attachment, and the quality of the experimental clamp.

As mentioned earlier we use a Galerkin's approximation, based on an orthonormal basis of functions $\{\varphi_1(x), \varphi_2(x), \cdots\}$ which satisfy the boundary conditions, to convert the above PDE (4) into ODEs. As a result, for each $j \in [1, N]$ we have

$$\sum_{i=1}^{N} M_{ji}\ddot{u}_i + \sum_{i=1}^{N} K_{ji}u_i = fP_j, \tag{5}$$

where the u_i are the generalized coordinates associated with the basis functions,

$$K_{ji} = \int_{x=0}^{x=L} EI(x)\varphi_j(x)\varphi_i^{(iv)}(x)dx, \tag{6}$$

$$M_{ji} = \int_{x=0}^{x=L} \rho\varphi_j(x)\varphi_i(x)dx + m_L\varphi_j(L)\varphi_i(L), \tag{7}$$

$$P_j = \varphi_j(L) + l\varphi'_j(L) - l\varphi_j(L), \tag{8}$$

and f represents the modulation of the friction force, which depends now on displacement quantities u_i and their time derivatives. Assuming $\mathbf{M}$ to be invertible, we can write equation (5) as

$$\ddot{\mathbf{u}} = \mathbf{M}^{-1}\mathbf{K}\mathbf{u} + \mathbf{M}^{-1}\mathbf{P}f(\mathbf{u},\dot{\mathbf{u}}). \tag{9}$$

Also if we assume that the modal viscous damping is proportional to the stiffness, then the only difference to the above equation would be a term $c\mathbf{M}^{-1}\mathbf{K}\dot{\mathbf{u}}(t)$ added to the right hand side. Here c would be proportional to a characteristic damping coefficient. Let us now write the system equation with such damping in the state space form, with the elements of the state $\mathbf{U}$ defined by elements $U_1 = u_1, \cdots, U_n = u_n, U_{n+1} = \dot{u}_1, \cdots U_{2n} = \dot{u}_n$. Thus we end up with the following state-space equation:

$$\dot{\mathbf{U}} = \begin{bmatrix} \mathbf{0} & \mathbf{I} \\ \mathbf{M}^{-1}\mathbf{K} & c\mathbf{M}^{-1}\mathbf{K} \end{bmatrix} \mathbf{U} + \begin{bmatrix} \mathbf{0} \\ \mathbf{M}^{-1}\mathbf{P} \end{bmatrix} f(\mathbf{U}). \tag{10}$$

For clarity let us denote the coefficient of the state $\mathbf{U}$ as $\mathbf{A}$ and the coefficient of the forcing term $f(\mathbf{U})$ as $\mathbf{B}$. Thus

$$\dot{\mathbf{U}} = \mathbf{A}\mathbf{U} + \mathbf{B}f. \tag{11}$$

These equations can be numerically integrated to find the values of $U_i(t)$, which in turn can be used to obtain the value of $y(x,t)$ and its derivatives at any point on the beam, provided we know f. The quantity f is the same as the frictional forces at the massless contact point, and is implicitly time dependent. When the contact elements are sliding the relative velocity between the contact elements at the point of contact is the same as the relative velocity between the slider and the tip of the beam. Hence in the sliding mode, we have f well defined by the kinetic friction force.

However when the two contact elements are stuck, from the dynamic equilibrium we can verify that the static friction force is in equilibrium with $K_C(Y_T(t) - Y_C(t))$, the elastic force due to contact compliance. Here, Y_C and Y_T are the displacements of the contact surface and the beam tip, and K_C is the contact stiffness. This in turn is the excitation force f. $Y_C(t) = a\sin(\omega t)$, the slider displacement, is imposed. The tip displacement $Y_T(t)$ can be calculated from the state $\mathbf{U}$. Thus we have

$$\begin{aligned} f &= K_C(Y_C(t) - Y_T(t)) \\ &= K_C(Y_C(t) - \mathbf{\Psi}^*\mathbf{U}), \end{aligned} \tag{12}$$

where

$$\mathbf{\Psi}^* = [\ \varphi_1(L) + l\varphi_1'(L) \ \cdots \ \varphi_{2n}(L) + l\varphi_{2n}'(L) \ 0 \ \cdots \ 0.\ ] \tag{13}$$

In the above force relation we have the effective force in terms of the state and input displacement which are all known. Thus we know the effective restraining force, which can also be used in comparison with the static friction bounds to determine when the contact elements slip.

8   Numerical Simulations and Validation

Numerical simulation of stick-slip systems needs to be carried out carefully. The presense of stick and slip leads to an ever changing number of active degrees of freedom, and hence an unsteady topology. General algorithms for handling such systems have been developed [34, 35, 36]. The simulation algorithm needs to keep track of kinematic and kinetic "indicators" of stick and slip and the contact forces, and also incorporate ways of determining the contact forces and reducing the system order as needed. In the particular case of a beam with friction applied at a point, Whiteman and Ferri [37] have outlined an effective method. The algorithm we apply has some similarities, and will be briefly discussed in terms of the formulation in the previous section.

The numerical simulation technique involved the usage of a Runge-Kutta ODE integrator in double precision with a variable-time-step. In our application we have used the full-order set of equations (11) for both stick and slip, and have defined the frictional force based on the kinetic friction coefficient when slipping, and based on equation (12) when stuck. We monitor the sticking friction force in this equation to determine when the static friction is exceeded, such that slip ensues. Using equation (11) for stick means that we are integrating a redundant set of equations. This has not adversely effect the results. It is similar to integrating, for example, the acceleration to obtain the (known) velocity for a stuck belt-driven single-degree-of-freedom friction oscillator.

For a given set of initial conditions we determine the relative velocity and the excitation force. These we use to define a flag for slip or stick. At each time step of integration we determine the relative velocity and effective force on the beam and compare their values for slip and stick criteria. In case of a cross-over (relative velocity crossing zero or the effective force crossing the static friction

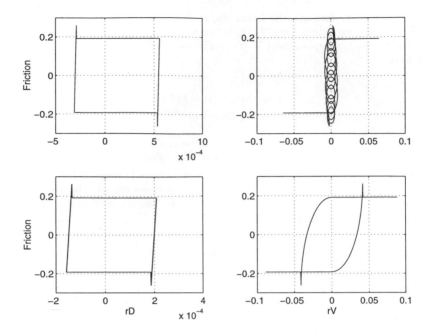

Fig. 10.   Simulated friction characteristics of the compliant contact model with $\mu_S > \mu_K$. The top row shows the case of stick-slip and the bottom row shows the case of micro-stick. The left column shows the friction vs. the relative displacement, and the right column depicts the friction vs. the relative velocity.

force bounds) we step back and use a simple interpolation to determine a smaller time step, and integrate again. This we continue until the cross-over is within a predefined tolerance zone. At this point after verifying the respective criterion we move to the subsequent mode. In this way we compute the state **U** at each time step which can in turn be used to obtain strain or displacement or velocity at any given point on the beam.

For verification of the friction model that was used in the simulations we illustrate the simulated friction force characteristics in Figure 10. We recorded these measurements using the 3-POMF reduced model (the POMFs derived from the 12 Hz dynamics). These match the characteristics of the compliant contact during stick and stick-slip [16] also obeserved in the experiment [38].

We simulate frequency sweeps in order to compare the model with the experiment [38]. In the experimental system, a close fit of the relationship between the shaker frequency and amplitude is

$$a = \exp(-4.4514 - 0.9037 \log \frac{\omega}{2\pi}). \tag{14}$$

Here we used the same frequency-amplitude relation in our numerical "frequency" sweeps. Each step in the sweep was carried for 600 cycles of which only the last 100 cycles were stored. Thus we are discarding 500 cycles to account for transients. Incidentally we applied a one-percent proportional damping, and the two-percent settling time is well within this transient time. We however note that the damping effects could be quite different in this nonlinear system. In each of the excitation cycles we computed the strain near the base of the beam, corresponding to the experimental strain gage closest to the clamp, whenever $v(t)$ crosses zero with a positive slope, so as to match the experimental sweeps [38]. Thus we simulated 100 strain values at each frequency step when the slider acceleration crossed its peak. We swept from $5 - 60$ Hz in steps of 0.1 Hz. We caution the reader that the above plots are not the amplitude-frequency response curves. These plots can be viewed as sweeps of Poincare sections along the curve described by the equation (14). in the configuration space, or as the real part of a nonlinear frequency response.

In Figure 11 we show the sweeps obtained from based on POMFs determined at 12.0 Hz in the physical experiment superimposed on the experimental frequency sweep.

From these plots we see that the most dominant mode alone is not able to predict the system's behavior. With two dominant modes we are able to retrieve the resonant dynamics around 12 Hz range (the POMs are extracted from the chaotic orbits at this frequency), but the high frequency dynamics are not realized. However with three POMs we are able to see the higher resonant frequencies, though the instabilities (indicated by the smear of points) are not as pronounced as in the experiment. These instabilities become more prominent as we include more POMs but the sweep characteristics (represented by the resonant frequencies) remain same. The inclusion of higher modes results in higher computing time, which grows exponentially with increasing number of modes.

On comparing these predicted sweeps with the ones from the experiment we realize that though the stick-slip instabilities around 12 and 41 Hz of the physical experiment do occur, the latter instablities occur at slightly higher frequencies. The dynamics in the unstable regions in the predicted sweeps is nonperiodic but is not as prominent as in the experiment, especially at the higher frequencies. However we note that the periodic orbits and also the zones of pure stick, which occupy broad regions between the resonances, are well predicted by the reduced model.

We expected this shift from the linear behavior of the numerical model. The reduction in order though introduces new constraints and in a way stiffens the system, it is not truly the cause for the shift in higher frequency dynamics to

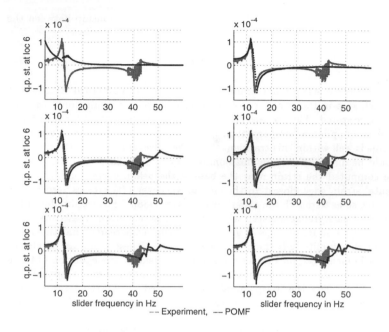

Fig. 11. "Frequency sweeps" of reduced models obtained at different levels of reduction using POMFs (extracted from experiment at 12.0 Hz and the continuous versions orthonormalized) superimposed on the experimental sweep. The grey curve is experimental, and the solid black curve is simulated.

higher frequencies. This we verified by carrying a frequency sweep using a numerical model with ten LNMs. The addition of more modes did not significantly improve the dynamic characteristics with reference to the frequency shift. Thus we attribute the shift to the fundamental modeling assumption of "lumped mass at the free end of a Euler-Bernoulli cantilever beam".

In order bypass the uncertainities involved in our modeling efforts, viz., beam model, damping, friction characteristics, etc., we considered a model similar to that of the POMF model, but with a large number of LNMs (ten) as a "truth set". We compared the frequency sweeps of reduced models with that of this "truth set" (see Figure 12). This comparison reveals that the resonant frequencies predicted by the POMF models are almost same as the ones predicted by the "truth set" model. The convergence of the reduced model frequency sweep with that of the "truth set" is acheived rather earlier just with three POMFs.

For the three POMF reduced model (from the 12.0 Hz chaotic dynamics in

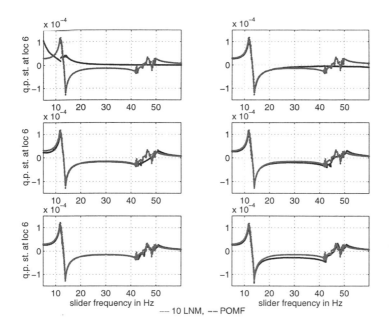

Fig. 12. "Frequency sweeps" of reduced models obtained at different levels of reduction using POMFs (extracted from experiment at 12.0 Hz and the continuous versions orthonormalized), which are the black curves, superimposed on that of the 10-LNM model ("truth set"), which are the grey curves.

the experiment) we show the phase portraits (displacement versus velocity of the contact point on the beam) in Figure 13. These are obtained at 11, 11.9 and 42.9 Hz respectively of the resulting dynamics in the sweeps. These may be compared with the experimental phase-portraits [38].

9 Discussion and Elaboration

We have taken a set of proper orthogonal modes from the response at a single parameter set and used them as the foundation for reduced order modeling of the beam equations. The resulting low-order model was then simulated for a parameter range (frequency sweep) and compared to the experiment. There is no reason to assume that the POMs found at 12 Hz should represent the dynamics at any other frequency, or at other choices of initial conditions. However, in this case the extrapolation of the POM-based model to other parameter values has drawn

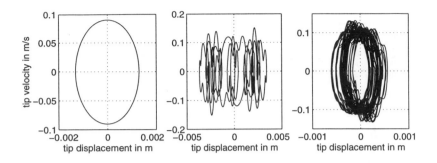

Fig. 13.   Phase-portraits (contact point displacement versus velocity) from the three-POMF reduced model at 11, 11.9 and 42.9 Hz.

out some of the dynamical features. It is conceivable, however, that a better match would be made between simulated and experimental frequency sweeps if this issue were overcome.

In our projections, we started with POMs obtained from discrete measurements, and used them to fit continous POMFs. While the POMs represent optimal energy distributions of the data, there is no such guarantee for the POMFs. We had seen, however, that as large numbers of redundant false sensors were added (in the conversion from strain to displacement, a set of basis functions were used, and these could be evaluated at any number of positions) [38, 26], the discrete POMs seemed to converge closely to the POMFs, indicating that the POMFs are close to optimal. In the reduced order modeling, continuous POMFs were integrated into the PDE to obtain the ODEs. We could have used discrete POMs, and applied them to the PDE by means of a numerical quadrature to obtain ODEs. Indeed, in some applications were the PDE is not so simply defined, this may be a more sensible approach.

We have also performed simulations which incorporated rigid contacts as opposed to compliant contacts. These are shown in Figure 14. The results with the rigid contacts did not agree as well with experiments as the simulations with compliant contacts. The rigid contacts were unable to generate any features that would suggest instabilities or resonances in the 40 Hz area. Previous work has shown that the dynamic hysteretic effects of compliant contacts are more significant as the frequency increases [16]. It might be interesting to investigate high-frequency applications such as squeak and squeal to find out whether contact compliance is significant to these phenomena, and whether it has a stabilizing or destabilizing effect.

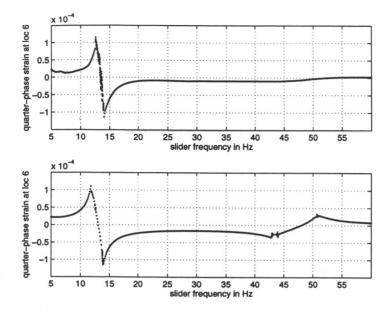

Fig. 14. Frequency sweeps of the model with a rigid contact, using the same POMF-based modal reduction as before. The black curve represents the simulted sweeps, and the grey curve represents experimental sweeps.

Finally, we address an issue mentioned earlier. Recall the static friction did not reveal itself in the dynamic friction plots, and that it was then estimated in a separate static experiment. It is possible that the static friction is itself dynamic [?]. To this end, we may ask what effect the static friction may have on the dynamics. To this end, we have simulated the i behavior at 12.5 Hz and 43.9 Hz in the POMF-reduced model. The results, shown in Figure 15 in terms of parameter

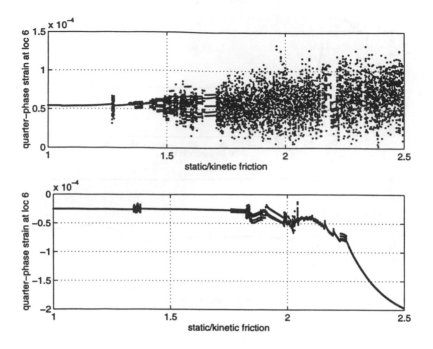

Fig. 15.   Sweeps of the ratio μ_s/μ_k in the model reduced by means of POMFs of the three dominant POMs obtained at 12.0 Hz). The top figure corresponds to a sweep at 12.5 Hz and the bottom plot represents a sweep at 43.9 Hz.

sweeps, indicate that the ratio between the static and kinetic friction coefficients can have a strong effect on the occurrence of instabilities.

10   Conclusions

We have studied the dynamics of an experimental beam excited by dry friction. We found the friction to have the characteristics of a massless compliant contact. Modeled as such, the force-displacement and force-velocity characteristics are mutually consistent, and support reference [16] in the features and applicability of this friction model.

The response of the beam included periodic and nonperiodic oscillations. We performed proper orthogonal decomposition on the nonperiodic data, and found the proper orthogonal modes. The chaotic 12-Hz data ensemble required three POMs to capture 99.99% of the signal power.

The number of POMFs required to span the spatial description is less than the number of linearized normal modes. Hence, it is expected that the POMFs would provide a more efficient basis for applying a Galerkin projection for reducing the order of the associated model. This is found to be the case in other works [5, 26].

We reduced the order of a nonlinear beam model by using the POMFs, and evaluated the model using frequency sweeps. Though we found some discrepancies between the physical and numerical experiments, we attribute these discrepancies to the original mathematical model of the system, such as the assumptions of Euler-Bernoulli model for the cantilever, lumped mass for the loading beam and friction law to be a simple Coulomb with contact compliance characteristic.

We used POMs obtained from the chaotic dynamics at 12.0 Hz, three of which captured 99.99 % of the signal power, and fit POMFs to these POMs using linear normal modes as basis functions. We then reduced the PDE model by using the POMFs, and compared simulations to experimental results, and to simulations based on LNMs. The projections based on POMFs converge to both the experimental sweep and the sweep from the "truth set", with just three POMFs. This is promising, since the experimental nature of the determination of POMFs makes it more advantageous as for complex systems the LNMs may not be readily available.

Finally, we have seen that frictional contact compliance may have significant effects on the dynamic response, especially at higher frequencies. The ratio of the coefficients of static and kinetic friction also has an impact on the dynamics.

11 Acknowledgements

Support for this research program comes gratefully from the National Science Foundation (CMS-9624347). We appreciate comments from Chuck McCluer, Clark Radcliffe, and Steve Shaw.

References

1. J. L. Lumley. The structure of inhomogeneous turbulent flow. In *Atmospheric Turbulence and Radio Wave Propagation*, pages 166–178. Nauka, Moscow, 1967.

2. G. Berkooz, P. Holmes, and J. L. Lumley. The proper orthogonal decomposition in the analysis of turbulent flows. In *Annual Review of Fluid Mechanics*, volume 25. Annual Reviews Inc., New York, 1993.

3. Philip M. Fitzsimons and Chunlei Rui. Determining low dimensional models of distributed systems. *Advances in Robust and Nonlinear Control Systems, ASME DSC*, 53:9 15, 1993.

4. K. D. Murphy. Using the Karhunen-Loève decomposition to examine chaotic snap-through oscillations of a buckled plate. In *Proceedings in Sixth Conference on Nonlinear Vibrations, Stability and Dynamics of Structures*, Virginia, June 1996.

5. M. F. A. Azeez and A. F. Vakakis. Using Karhunen-Loève decomposition to analyze the vibroimpact response of a rotor. In *Proceedings in Seventh Conference on Nonlinear Vibrations, Stability and Dynamics of Structures*, Blacksburg, July 1998.

6. X. Ma, M. A. F. Azeez, and A. F. Vakakis. Nonparametric nonlinear system identification of a nonlinear flexible system using proper orthogonal mode decomposition. In *Proceedings in Seventh Conference on Nonlinear Vibrations, Stability and Dynamics of Structures*, Blacksburg, July 1998.

7. J. P. Cusumano and B. Y. Bai. Period-infinity periodic motions, chaos and spatial coherence in a 10 degree of freedom impact oscillator. *Chaos, Solitons and Fractals*, 3(5):515–535, 1993.

8. M. A. Davies and F. C. Moon. In *Nonlinear Dynamics: the Richard Rand 50th Anniversary Volume*. World Scientific, Singapore, 1997.

9. S. R. Sipcic, A. Benguedouar, and A. Pecore. Karhunen-Loève decomposition in dynamical modeling. In *Proceedings in Sixth Conference on Nonlinear Vibrations, Stability and Dynamics of Structures*, Virginia, June 1996.

10. E. Kreuzer and O. Kust. Proper orthogonal decomposition—an efficient means of controlling self-excited vibrations of long torsonal strings. In *Nonlinear Dynamics and Controls*, volume ASME DE-Vol. 91, pages 105–110, 1996.

11. I. T. Georgiou and I. B. Schwartz. A proper orthogonal decomposition approach to coupled mechanical systems. In *Nonlinear Dynamics and Controls*, volume ASME DE-Vol. 91, pages 7–12, 1996.

12. B. Feeny and R. Kappagantu. On the physical interpretation of proper orthogonal modes in vibrations. *Journal of Sound and Vibration*, 1996.

13. C. Canudas de Wit, H. Olsson, K. J. Astrom, and P. Lischinsky. A new model for control of systems with friction. *IEEE Transactions on Automatic Control*, 40(3):419–425, 1995.

14. P. Dahl. A solid friction model. El Segundo, CA, 1968. Tech. Rep. TOR-0158 (3197-18)-1.

15. J. S. Courtney-Pratt and E. Eisner. The effect of a tangential force on the contact of metallic bodies. A238:529–550, 1957.

16. J.-W. Liang and B. F. Feeny. Dynamical friction behavior in a forced oscillator with a compliant contact. *Journal of Applied Mechanics*, 65(1):250–257, 1998.

17. J. T. Oden and J. A. C. Martins. Models and computational methods for dynamic friction phenomena. *Computer Methods in Applied Mechanics and Engineering*, 52:527–634, September 1985.

18. A. Harnoy, B. Friedland, and H. Rachoor. Modeling and simulations of elastic and friction forces in lubricated bearing for precise motion control. *Wear*, 172:155–165, 1994.

19. D. C. Threlfall. The inclusion of Coulomb friction in mechanisms programs with particular reference to DRAM. *Zeitschrift für angewandte Mathematik und Mechanik*, 76(8):475–483, 1978.

20. K. Y. Sanliturk and D. J. Ewins. Modeling two-dimensional friction contact and its application using harmonic balance method. *Journal of Sound and Vibration*, 193(2):511–524, 1996.

21. J. H. Griffin. Friction damping of resonant stresses in gas turbine engine airfoils. *Journal of Engineering for Power*, 102:329–333, 1980.

22. A. A. Ferri and B. S. Heck. Vibration analysis of dry damped turbine blades using singular perturbation theory. *Journal of Vibration and Acoustics*.

23. J. Guillen and Ch. Pierre. Analysis of the forced response of dry-friction dampined structural systems using and efficient hyprid frequency-time method. In *Nonlinear Dynamics and Controls*, volume ASME DE-Vol. 91, pages 41–50, 1996.

24. J.-H. Wang. Design of a friction damper to control vibration of turbine blades. In F. Pfeiffer A. Guran and K. Popp, editors, *Dynamics and Friction: Modeling, Analysis, and Experiment*. World Scientific, Singapore, 1996.

25. A. A. Ferri. Friction damping and isolation systems. *Journal of Mechanical Design*, 117(B):196–206, 1995.

26. R. V. Kappagantu. *An "Optimal" Modal Reduction of Systems with Frictional Excitation*. Ph.D. thesis, Michigan State University, East Lansing, 1997.

27. A. M. Fraser and H. L. Swinney. Information and entropy in strange attractors. *Physical Rev*, 33A:1134, 1986.

28. M. B. Kennel, R. Brown, and H. D. I. Abarbanel. Determining embedding dimension for phase-space reconstruction using a geometrical construction. *Physics Letters A*, 45(6):3403–

3411, 1992.

29. A. Wolf, J. B. Swift, H. L. Swinney, and J. A. Vastano. Determining Lyapunov exponents from a time series. *Physica D*, 16:285–317, 1985.

30. J. P. Cusumano, M. T. Sharkady, and B. W. Kimble. Spatial coherence measurements of a chaotic flexible-beam impact oscillator. In *Aerospace Structures: Nonlinear Dynamics and System Response*, volume ASME AD-Vol. 33.

31. D. P. Broomhead and G. P. King. Extracting qualitative dynamics from experimental data. *Physica D*, 20:217–226, 1986.

32. B. Ravindra. Comments on on the physical interpretation of proper orthogonal modes in vibrations. *Journal of Sound and Vibration*.

33. Leonard Meirovitch. *Analytical Methods in Vibrations*. Macmillan, New York, 1967.

34. F. Pfeiffer and Ch. Glocker. *Multibody dynamics with unilateral contacts*. Wiley, New York, 1996.

35. M. Wosle and F. Pfeiffer. Dynamics of multibody systems containing dependent unilateral constraints with friction. *Journal of Vibration and Control*, 2(2):161–192, 1996.

36. Ch. Glocker and F. Pfeiffer. Dynamical systems with unilateral constraints. *Nonlinear Dynamics*, 9(3):245–259, 1992.

37. W. E. Whiteman and A. A. Ferri. Multi-mode analysis of beam-like structures subjected to displacement-dependent dry-friction damping. *Journal of Sound and Vibration*, 207(3):403–418, 1997.

38. R. V. Kappagantu and B. F. Feeny. *Proper orthogonal modal analysis of a frictionally excited beam (part 1)*. submitted to Nonlinear Dynamics, 1998.

39. Brian Armstrong-Helouvry, Pierre Dupont, and Carlos Canudas de Wit. A survey of models, analysis tools and compensation methods for the control of machines with friction. *Automatica*, 30(7):1083–1138, 1994.

Dynamics with Friction: Modeling, Analysis and Experiment, Part II, pp. 155–189
edited by A. Guran, F. Pfeiffer and K. Popp
Series on Stability, Vibration and Control of Systems, Series B, Vol. 7
© World Scientific Publishing Company

TRANSIENT WAVES
IN LINEAR VISCOELASTIC MEDIA

FRANCESCO MAINARDI

Department of Physics, University of Bologna
46 Via Irnerio, Bologna 40126, Italy

ABSTRACT

In this chapter we review the main aspects of linear wave propagation in homogeneous, semi-infinite, viscoelastic media. In particular, we consider the so-called transient waves, thus named since they are generated by impact on an initially quiescent medium. The utility of the Laplace transform technique is pointed out in dealing with the evolution of these waves. We also discuss the concept of signal velocity and we provide wave-front expansions suitable for numerical computation.

Introduction

Transient waves in linear viscoelastic media are a noteworthy example of on linear dispersive waves in the presence of dissipation. They are obtained from a one-dimensional initial-boundary value problem, that we are going to solve using the Laplace transform technique. We already know from our chapter on *Linear Viscoelasticity*[10], henceforth referred to as *LV*, that the Laplace transform technique is suited to treat the various constitutive equations for linear viscoelastic bodies. Our presentation is mostly based on our papers, see *e.g.* Refs [11−26] .

We first consider the general problem of transient waves introducing the Laplace transform. This approach allows us to obtain the integral representation of the solution and leads in a natural way to the concept of wave-front velocity.

In Sect. 2. we consider the structure of the wave equations in the original space-time domain after inversion from the Laplace domain; in particular, we provide the explicit wave equations for the most used viscoelastic models.

In Sect. 3. we introduce the Fourier integral representation of the solution which leads to the notion of the complex index of refraction. The dispersive and dissipative properties of the viscoelastic waves are then investigated by considering the normal-mode solutions: the notions of phase/group velocity and attenuation coefficient are introduced and the anomalous character of the dispersion is highlighted.

In Sect. 4. we deal with the problem of finding a suitable definition for the signal velocity of viscoelastic waves. In fact, because of the anomalous dispersion and far from negligible dissipation, the identification of the group velocity with the above velocity is lost and the subject matter must be revisited. This argument allows us to consider the saddle-point method which provides approximate solutions, which are valid for long times elapsed from the wave-front.

Finally, in the subsequent Sections 5. and 6., we look at series recurrence methods which provide wave-front expansions for the solutions. In Sect. 5. we prove the convergence of the series in any space-time domain for the so-called *analytical models* of viscoelasticity However, the convergence is slow for long times elapsed from the wave front, so that, in order to obtain the match with the previous saddle-point approximation, the acceleration technique of rational Padè approximants is suggested. For the other models of viscoelasticity, treated in Sect. 6, we point out the asymptotic character of our expansions as the time elapsed from the wave front tends to zero; in this respect our method can be seen as a generalization of the ray-series method known for electro-magnetic waves.

1. Statement of the Problem by Laplace Transform

Problems of transient viscoelastic waves essentially concern the response of a long viscoelastic rod of uniform small cross-section to dynamical (uniaxial) loading conditions.

According to the elementary theory (see *e.g.* Bland[27], Hunter[28], Caputo & Mainardi[11], Christensen[29], Pipkin[30], Chin[31], Graffi[32]) the rod is taken to be homogeneous (of density ρ), semi-infinite in extent ($x \geq 0$), and undisturbed for $t < 0$. For $t \geq 0$ the end of the rod (at $x = 0$) is subjected to a disturbance (input) denoted by $r_0(t)$. The response variable (output) denoted by $r(x, t)$ may be either the displacement $u(x, t)$, the particle velocity $v(x, t) = u_t(x, t)$, the stress $\sigma(x, t)$, or the strain $\epsilon(x, t)$, where subscripts denote partial derivatives.

The mathematical problem consists of finding a solution for $r(x, t)$ in the region $x > 0$ and $t > 0$ which satisfies the following field equations,
the *equation of motion*

$$\boxed{\sigma_x = \rho u_{tt}}, \tag{1.1}$$

the *kinematic equation*

$$\boxed{\epsilon(x, t) = u_x(x, t)}, \tag{1.2}$$

and a linear constitutive equation (the *stress–strain relation*)

$$\boxed{\sigma(x, t) = F[\epsilon(x, t)]}, \tag{1.3}$$

under the boundary conditions

$$\boxed{r(0, t) = r_0(t), \quad \lim_{x \to \infty} r(x, t) = 0, \quad t > 0}, \tag{1.4}$$

and the homogeneous initial conditions

$$r(x,0) = r_t(x,0) = 0, \quad x > 0.$$ (1.5)

The stress–strain relation is the equation describing the mechanical properties of the rod and, therefore, it is the constitutive equation for the assumed viscoelastic model. It is most conveniently treated using the Laplace transform as shown in *LV* [see eqs (7a) and (7b)]. Recalling our notation for the Laplace transform of a generic (locally summable) function $f(t)$,

$$f(t) \div \mathcal{L}[f(t)] = \overline{f}(s) := \int_0^\infty e^{-st} f(t)\,dt, \quad s \in \mathbb{C},$$

we obtain

$$\overline{\sigma}(x,s) = s\overline{G}(s)\,\overline{\epsilon}(x,s),$$ (1.6a)

or

$$\overline{\epsilon}(x,s) = s\overline{J}(s)\,\overline{\sigma}(x,s),$$ (1.6b)

where $\overline{G}(s)$ and $\overline{J}(s)$ denote the Laplace transform of the *relaxation modulus* $G(t)$ and *creep compliance* $J(t)$, respectively. Henceforth, eqs (1.6) are referred to as the *relaxation representation* and *creep representation* of the stress–strain relation (1.3).

Applying the Laplace transform to the other field equations and using (1.4) and (1.5), we obtain

$$\overline{r}_{xx}(x,s) - [\gamma(s)]^2\,\overline{r}(x,s) = 0 \implies \overline{r}(x,s) = \overline{r}_0(s)\,e^{-\gamma(s)x},$$ (1.7)

where

$$\gamma(s) = \begin{cases} \sqrt{\rho}\,s\,[s\overline{J}(s)]^{1/2} \\ \sqrt{\rho}\,s\,[s\overline{G}(s)]^{-1/2} \end{cases},$$ (1.8)

with

$$\gamma(s) \geq 0 \quad \text{for} \quad s \geq 0, \qquad \widetilde{\gamma(s)} = \gamma(\tilde{s}) \quad \text{for} \quad s \in \mathbb{C}.$$ (1.9)

Here, the $\tilde{\ }$ denotes the complex conjugate.

The solution $r(x,t)$ is therefore given by the Bromwich formula,

$$r(x,t) = \frac{1}{2\pi i} \int_{Br} ds\,\overline{r}_0(s)\,e^{st\,-\,\gamma(s)x},$$ (1.10)

in which Br denotes the Bromwich path, *i.e.* a vertical line lying to the right of all singularities of $\overline{r}_0(s)$ and of $\gamma(s)$. Because of (1.8), the singularities of $\gamma(s)$ result from the explicit expressions of the functions $s\bar{J}(s)$, $s\bar{G}(s)$ and, thus, of the material functions $J(t)$ and $G(t)$, respectively.

From the analysis carried out in Sections 3 and 4 of LV, we obtain the expressions valid for viscoelastic models with *discrete* and *continuous* distributions of retardation/relaxation times, respectively. We recognize that, in the *discrete* case $s\bar{J}(s)$ and $s\bar{G}(s)$ are rational functions with zeros and poles alternating in the negative real axis, while in the *continuous* case they are sectionally analytic functions, with the cut in the negative real axis. Consequently, $\gamma(s)$ turns out to be analytic over the entire s-plane cut along the negative real axis, where $s\bar{J}(s)$ and $s\bar{G}(s)$, if not sectionally analytic, attain their zeros and poles.

It may be convenient in (1.10) to introduce the so-called *impulse response*

$$\mathcal{G}(x,t) \div \bar{\mathcal{G}}(x,s) := e^{-\gamma(s)x}, \tag{1.11}$$

which is the solution corresponding to $r_0(t) = \delta(t)$. Consequently, we can write

$$r(x,t) = \int_0^t \mathcal{G}(x, t-\tau) r_0(\tau)\, d\tau = \mathcal{G}(x,t) * r_0(t), \tag{1.12}$$

where $*$ denotes the (unilateral) time convolution.

A general result, which can be easily obtained from the Laplace representation (1.10), concerns the velocity of propagation of the head (or *wave-front velocity*) of the disturbance[33]. This velocity, denoted by c, is readily obtained by considering the following limit in the complex s-plane

$$\frac{\gamma(s)}{s} \to \frac{1}{c} > 0 \quad \text{as} \quad Re[s] \to +\infty. \tag{1.13}$$

According to the analysis carried out in LV [see Sect. 2], the limit in (1.13) is true for viscoelastic models of types I and II, which exhibit an instantaneous elasticity, namely $0 < G(0^+) = 1/J(0^+) < \infty$, for which

$$\begin{cases} s\bar{G}(s) \to G(0^+), \\ s\bar{J}(s) \to J(0^+), \end{cases} \quad \text{as} \quad Re[s] \to +\infty. \tag{1.14}$$

From (1.8)-(1.9) and (1.13)-(1.14) we easily recognize that, setting $G_0 = G(0^+)$ and $J_0 = J(0^+)$,

$$c = \begin{cases} 1/\sqrt{\rho J_0} \\ \sqrt{G_0/\rho} \end{cases}, \tag{1.15}$$

and, from the application of Cauchy's theorem in (1.10), $r(x,t) \equiv 0$ for $t < x/c$. Therefore, c represents the wave-front velocity, *i.e.* the maximum velocity exhibited by the wave precursors. On the other hand, the viscoelastic bodies of types III-IV, for which $J_0 = 0$, exhibit an infinite wave-front velocity. Because it is difficult to conceive of a body admitting an infinite propagation velocity in this respect, the quantity c is assumed finite throughout the remainder of the present analysis, with the exception of certain isolated cases.

Using (1.8) and (1.13)-(1.15), we find it convenient to set

$$\gamma(s) = \frac{s}{c} n(s),$$ (1.16)

where

$$n(s) := \left[\frac{s\overline{J}(s)}{J_0}\right]^{1/2} = \left[\frac{G_0}{s\overline{G}(s)}\right]^{1/2} \to 1 \quad \text{as} \quad Re[s] \to +\infty.$$ (1.17)

Of course, $n(s)$ takes on itself the multivalued nature of $\gamma(s)$, exhibiting the same branch cut on the negative real axis and the same positivity and crossing-symmetry properties, *i.e.*

$$n(s) \geq 0 \quad \text{for} \quad s \geq 0, \qquad \widetilde{n(s)} = n(\tilde{s}), \quad s \in \mathbb{C}.$$ (1.18)

2. The Structure of Wave Equations in the Space–Time Domain

After having reported on the integral representation of transient waves in the Laplace domain, we devote this section to deriving the evolution equations in the original space-time domain and to discussion of their mathematical structure. For this purpose, we introduce the following non-dimensional functions, related to the material functions $J(t)$ and $G(t)$,

a) Rate of Creep

$$\Psi(t) := \frac{1}{J_0} \frac{dJ}{dt} \geq 0,$$ (2.1a)

b) Rate of Relaxation

$$\Phi(t) := \frac{1}{G_0} \frac{dG}{dt} \leq 0.$$ (2.1b)

The Laplace transforms of these functions turn out to be related to $n(s)$ through (1.17); we obtain

$$[n(s)]^2 := \frac{s\overline{J}(s)}{J_0} = 1 + \overline{\Psi}(s),$$ (2.2a)

and

$$[n(s)]^{-2} := \frac{s\overline{G}(s)}{G_0} = 1 + \overline{\Phi}(s).$$ (2.2b)

As a consequence of (1.7), (1.16) and (2.2), we obtain the creep and relaxation representation of the wave equations in the Laplace domain as follows

$$\overline{r}_{xx}(x,s) - \frac{s^2}{c^2}\left[1 + \overline{\Psi}(s)\right]\overline{r}(x,s) = 0 \,, \tag{2.3a}$$

$$\left[1 + \overline{\Phi}(s)\right]\overline{r}_{xx}(x,s) - \frac{s^2}{c^2}\overline{r}(x,s) = 0 \,. \tag{2.3b}$$

Thus, the required wave equations in the space-time domain can be obtained by inverting (2.3a) and (2.3b), respectively. We get the following integro-differential equations of convolution type

a) Creep Representation

$$\left[1 + \Psi(t) *\right]\frac{\partial^2 r}{\partial t^2} = c^2 \frac{\partial^2 r}{\partial x^2}, \quad r = r(x,t) \,, \tag{2.4a}$$

b) Relaxation Representation

$$\frac{\partial^2 r}{\partial t^2} = c^2 \left[1 + \Phi(t) *\right]\frac{\partial^2 r}{\partial x^2}, \quad r = r(x,t) \,. \tag{2.4b}$$

We recall that, in view of their meaning, the kernel functions $\Psi(t)$ and $\Phi(t)$ are usually referred to as the *memory functions* of the creep and relaxation representations, respectively.

Let us now consider the case of mechanical models treated in *LV*, Sect. 3. For them $s\overline{J}(s)$ and $s\overline{G}(s)$ turn out to be *rational* functions in $\mathbb{C}$ with simple poles and zeros on the negative real axis and, possibly, with a simple pole or a simple zero at $s = 0$, respectively. In particular, we recall

$$s\overline{J}(s) = \frac{1}{s\overline{G}(s)} = \frac{P(s)}{Q(s)}, \quad \text{where} \quad \begin{cases} P(s) = 1 + \sum_{k=1}^{p} a_k\, s^k, \\ Q(s) = m + \sum_{k=1}^{q} b_k\, s^k, \end{cases} \tag{2.5}$$

with $q = p$ for models exhibiting instantaneous elasticity ($J_0 = a_p/b_p$) and $q = p+1$ for the others. In this case the evolution equation (1.7) in the Laplace domain can be easily inverted and reads

$$\left[m + \sum_{k=1}^{q} b_k \frac{\partial^k}{\partial t^k}\right]\frac{\partial^2 r}{\partial x^2} = \rho \left[1 + \sum_{k=1}^{p} a_k \frac{\partial^k}{\partial t^k}\right]\frac{\partial^2 r}{\partial t^2}, \quad r = r(x,t) \,. \tag{2.6}$$

This is to say that for the mechanical models the integral convolutions entering the evolution equations (2.4a)-(2.4b) can be eliminated to yield time derivatives.

We now turn to the most elementary mechanical models, reporting for each of them the corresponding evolution equation with $r = r(x,t)$.

$Newton :$ $\qquad \sigma(t) = b \dfrac{d\epsilon}{dt}$,

$$\boxed{\dfrac{\partial r}{\partial t} = D \dfrac{\partial^2 r}{\partial x^2}, \qquad D = \dfrac{b}{\rho}.} \tag{2.7}$$

In this case we obtain the classical *diffusion equation*, which is of parabolic type.

$Voigt :$ $\qquad \sigma(t) = m\,\epsilon(t) + b\dfrac{d\epsilon}{dt}$,

$$\boxed{\dfrac{\partial^2 r}{\partial t^2} = c_0^2 \left(1 + \tau_\epsilon \dfrac{\partial}{\partial t}\right) \dfrac{\partial^2 r}{\partial x^2}, \qquad c_0^2 = \dfrac{m}{\rho}, \qquad \tau_\epsilon = \dfrac{b}{m}.} \tag{2.8}$$

Also in this case we obtain a parabolic equation, but of the third order.

$Maxwell :$ $\qquad \sigma(t) + a\dfrac{d\sigma}{dt} = b\dfrac{d\epsilon}{dt}$,

$$\boxed{\dfrac{\partial^2 r}{\partial t^2} + \dfrac{1}{\tau_\sigma}\dfrac{\partial r}{\partial t} = c^2 \dfrac{\partial^2 r}{\partial x^2}, \qquad c^2 = \dfrac{b}{a\rho}, \qquad \tau_\sigma = a.} \tag{2.9}$$

This is the so-called *telegraph equation*, which is of hyperbolic type.

$S.L.S. :$ $\qquad \left[1 + a\dfrac{d}{dt}\right]\sigma(t) = \left[m + b\dfrac{d}{dt}\right]\epsilon(t), \qquad 0 < m < \dfrac{b}{a}$,

$$\boxed{\dfrac{\partial}{\partial t}\left(\dfrac{\partial^2 r}{\partial t^2} - c^2 \dfrac{\partial^2 r}{\partial x^2}\right) + \dfrac{1}{a}\left(\dfrac{\partial^2 r}{\partial t^2} - c_0^2 \dfrac{\partial^2 r}{\partial x^2}\right) = 0, \qquad c^2 = \dfrac{b}{a\rho}, \qquad c_0^2 = \dfrac{m}{\rho}.} \tag{2.10}$$

We recall that $\tau_\epsilon = b/m$, $\tau_\sigma = a$, so that $c_0/c = \tau_\sigma/\tau_\epsilon = a\,m/b$. This common ratio will be denoted by χ, where $0 < \chi < 1$. The Maxwell model is recovered as a limit case when $\chi \to 0$. Here we have an hyperbolic equation of the third order with characteristics (related to c) and subcharacteristics (related to c_0), as noted by Chin[31].

3. The Complex Index of Refraction: Dispersion and Attenuation

In alternative to (1.10), there is another (perhaps more common) integral representation of the solution, which is based on the Fourier transform of causal functions (*i.e.* vanishing for $t < 0$). The Fourier integral representation can be formally derived from the Laplace representation setting $s = i\omega$ in (1.10). We recall our notation for the Fourier transform of a generic (summable) function $f(t)$,

$$f(t) \div \mathcal{F}[f(t)] = \hat{f}(\omega) := \int_{-\infty}^{+\infty} e^{-i\omega t} f(t)\, dt, \quad \omega \in \mathbb{R}.$$

Thus we can re-write (1.10) as

$$r(x,t) = \frac{1}{2\pi} \int_{-\infty}^{+\infty} d\omega\, \hat{r}(0,\omega)\, e^{i\,[\omega t - \kappa(\omega)x]}, \tag{3.1}$$

where $\kappa(\omega) = \gamma(i\omega)$ is referred to as the *complex wave number*. To ensure that $r(x,t)$ is a real function of x and t, the following *crossing-symmetry relation* holds

$$\widetilde{\kappa(\omega)} = -\kappa(-\omega), \quad \omega \in \mathbb{R}, \tag{3.2}$$

which shows that the real/imaginary part of $\kappa(\omega) = \kappa_r(\omega) + i\,\kappa_i(\omega)$ is odd/even, respectively. We recognize that the *complex wave number* turns out to be related to the *complex modulus* $G^*(\omega)$ and to the *complex compliance* $J^*(\omega)$ [defined in *LV*, see (69)] by the following relations

$$\kappa(\omega) = \sqrt{\rho}\,\omega\,[G^*(\omega)]^{-1/2} = \sqrt{\rho}\,\omega\,[J^*(\omega)]^{1/2}. \tag{3.3}$$

The integral representations (1.10) and (3.1) are to be used according to convenience. The Fourier representation is mostly used to show the dispersive nature of the wave motion, which is related to the *dispersion equation* of the problem:

$$\mathcal{D}(\omega, \kappa) = 0, \; \omega \in \mathbb{R}, \; \kappa \in \mathbb{C}. \tag{3.4}$$

After this, the complex wave number $\kappa = \kappa(\omega)$ can be obtained as a specific branch (*normal mode*) of the *dispersion equation* and the quantity

$$e^{i\,[\omega t - \kappa(\omega)x]} = e^{\kappa_i(\omega)x}\, e^{i[\omega t - \kappa_r(\omega)x]} \tag{3.5}$$

is referred to as the corresponding *normal-mode solution*.

Also the function $\gamma(s)$ can be found from the *dispersion equation*: setting in (3.4) $\omega = -is$ and solving for $\kappa = \kappa(-is)$, we have $\gamma(s) = \pm\kappa(-is)$, where the choice of sign is dictated by the condition that the real part of γ should be positive when s is real and positive[34].

The Fourier representation allows us to point out the importance of the analytic function $n(s)$ for problems of wave propagation. For this purpose let us consider $n(s)$ on the imaginary axis (the frequency axis) and put it in relation with the complex wave number. Then, using (3.3) with (1.16)-(1.17), we write

$$\boxed{\kappa(\omega) = \frac{\omega}{c} \, n^*(\omega)}, \qquad (3.6)$$

where

$$\boxed{n^*(\omega) := n(i\omega) = \left[\frac{G_0}{G^*(\omega)}\right]^{1/2} = \left[\frac{J^*(\omega)}{J_0}\right]^{1/2} \to 1 \quad \text{as} \quad \omega \to \infty}. \qquad (3.7)$$

The quantity $n^*(\omega)$ is referred to as the *complex index of refraction* of the viscoelastic medium with respect to mechanical waves, in analogy with the optical case.

From (3.2) and (1.17)-(1.18) we obtain the *crossing-symmetry relation* for $n^*(\omega)$, namely

$$\boxed{\widetilde{n^*(\omega)} = n^*(-\omega) \quad \omega \in \mathbb{R}}, \qquad (3.8)$$

which shows that the real/imaginary part of the complex refraction index is even/odd, respectively. We note the property

$$\boxed{\begin{cases} n_r(\omega) := Re\left[n^*(\omega)\right] \geq 0 & \text{for} \quad \omega \geq 0, \\ n_i(\omega) := Im\left[n^*(\omega)\right] \leq 0 & \text{for} \quad \omega \geq 0, \end{cases}} \qquad (3.9)$$

which derives from (3.7) recalling that the same property holds for $J^*(\omega)$ [see *LV*, Sect. 6].

We also remember the so-called *Krönig-Kramers* or $K - K$ *relations*, which hold between the real and imaginary parts of the complex index of refraction $n^*(\omega)$ as a consequence of the *causality*,

$$\boxed{\begin{cases} n_r(\omega) = 1 - \dfrac{2}{\pi} \displaystyle\int_0^\infty \dfrac{\omega' \, n_i(\omega') - \omega \, n_i(\omega)}{\omega'^2 - \omega^2} \, d\omega' , \\ n_i(\omega) = \dfrac{2\omega}{\pi} \displaystyle\int_0^\infty \dfrac{n_r(\omega') - n_r(\omega)}{\omega'^2 - \omega^2} \, d\omega' , \end{cases}} \qquad (3.10)$$

where the integrals are intended as Cauchy principal values. Similar relations are expected to hold also between the real and imaginary parts of the complex wave number $\kappa(\omega)$ and the dynamical functions $J^*(\omega)$, $G^*(\omega)$, provided that they refer to *causal* models of viscoelasticity, *i.e.* to viscoelastic models of type I and II. For applications of the $K - K$ *relations* in propagation problems of viscoelastic waves, we refer the interested reader to Futterman[35], Strick[36], Chin[31], Aki & Richards[37] and Ben-Menahem & Singh[38].

We now go on to point out that, for a given linear viscoelastic medium, $n^*(\omega)$ characterizes the relative *dispersion* and *attenuation*. To this end, following the pioneering analysis of Sommerfeld and Brillouin[39] carried out in 1914 for electromagnetic waves propagating in a dielectric, we consider the viscoelastic response to a sinusoidal excitation of a given frequency Ω provided at $x = 0$. Adopting the Laplace representation, for which

$$r_0(t) = R \cos \Omega t \div \bar{r}_0(s) = R \frac{s}{s^2 + \Omega^2},$$

(3.11)

where R is a constant, we obtain

$$r(x,t) = \frac{R}{2\pi i} \int_{Br} \frac{s}{s^2 + \Omega^2} e^{s[t - (x/c)\, n(s)]} \, ds.$$

(3.12)

The singularities of the integrand in (3.12) are given by the branch cut of $n(s)$ and the two simple poles on the imaginary axis, $s_{\pm} = \pm i\Omega$, exhibited by the source.

By applying the Cauchy theorem, one gets the following picture of the course of the mechanical disturbance at a distance $x > 0$. Up to the time $t = x/c$ no motion occurs; at $t = x/c$ the wave motion starts from a certain amplitude; for $t > x/c$ it consists of two parts: the *transient state* and the *steady state*. The *transient state* is the contribution due to the branch cut: it is expected to vanish at any x as $t \to \infty$; the *steady state* is the sum contribution from the poles: it gives the limiting value of the response variable as $t \to \infty$. Recognizing that

$$n(\pm i\Omega) = n_r(\Omega) \pm i\, n_i(\Omega) = \frac{c}{\Omega} [\kappa_r(\Omega) \pm i\, \kappa_i(\Omega)],$$

(3.13)

the *steady state* turns out

$$r_S(x,t) = e^{-\alpha(\Omega)\, x} \cos \Omega [t - x/V(\Omega)],$$

(3.14)

where

$$\begin{cases} V(\Omega) = \dfrac{c}{n_r(\Omega)} = \dfrac{\Omega}{\kappa_r(\Omega)} \geq 0, \\[2mm] \alpha(\Omega) = -\Omega \dfrac{n_i(\Omega)}{c} = -\kappa_i(\Omega) \geq 0. \end{cases}$$

(3.15)

This is to say that the *steady-state* response to a sinusoidal impact of frequency Ω is a particular *normal-mode solution*, whose complex wave number and real frequency satisfy the *dispersion equation* (3.4) with $\omega = \Omega$. We refer to (3.14) as a *pseudo-monochromatic* wave, since it is strictly sinusoidal only in time (with frequency Ω). The phase of this wave propagates into the medium with *phase velocity* $V(\Omega)$ while its amplitude exponentially decays in space with *attenuation coefficient* $\alpha(\Omega)$. The quantities V and α are related to each other because of the *Krönig-Kramers relations* (3.10).

The main properties of any linear dispersive motion are known to be related to the analysis of the *dispersion equation* and *normal mode solutions*. In particular, in the absence of dissipation (ω and κ are both real), the normal-mode solutions are mono-chromatic waves and the concept of *group velocity* U plays the same fundamental role, both versus κ and versus ω, see for example Lighthill[40], Whitham[41], Thau[34], Baldock & Bridgeman[42].

In the presence of dissipation we need to distinguish which of the variables κ, ω is real and which one is complex. Since we are interested in the impact problem, see (3.1), we must take ω as real, and consequently define the *phase velocity* $V(\omega)$ and the *group velocity* $U(\omega)$ as follows

$$
\begin{cases}
V(\omega) := \dfrac{\omega}{\kappa_r(\omega)} = \dfrac{c}{n_r(\omega)} \\[4mm]
U(\omega) := \left[\dfrac{d}{d\omega}\,\kappa_r(\omega) \right]^{-1} = \dfrac{c}{n_r(\omega) + \omega\,\dfrac{dn_r}{d\omega}}
\end{cases}
\tag{3.16}
$$

Of course, these definitions reduce to the ordinary ones when κ is real. We note that, while the concept of phase velocity retains its kinematic meaning of the phase speed in the presence of dissipation [see (3.14)], the above definition of group velocity is expected to lose its usual kinematic meaning of the group (*wave packet*) speed. In fact for dispersive but non-dissipative waves the concept of group velocity arises from the consideration of a superposition of two monochromatic waves of equal amplitude and nearly equal frequency and wavelength. In the presence of dissipation two such waves cannot exist at all times because they are attenuated by different amounts due to the imaginary part of κ or of ω, see Bland[27].

The relation between the phase and the group velocities turns out to be

$$
\boxed{\dfrac{1}{U} = \dfrac{1}{V} + \omega\,\dfrac{d}{d\omega}\left(\dfrac{1}{V}\right).}
\tag{3.17}
$$

As usual, we refer to the case $0 < U < V$ as *normal dispersion*, while the other cases ($U > V > 0$ and $U < 0 < V$) are referred to as *anomalous dispersion*. It is easy to recognize from (3.16)-(3.17) that the dispersion is normal or anomalous when $dn_r/d\omega > 0$ or $dn_r/d\omega < 0$, respectively; in other words, when the phase velocity is a decreasing or increasing function of ω.

It is possible to prove[19] that the mathematical structure of the *Linear Viscoelasticity* requires the dispersion of viscoelastic waves to be *completely anomalous, i.e.* anomalous throughout the full frequency range, with $0 < V(\omega) \le U(\omega)$. The sign of equal holds exclusively for $\omega \to 0$ and $\omega \to \infty$, which represent the non-dispersive limits.

As a result, for $0 < \omega < \infty$ the phase velocity $V(\omega)$ turns out to be a increasing function of ω, less than the wave-front velocity $c = V(\infty)$. As far as the group velocity $U(\omega)$ is concerned, we expect that a value ω_0 of the frequency can be found such that for $\omega > \omega_0$ it turns out that $U(\omega) \geq c$. Since the group velocity of viscoelastic waves may attain non-physical values (*i.e.* greater than the wave-front velocity), we need to revisit the concept of *signal velocity* usually identified with the group velocity when the dispersion is normal and the dissipation is absent or negligible. We shall treat this interesting topic in the next Section.

In order to show the dispersion and attenuation for viscoelastic waves, Fig. 3.1 shows the plots of the phase/group velocity and attenuation coefficient versus frequency corresponding to the two simplest viscoelastic models of type I and II, *i.e.* the SLS and the *Maxwell* model, respectively.

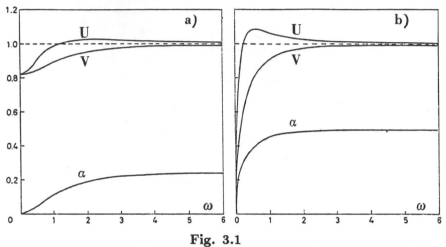

Fig. 3.1

Phase Velocity V, Group Velocity U and Attenuation Coefficient α versus frequency ω for a) SLS model, b) Maxwell model.

For plotting we refer to the constitutive equations of the two models, see eqs (2.9) and (2.10), and we consider non-dimensional variables, scaling the frequency with $\omega_* = 1/\tau_\sigma$, the velocities V and U with c, and the coefficient α with $\alpha_* = 1/(c\tau_\sigma)$. For the SLS we assume $\tau_\epsilon = 1.5\,\tau_\sigma$. In practice this is equivalent to assuming $c = 1$ and

$$n(s) = \left[\frac{s + 1/\tau_\sigma}{s + 1/\tau_\epsilon}\right]^{1/2}, \tag{3.18}$$

with $\tau_\sigma = 1$, $\tau_\epsilon = 1.5$ for the S.L.S. model and $\tau_\sigma = 1$, $\tau_\epsilon = \infty$ for the *Maxwell* model.

4. The Signal Velocity and the Saddle-Point Approximation

The phenomenon by which the group velocity attains non-physical values and thus must not be identified with the *signal velocity* led Sommerfeld and Brillouin[39] to investigate the subject matter more carefully, by considering the propagation of an electro-magnetic signal of type (3.12) in a dielectric. Assuming a *Lorentz-Lorenz dispersion equation*, that is appropriate to electromagnetic waves in a dielectric, the above authors introduced a suitable definition of signal velocity in order to meet the physical requirements also in the frequency range where the dispersion is anomalous and the absorption is high. The equivalence with the group velocity was effectively proved except in that frequency range, where some estimations of the appropriate signal velocity was provided by Brillouin[43] and then improved by Baerwald[44], see also Elices & García-Moliner[45].

For viscoelastic waves, which exhibit only anomalous dispersion and relatively high absorption, the problem of such an identification turns out to be of some relevance. Therefore, extending the classical arguments of Ref[39], the Author has in the past considered the general problem of the signal velocity[18-20]. Here we summarize the main results applicable to the class of viscoelastic waves.

The Author has also considered the problem of a correct definition of the energy velocity[21-26], related to dispersive motions in the presence of dissipation. For viscoelastic waves we refer to the treatise by Bland[27].

As far as the *signal velocity* is concerned, let us start from the Laplace integral representation (3.12) valid for the sinusoidal signalling problem (3.11) *

For his definition Brillouin used the path of steepest descent to evaluate the Bromwich integral in (3.12). For this purpose, let us change the original variables x, t into $\xi = x/c$, $\theta = ct/x > 1$, so that the solution reads

$$r(\xi, \theta) = \frac{R\,\Omega}{2\pi i} \int_{Br} \frac{e^{\xi\,F(s; \theta)}}{s^2 + \Omega^2}\, ds \qquad (4.1)$$

where

$$F(s; \theta) = s\,[\theta - n(s)]. \qquad (4.2)$$

After this , the original fixed path Br is to be deformed into the new, moving path $L(\theta)$, that is of steepest descent for the real part of $F(s; \theta)$ in the complex s-plane. For any given $\theta > 1$ this path L turns out to be defined by the following properties:
(i) L passes through the saddle point s_0 of $F(s; \theta) = \Phi(s, \theta) + i\,\Psi(s, \theta)$, *i.e.*

$$\left.\frac{dF}{ds}\right|_{s=s_0} = 0 \iff n(s) + s\frac{dn}{ds} = \theta \quad \text{at} \quad s = s_0(\theta); \qquad (4.3)$$

---

* The original analysis by Brillouin was made by using the Fourier integral and deforming the path of integration into the complex domain. It may be shown by a simple change of variable that the complex Fourier integral employed by Brillouin is exactly the Bromwich integral (3.12).

(ii)   the imaginary part of $F(s; \theta)$ is everywhere constant on L, *i.e.*

$$\Psi(s; \theta) = \Psi(s_0; \theta), \quad s \in L; \tag{4.4}$$

(iii)   the real part of $F(s; \theta)$ attains an absolute maximum at s_0 along L, *i.e.*

$$\Phi(s; \theta) \leq \Phi(s_0; \theta), \quad s \in L; \tag{4.5}$$

(iv)   the integral on L is equivalent either to the original one or differs from it by a finite residue contribution.

In our case the integral on $L(\theta)$ turns out to be equal to the Bromwich integral, or different from it by the steady state solution (2.15), according to $1 \leq \theta < \theta_s(\Omega)$ or $\theta > \theta_s(\Omega)$, respectively, where $\theta_s(\Omega)$ is the value of θ for which $L(\theta)$ intersects the imaginary axis at the frequency $\pm\Omega$.

Therefore, the representation of the wave motion by the integral on $L(\theta)$, which we refer to as the *Brillouin representation*, allows one to recognize the arrival of the steady state and, following Brillouin, to define the *signal velocity* as

$$\boxed{S(\Omega) := \frac{c}{\theta_s(\Omega)}}. \tag{4.6}$$

The condition $\theta \geq 1$ ensures that, in any dispersive motion where the representation (4.1) holds, the velocity S is always less than c, the wave-front velocity; this property is independent of the fact that the phase velocity V and/or the group velocity U can *a priori* be greater than c.

We note that the analytical determination of the path L and its evolution is a difficult task even for simple models of viscoelasticity as the SLS, see for example Rubin[43], Mainardi[12], Chin[31]. Consequently, the exact evaluation of the signal velocity S may appear only as a numerical achievement. In other words, for a general linear dispersive motion, the definition of signal velocity by Brillouin appears as a computational prescription, and its evaluation, in general depending on the explicit knowledge of $n(s)$ in the complex s-plane, appears more difficult than the evaluation of the phase velocity V and group velocity U, which both depend on the values of $n(s)$ only in the imaginary axis. However, while in the absence of dissipation one can prove the identification of signal velocity with group velocity, *i.e.* $S(\Omega) = U(\Omega)$, see Refs [39,18], in the presence of dissipation and full anomalous dispersion (as for linear viscoelastic waves), the identification of signal velocity with phase velocity, *i.e.* $S(\Omega) = V(\Omega)$ turns out to hold.

This simple and surprising result is essentially based on the property that, because of the completely anomalous dispersion, the relevant saddle point $s_0(\theta)$ remains located on the real axis on which $F(s; \theta)$ attains only real values, as proved by the Author[19]. We easily recognize that the saddle point moves from $+\infty$ (at the wave-front, $\theta = 1$) up to the largest branch point of $n(s)$ (for $\theta = \infty$), and that the path L, of steepest descent through s_0, is a curve that encloses the branch cut of

$n(s)$. Thus, when the path $L(\theta)$ intersects the imaginary axis at the poles $s = \pm i\Omega$, we obtain

$$Im\ [F(s_0, \theta_S)] = Im\ [F(\pm i\Omega, \theta_S)] = \pm \Omega\ [\theta_S - n_r(\Omega)] = 0\,. \qquad (4.7)$$

This means that

$$\boxed{S(\Omega) := \frac{c}{\theta_s(\Omega)} = \frac{c}{n_r(\Omega)} := V(\Omega)}\,. \qquad (4.8)$$

A simple but relevant example of anomalous dispersion is provided by the Maxwell model of viscoelasticity, which is equivalent to the telegraph equation (2.9). Carrier & al[47] and Thau[34] considered this equation in order to evaluate the Brillouin signal velocity but overlooked the identification with the phase velocity. In Ref[20] the author resumed this interesting case and, in addition, provided numerical results. Here we note that in such a case the forerunner turns out to be not oscillating but less damped than the (oscillating) steady state; they match only after a sufficiently long time. The identification of signal velocity with phase velocity is to be interpreted differently from the identification with group velocity valid for normal dispersion. We observe the existence of a substantial forerunner which is the source of great disturbance for transmissions. In other words, the role of the main signal is played by this forerunner which becomes appreciable and of the same order as the steady state when the time elapsed from the wave-front corresponds to the phase velocity, i.e. $\tau = x/V(\Omega) - x/c$.

In Fig. 4.1 we show the evolution of the path $L(\theta)$ in the s-complex plane for the Maxwell model. In this case the analytical determination of the path $L(\theta)$ is straightforward, as shown later. In the figure we find it convenient to introduce the parameter $\mu := 1/\theta$ and exhibit the path for the following five values of μ : $\mu_1 = 0.85$, $\mu_2 = 0.75$, $\mu_3 = 0.50$, $\mu_4 = 0.25$, $\mu_5 = 0.15$.

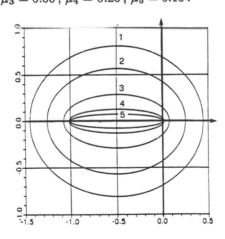

Fig. 4.1

The evolution of the steepest-descent path $L(\theta)$ for the Maxwell model.

When the integration is carried out on the *entire* curve L, the exact result is expected to be found for any x and t, without any problem related to an oscillating integrand since on L the imaginary part of $F(s, \theta)$ is constant.

From the above one can argue that any numerical technique of integration on L can successfully be used, provided that the path itself may be determined with sufficient accuracy, starting from the relevant saddle point s_0. In this respect Mainardi & Vitali[24] found it convenient to use their specific routine that enabled them to evaluate the path L in the complex s-plane by solving the equation (4.4) numerically. It is based on rational approximations (Padé Approximants of type II), according to a simple algorithm, that provides an exponential convergence

For large x and t we can presume that the dominant contribution to the integral along L comes from a neighborhood of s_0: this yields the leading term in the asymptotic expansion of the wave form, that can easily be found by the so called *saddle-point method*, see e.g. Refs[30,31,34,39-49]. For this method a detailed knowledge of the path L is not usually required, while the location of the relevant saddle point with respect to the singularities of $\bar{r}_0(s)$ is important to get a uniform approximation, see Bleistein & Handelsman[49].

We can easily check the validity of the approximation provided the saddle-point method for the so-called *Lee – Kanter problem*[50], in the Maxwell model of viscoelasticity where the exact solution is known in a closed form. In this case, using suitable non-dimensional variables, we have $\bar{r}_0(s) = 1/[s\, n(s)]$ where $n(s)$ is provided by (3.18) with $\tau_\sigma = 1$, $\tau_\epsilon = \infty$. Consequently

$$\bar{r}(x, s) = \frac{e^{\,st - s(1 + 1/s)^{1/2}\, x}}{s\,(1 + 1/s)^{1/2}} \div r(x, t) = e^{-t/2}\, I_0\left[\frac{\sqrt{t^2 - x^2}}{2}\right], \quad t \ge x, \quad (4.9)$$

where I_0 denotes the modified Bessel function of order 0. The equation (4.3) for the saddle points reads

$$\left(1 + \frac{1}{s}\right)^{1/2}\left[1 - \frac{1}{2(1 + s)}\right] = \theta, \quad (4.10)$$

which, when rationalized, is a quadratic equation in s with two real solutions. These solutions are the required saddle points, which read

$$s^\pm = -\frac{1}{2}\left[1 \mp \frac{\theta}{\sqrt{\theta^2 - 1}}\right]. \quad (4.11)$$

Therefore, these saddle points move on the real axis from $\pm\infty$ (for $\theta = 1$) to the branch points of $n(s)$: $s_*^+ = 0$, $s_*^- = -1$ (for $\theta = \infty$). The path of steepest descent through $s^\pm$ is defined by the condition (4.4), that now reads

$$\Psi(s; \theta) = Im\left\{s\theta - [s(s + 1)]^{1/2}\right\} = 0. \quad (4.12)$$

Setting $s = \xi + i\eta$, this condition leads to the following algebraic equation

$$\frac{(\xi + 1/2)^2}{a^2} + \frac{\eta^2}{b^2} = 1, \quad a = \frac{\theta}{2\sqrt{\theta^2 - 1}}, \quad b = \frac{1}{2\sqrt{\theta^2 - 1}}. \qquad (4.13)$$

Thus, the curve L represented by (4.13) is an ellipse, with axes $2a$, $2b$ and foci in the branch points $s_*^{\pm}$, that intersects the real axis in the saddle points $s^{\pm}$.

It is easy to prove that s^+ is the relevant saddle point s_0, being a maximum of $Re\{F(s; \theta)\}$ on L, and that the original Bromwich path Br can be deformed into the ellipse L. In Fig. 4.1 the curve L has been shown in the complex s–plane for different values of the parameter θ; from the figure we easily understand the evolution of the ellipse from a big circle at infinity (for $\theta = 1$) to the segment of the branch cut (for $\theta = \infty$).

The saddle-point method provides the following approximate representation (that is expected to be good for large values of time)

$$r(\theta, t) \sim \sqrt{\frac{2\theta}{\pi t}} \, [s_0(s_0 + 1)]^{1/4} \, e^{t\{s_0 - [s_0(s_0 + 1)]^{1/2}/\theta\}}. \qquad (4.14)$$

This approximation is of course not uniform on the wave-front, since for $\theta \to 1$ the two saddle points coalesce at infinity. However, it can be proved to be regular for $\theta \to \infty$; for $\theta = \infty$ (*i.e.* at $x = 0$) it yields $r(x = 0, t) \sim 1/\sqrt{\pi t} \leq \exp(-t/2) I_0(t/2)$, where in the R.H.S. the exact value is quoted from (4.9).

In Fig. 4.2 we compare the exact solution given by (4.9) [continuous line] with the approximate solution given by (4.14) [dashed line] for some fixed values of time $[t = 2, 4, 6, 8, 10]$. We recognize that the saddle-point method provides a satisfactory approximation (*i.e.* a discrepancy of less than one per cent) for any x, only if $t \geq 8$. For smaller times the contribution to the integral from a neighborhood of the relevant saddle point s_0 is thus inadequate to represent the solution.

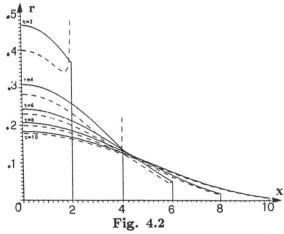

Fig. 4.2

The pulse response for the Lee   Kanter problem depicted versus x.

5. The Regular Wave-Front Expansion

Let us consider viscoelastic models for which the *material functions* $J(t)$, $G(t)$ and consequently the respective *memory functions* $\Psi(t)$, $\Phi(t)$ defined in (2.1a)-(2.1b) are *entire functions of exponential type.* * Henceforth, we shall refer to these models as *full analytical models* of viscoelasticity.

Assuming the *creep representation* we can write

$$J(t) = J_0 \left[1 + \sum_{k=1}^{\infty} \psi_k \frac{t^k}{k!} \right], \tag{5.1a}$$

where

$$J_0 := J(0^+), \quad \psi_k := \frac{1}{J_0} \left[\frac{d^k J}{dt^k} \right]_{t=0^+}. \tag{5.2a}$$

Thus,

$$[n(s)]^2 := \frac{s\overline{J}(s)}{J_0} = 1 + \overline{\Psi}(s) = 1 + \sum_{k=1}^{\infty} \frac{\psi_k}{s^k}, \quad |s| > \delta_\psi. \tag{5.3a}$$

Similarly, assuming the *relaxation representation* we write

$$G(t) = G_0 \left[1 + \sum_{k=1}^{\infty} \phi_k \frac{t^k}{k!} \right], \tag{5.1b}$$

where

$$G_0 := G(0^+), \quad \phi_k := \frac{1}{G_0} \left[\frac{d^k G}{dt^k} \right]_{t=0^+}. \tag{5.2b}$$

Thus,

$$[n(s)]^{-2} := \frac{s\overline{G}(s)}{G_0} = 1 + \sum_{k=1}^{\infty} \frac{\phi_k}{s^k}, \quad |s| > \delta_\phi. \tag{5.3b}$$

---

* We recall that an entire function $f(z)$, $z \in \mathbf{C}$, is said to be of exponential type if, roughly speaking, it is bounded by $M \exp(\delta z)$, see Doetsch[51]. The definition is more precisely stated on the Taylor series of $f(z)$ using the notation $f(z) \in \{1, \delta\}$ with $\delta > 0$, see Widder[52]. Then we have

$$f(z) = \sum_{k=0}^{\infty} \frac{a_k}{k!} z^k. \qquad f(z) \in \{1, \delta\} \iff \varlimsup_{k \to \infty} |a_k|^{1/k} \leq \delta.$$

For greater detail we refer to Refs[51,52].

The material functions (and their derivatives) are known to be defined as causal functions. Here, of course, the analytic continuation from $t \in \mathbf{R}^+$ to $t \in \mathbf{C}$ is understood.

For a known theorem on Laplace transforms, see *e.g.* Ref[52], we have

$$\boxed{\Psi(t) \in \{1, \delta_\psi\} \iff \bar{\Psi}(s) \in \mathcal{A} \quad \text{for} \quad |s| > \delta_\psi, \quad \bar{\Psi}(\infty) = 0}, \tag{5.4a}$$

and

$$\boxed{\Phi(t) \in \{1, \delta_\phi\} \iff \bar{\Phi}(s) \in \mathcal{A} \quad \text{for} \quad |s| > \delta_\phi, \quad \bar{\Phi}(\infty) = 0}, \tag{5.4b}$$

where $\mathcal{A}$ denotes the class of analytic functions. Consequently also $[n(s)]^{\pm 2}$ turn out to be analytic functions, regular at infinity, where they assume the value 1. As a matter of fact, to ensure that $\bar{\Psi}(s)$ and $\bar{\Phi}(s)$ are analytic and vanishing at infinity and, hence, $\Psi(t)$ and $\Phi(t)$ are entire functions of exponential type, we have to assume that the retardation/relaxation spectra of the viscoelastic body are either *discrete* (mechanical models) or *continuous* with a finite branch cut on the negative real axis.

In view of the mathematical structure of *Linear Viscoelasticity* also the function $n(s)$ turns out to be an analytic function, regular at infinity but with unitary value. This means that a positive number δ can be found so that $n(s)$ is represented by the following power series, which is absolutely convergent for $|s| > \delta$,

$$\boxed{n(s) = 1 + \frac{n_1}{s} + \frac{n_2}{s^2} + \dots, \quad |s| > \delta.} \tag{5.5}$$

In principle, the series coefficients n_k ($k \in \mathbb{N}$) can be obtained from the coefficients ψ_k and ϕ_k of the memory functions, that we shall refer to as the *creep* and *relaxation* coefficients, respectively. Taking from now on $c = 1$ for convenience, we easily obtain the first few coefficients of $n(s)$, *i.e.*

$$n_1 = \frac{1}{2}\psi_1, \quad n_2 = \frac{1}{2}\psi_2 - \frac{1}{8}\psi_1^2, \quad n_3 = \frac{1}{2}\psi_3 - \frac{1}{4}\psi_1\psi_2 + \frac{1}{16}\psi_1^3, \quad \dots \tag{5.6a}$$

and

$$n_1 = -\frac{1}{2}\phi_1, \quad n_2 = -\frac{1}{2}\phi_2 + \frac{3}{8}\phi_1^2, \quad n_3 = -\frac{1}{2}\phi_3 + \frac{3}{4}\phi_1\phi_2 - \frac{5}{16}\phi_1^3, \quad \dots \tag{5.6b}$$

Recalling that the *creep* and *relaxation* functions are *completely monotonic* [see (31) in *LV*], we point out that $(-1)^n\psi_n \leq 0$ and $(-1)^n\phi_n \geq 0$; in particular, we note that $n_1 \geq 0$, $n_2 \leq 0$, $n_3 \geq 0$.

In this section we wish to present an efficient recursive method which allows us to obtain a *convergent wave-front expansion* for the solution $r(x, t)$, starting from the *creep* coefficients, as proposed by Mainardi and Turchetti[14].

For this purpose, we start from the Laplace image of the solution, $\bar{r}(x, s) = \exp[-\gamma(s)x] = \exp[-s(n(s)x]$, see (1.7), (1.16), writing

$$\boxed{\gamma(s) = s + \alpha - \frac{\beta}{s} + \dots}, \tag{5.7}$$

where, for convenience, we have put

$$\boxed{\alpha = n_1 > 0, \quad \beta = -n_2 \geq 0}. \tag{5.8}$$

Let us re-write the transform solution as follows

$$\bar{r}(x,s) := \bar{r}_0(s)\,e^{-\gamma(s)x} = e^{-\alpha x}\left[e^{-sx}\,\bar{R}(x,s)\right],\tag{5.9}$$

which yields in the time domain

$$r(x,t) = e^{-\alpha x}\,R(x,t-x).\tag{5.10}$$

The purpose of the exponentials in (5.9) is to isolate the wave front propagating with velocity $c=1$ and with amplitude-decay α.

The equation satisfied by $\bar{R}(x,s)$ is, from (1.7) and (5.9),

$$\left\{\frac{d^2}{dx^2} - 2(s+\alpha)\frac{d}{dx} - \left[\gamma^2(s) - (s+\alpha)^2\right]\right\}\bar{R}(x,s) = 0,\tag{5.11}$$

subject to the initial condition $\bar{R}(0,s) = \bar{r}_0(s)$. After simple manipulation we obtain

$$\left[\gamma^2(s) - (s+\alpha)^2\right] = -2\beta + \sum_{j=1}^{\infty}\frac{\psi_{j+2}}{s^j}.\tag{5.12}$$

This shows that the differential operator acting on $\bar{R}(x,s)$ in (5.11), apart from the first 2 coefficients α and β, does not depend on the series coefficients n_j of (5.5) but only on the creep coefficients ψ_j.

For $\bar{R}(x,s)$ we now seek a series expansion in integer powers of $1/s$, which is expected to depend explicitly on α, β and creep coefficients. Based on the theory of Laplace transforms, see *e.g.* Ref[51], one is led to think that the term by term inversion of such series provides in the time domain an expansion for $R(x,t)$ *asymptotic* as $t \to 0^+$. Consequently, a formal *wave-front* expansion for $r(x,t)$ is expected to be of the kind considered by Achenbach and Reddy[53] and Sun[54], using the theory of propagating surfaces of discontinuity, and by Buchen[55], using the ray-series method. Here, however, we show that a term by term inversion yields an expansion which is *convergent* in any space-time domain. For this purpose we use the following theorem, stated in Ref[14].

THEOREM - *Let* $\bar{R}(x,s)$ *be analytic for* $|s| > \delta$, *uniformly continuous in x for* $0 \le x \le X$, *with* $\bar{R}(x,\infty) = 0$, *whose expansion reads*

$$\bar{R}(x,s) = \sum_{k=0}^{\infty}\frac{w_k(x)}{s^{k+1}},\qquad |s| > \delta,\tag{5.13}$$

then the inverse Laplace transform $R(x,t)$ *is an entire function of t of exponential type, whose expansion reads*

$$R(x,t) = \sum_{k=0}^{\infty} w_k(x)\frac{t^k}{k!},\qquad \text{uniformly in x},\tag{5.14}$$

and viceversa.

We easily recognize that the conditions of the above theorem are fulfilled if
(i)   the creep function is an entire function of exponential type,
(ii) the input $r_0(t)$ is an entire function of exponential type, *i.e.*

$$r_0(t) \in \{1, \delta_0\} \iff \bar{r}_0(s) = \sum_{k=0}^{\infty} \frac{\rho_k}{s^{k+1}}, \quad |s| > \delta_0 \, ; \qquad (5.15)$$

While assumption (i) has been requested since the beginning, assumption (ii) can be released by considering the *impulse response* and performing a suitable convolution with $r_0(t)$.

We now illustrate the procedure to find the coefficients $w_k(x)$ by recurrence from the creep coefficients. The required expansion can be determined by substituting the expansions (5.12-13) into eqn (5.11) and collecting like powers of s. For this purpose let us expand the operator in the brackets at the L.H.S of (5.11) in power series of s. Dividing by the highest power of s, *i.e.* $-2s$, and defining this operator $\hat{L}(x, s)$, we obtain from (5.11) and (5.12)

$$\hat{L}(x, s) = \sum_{i=0}^{\infty} \hat{L}_i(x) \, s^{-i} \, , \qquad (5.16)$$

where

$$\begin{cases} \hat{L}_0 = \dfrac{d}{dx} \, , \\[2mm] \hat{L}_1 = -\dfrac{1}{2} \dfrac{d^2}{dx^2} + \alpha \dfrac{d}{dx} - \beta \, , \\[2mm] \hat{L}_i = \dfrac{1}{2} \psi_{i+1} \, , \quad i \geq 2 \, . \end{cases} \qquad (5.17)$$

With a minimum effort the coefficients $w_k(x)$ in (5.13) prove to be solutions of a recursive system of linear first-order differential equations with initial conditions $w_k(0) = \rho_k$ ($k = 0, 1, 2, \ldots$). In fact, since the expansion of $\hat{L}\{R(x, s)\}$ is given by termwise application of the L_i to the $w_k(x)$, the coefficients $w_k(x)$ satisfy the recursive system of equations

$$\begin{cases} \hat{L}_0 \, w_0 = 0 \, , \\[2mm] \hat{L}_0 \, w_k = -\sum \hat{L}_i \, w_j \, , \quad k \geq 1 \, , \end{cases} \qquad (5.18)$$

where the summation is taken over values of i, j for which $i + j = k + 1$. The solutions of this system are easily seen to be polynomials in x of order k, which we write in the form

$$w_k(x) = \sum_{h=0}^{k} A_{k,h} \frac{x^h}{h!} \, , \qquad (5.19)$$

where the $A_{k,h}$ are obtained from the initial data $A_{k,0} = \rho_k$ by the following recurrence relations, with $1 \leq h \leq k$,

$$A_{k,h} = \frac{1}{2} A_{k-1,h+1} - \alpha A_{k-1,h} + \beta A_{k-1,h-1} - \frac{1}{2} \sum_{j=2}^{k-h+1} \psi_{j+1} A_{k-j,h-1}. \quad (5.20)$$

Finally, we obtain the following series representation

$$R(x,t) = \sum_{k=0}^{\infty} \sum_{h=0}^{k} A_{k,h} \frac{x^h}{h!} \frac{t^k}{k!}, \quad (5.21)$$

that with (5.10) provides the requested solution.

The obtained series solution is easy to handle for numerical computations since it is obtained in a recursive way. Although the "mathematical" convergence of the series is ensured uniformly in any finite space-time domain, its "numerical" convergence is expected to slow down by increasing of the time elapsed from the wave-front, τ, with a rate depending on x. Since the result is an entire function of τ of exponential type which must be bounded at infinity, we are faced with the same difficulty as in the evaluation of $\exp(-\tau)$ using its Taylor expansion when τ is large. Mainardi & Turchetti [14-15] have proposed to accelerate the numerical convergence with the diagonal Padé approximants, henceforth referred to as PA, which are a noteworthy rational approximation. This technique allows one to use a reasonable number of series terms (no more than 20) in order to have a representation of the wave phenomenon in any space-time domain of physical interest, possibly with a matching with the long-time asymptotic solution, obtained by the saddle-point method.

By way of example, let us consider the SLS for which $n(s)$ is recalled in (3.18). Taking $\tau_\sigma = 1$ and $\tau_\epsilon = 1/\chi > 1$, we have $\delta = 1$ in (5.5) and

$$\Psi(t) = (1 - \chi) e^{-\chi t} \implies \psi_k = (1 - \chi)(-\chi)^{k-1}, \quad k = 1, 2, \dots . \quad (5.22)$$

In particular, we obtain $\alpha = (1 - \chi)/2$ and $\beta = (1 - \chi)(1 + 3\chi)/8$.

For $\chi = 0$ the SLS reduces to the $Maxwell$ model for which we simply get

$$\Psi(t) = 1, \implies \psi_k = \delta_{k1}, \quad k = 1, 2, \dots . \quad (5.23)$$

Consequently, for the $Maxwell$ model the recurrence relation (5.20) simplifies to

$$A_{k,h} = \frac{1}{2} A_{k-1,h+1} - \frac{1}{2} A_{k-1,h} + \frac{1}{8} A_{k-1,h-1}. \quad (5.24)$$

The computations carried out by Mainardi & Turchetti have concerned two typical signalling problems for the SLS, the $unit\ step\ problem$, $\bar{r}_0(s) = 1/s$ and the $Lee\text{-}Kanter\ problem$, $\bar{r}_0(s) = 1/[s\, n(s)]$.

In the Figures 5.1 and 5.2 we show the evolution of the transient waves versus x at fixed times, corresponding to the above problems, taking non-dimensional variables and $\chi = 0.5$.

At the left panel of each figure, we consider short times, where the wave-front discontinuity is still appearing, while at the right one we consider long times, where smooth solutions start to appear. Of course, in the cases at right, the matching with the saddle-point solution has been considered, in order to use not too many terms of the series for the PA. We note that in the actual computations the numerical convergence of the series is lost beyond a critical value of τ, no matter how many terms are computed (we work with a fixed number of digits), so that the P.A. technique turns out to be an indispensible complement to achieve the matching between the wave-front expansion and the saddle-point solution.

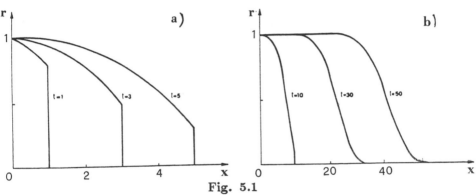

Fig. 5.1

The pulse response for the unit step problem in SLS depicted versus x for a) short times, b) long times.

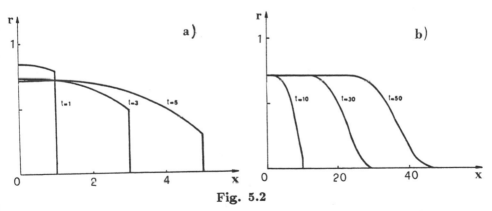

Fig. 5.2

The pulse response for the Lee-Kanter problem in SLS depicted versus x for a) short times, b) long times.

6. The Singular Wave-Front Expansion.

We now consider viscoelastic models for which the *material functions* $J(t)$, $G(t)$ and consequently the respective *memory functions* $\Psi(t)$, $\Phi(t)$ are no longer *entire functions of exponential type*. In this section we report the method developed by Buchen and Mainardi[13], which provides an asymptotic series solution when the creep compliance $J(t)$ of the viscoelastic medium exhibits at the time origin a behaviour of the following form

$$J(t) = J_0 + O(t^\alpha) \qquad \text{as} \quad t \to 0^+ , \tag{6.1}$$

where $J_0 \geq 0$ is the instantaneous compliance and $0 < \alpha \leq 1$.

Creep compliances which conform to this representation cover a wide class of viscoelastic materials including the models for which $J(t)$ has a well defined expansion about $t = 0^+$ in fractional non-negative powers of t with $J_0 \geq 0$.

The starting point of the Buchen-Mainardi method is the asymptotic behaviour of $\gamma(s) := \sqrt{\rho}\, s\, [s\bar{J}(s)]^{1/2}$, see (1.8), as $s \to \infty$. In general we have an expansion in decreasing powers of s of the form

$$\gamma(s) \sim \sum_{k=0}^{\infty} b_k s^{1-\beta_k} , \qquad 0 = \beta_0 < \beta_1 < \cdots . \tag{6.2}$$

Denoting by $\gamma_+(s)$ the sum of the first $(m+1)$ terms with $\beta_k \leq 1$ $(k = 0, 1, \ldots, m)$ and by $\gamma_-(s)$ the remainder of the series, we can write the transform solution as

$$\bar{r}(x, s) = \bar{r}_0(s)\, e^{-x\,\gamma(s)} = \bar{r}_0(s)\, e^{-x\,\gamma_+(s)}\, e^{-x\,\gamma_-(s)} . \tag{6.3}$$

Let us now put

$$\bar{R}(x, s) := e^{-x\,\gamma_-(s)} , \tag{6.4}$$

and, without loosing in generality, we agree to take as input the unit step Heaviside function, $r_0(t) = H(t)$, so that

$$\bar{r}_0(s) := 1/s . \tag{6.5}$$

The function $\gamma_+(s)$ is referred to as the principal part of the expansion of $\gamma(s)$; it can be obtained with a minimum effort from the first few terms of the expansion of $J(t)$ as $t \to 0^+$. From (1.8), (1.14-15) and (6.1) we easily infer

$$\begin{cases} J_0 = 0 \Longrightarrow b_0 = 0, \quad \beta_1 = \alpha/2, \\ J_0 > 0 \Longrightarrow b_0 = \sqrt{\rho J_0} = 1/c, \quad \beta_1 = \alpha. \end{cases} \tag{6.6}$$

For $\bar{R}(x, s)$ we seek an asymptotic expansion in negative powers of s as $s \to \infty$ of the kind

$$\bar{R}(x, s) \sim \sum_{k=0}^{\infty} w_k(x) s^{-\lambda_k}, \qquad 0 \le \lambda_0 < \lambda_1 < \dots. \tag{6.7}$$

Furthermore, let us set

$$\bar{\bar{\Phi}}_k(x, s) := s^{-(\lambda_k + 1)} e^{-x \left[\gamma_+(s) - s/c \right]}, \tag{6.8}$$

with the convention $1/c = 0$ when $J_0 = 0$. Then, from (6.7)-(6.8) the transform solution admits the following asymptotic expansion as $s \to \infty$

$$\bar{r}(x, s) \sim e^{-xs/c} \sum_{k=0}^{\infty} w_k(x) \, \bar{\bar{\Phi}}_k(x, s). \tag{6.9}$$

The purpose of the exponential function in (6.9) is to isolate, upon inversion, the wave front propagating with velocity c $(0 < c \le \infty)$.

From (6.2), (6.8) two things are evident for $k = 0, 1, \dots$, as $s \to \infty$

$$\begin{cases} \bar{\bar{\Phi}}_{k+1}(x, s) = o\left(\bar{\bar{\Phi}}_k(x, s) \right), \\ e^{\delta s} \, \bar{\bar{\Phi}}_k(x, s) \to \infty, \quad \forall \delta > 0. \end{cases} \tag{6.10}$$

Because of a Lemma from Erdélyi[56], we expect that the conditions (6.10) will allow a term by term inversion of (6.9), which provides the required asymptotic solution in the time-domain, as $t \to (x/c)^+$. Therefore, our asymptotic solution reads

$$r(x, t) \sim \sum_{k=0}^{\infty} w_k(x) \, \Phi_k(x, t - x/c), \quad \text{as} \quad t \to (x/c)^+. \tag{6.11}$$

We now discuss the recursive methods to determine
a) the functions $w_k(x)$ and the exponents λ_k ;
b) the functions $\Phi_k(x, t)$, the Laplace inverse of (6.8).

As far as the first goal is concerned, we note that $\bar{R}(x, s)$ formally satisfies the following differential equation obtained from (1.7) and (6.3),

$$\left\{ \frac{d^2}{dx^2} - 2 \left[\gamma_+(s) \right] \frac{d}{dx} - \left[\gamma^2(s) - \gamma_+^2(s) \right] \right\} \bar{R}(x, s) = 0, \tag{6.12}$$

subject to the initial condition

$$\bar{R}(0, s) = 1. \tag{6.13}$$

In order to obtain the coefficients and exponents in the asymptotic expansion (6.7), we shall use an argument which generalizes the one followed in the previous Section in order to allow non-integer powers of s. The argument is based a theorem stated by Friedlander & Keller[57], which we report for convenience.

THEOREM - *Let $\hat{L}$ be a linear operator with the asymptotic expansion with respect to a parameter ϵ, as $\epsilon \to 0$,*

$$\hat{L} \sim \sum_{i=0}^{\infty} \epsilon^{\nu_i} \hat{L}_i , \qquad 0 = \nu_0 < \nu_1 < \dots$$

and v a solution of
$$\hat{L} v = 0 ,$$

with the asymptotic expansion as $\epsilon \to 0$,

$$v \sim \sum_{k=0}^{\infty} \epsilon^{\lambda_k} v_k , \qquad \lambda_0 < \lambda_1 < \dots$$

If the asymptotic expansion of $\hat{L} v$ is given by termwise application of the $\hat{L}_i$ to the v_j and if $\hat{L}_0 v_k \neq 0$ for $k > 0$, then the coefficients v_k satisfy the recursive system of equations

$$\begin{cases} \hat{L}_0 v_0 = 0 , \\ \hat{L}_k v_k = - \sum \hat{L}_i v_j , \qquad k = 1, 2, \dots , \end{cases}$$

where the summation is taken over values of i, j for which $\nu_i + \lambda_j = \lambda_k$. The constant λ_0 is arbitrary but λ_k for $k > 0$ is the $(k+1)$-st number in the increasing sequence formed from the set of numbers $\lambda_0 + \sum_{i=1}^{\infty} m_i \nu_i$ where the m_i are any non-negative integers.

In the application of this theorem to our problem, the $\hat{L}_i$ are differential operators with respect to x, $\epsilon = s^{-1}$, $v = \bar{R}(x, s)$ and $v_k = w_k(x)$. If in (6.12) we expand the coefficients and divide the L.H.S. by the term containing the highest power of s (*i.e.* $-2b_0 s$ if $J_0 > 0$ or $-2b_1 s^{1-\beta_1}$ if $J_0 = 0$), we obtain quite generally:

$$\begin{cases} \hat{L}_0 = \dfrac{d}{dx} , \\ \hat{L}_i = p_i \dfrac{d^2}{dx^2} + q_i \dfrac{d}{dx} + r_i , \qquad i = 1, 2, \dots , \end{cases} \tag{6.14}$$

where p_i, q_i, r_i, like the exponents ν_i, can be determined from the behaviour of $J(t)$ as $t \to 0^+$, by the previous considerations. We point out that these quantities do not depend directly on the expansion of $\gamma_-(s)$, which would be prohibitive. Furthermore, comparing (6.7) and (6.13), we see that

$$\boxed{\lambda_0 = 0, \quad w_k(0) = \delta_{k0}} . \tag{6.15}$$

Then, because of the above theorem, the λ_k $(k > 0)$ are determined by the rule

$$\lambda_k = \sum_{i=1}^{\infty} m_i \, \nu_i \,, \tag{6.16}$$

and the coefficients $w_k(x)$ by the recursive system of differential equations

$$\begin{cases} \dfrac{dw_0}{dx} = 0 \,, \\[2mm] \dfrac{dw_k}{dx} = -\displaystyle\sum_{i,j} \left(p_i \dfrac{d^2}{dx} + q_i \dfrac{d}{dx} + r_i \right) w_j(x) \,, \quad k = 1, 2, \dots \,. \end{cases} \tag{6.17}$$

The solutions of this system are easily seen to be polynomials in x of order k, which we write in the form

$$w_k(x) = \sum_{h=0}^{k} A_{k,h} \, \frac{x^k}{k!} \,, \qquad k = 0, 1, \dots \,. \tag{6.18}$$

From (6.15) and (6.17) the coefficients are given by

$$\begin{cases} A_{k,0} = \delta_{k\,0} \,, \qquad h = 0 \,, \\[2mm] A_{k,h} = -\displaystyle\sum_{i,j} \left(p_i \, A_{j,h+1} + q_i \, A_{j,k} + r_i \, A_{j,h-1} \right) \,, \qquad 1 \le h \le k \,, \\[2mm] A_{k,h} = 0 \,, \qquad h > k \,. \end{cases} \tag{6.19}$$

Let us now consider the functions $\Phi_k(x,t)$. From (6.2), (6.6) and (6.8) we can write

$$\bar{\Phi}_k(x,s) = s^{-(\lambda_k+1)} \, \bar{E}_{\alpha_1}(y_1, s) \, \bar{E}_{\alpha_2}(y_2, s) \, \dots \, \bar{E}_{\alpha_m}(y_m, s) \,, \tag{6.20}$$

where, for $i = 1, 2, \dots, m$,

$$y_i = x \, b_i \,, \qquad \alpha_i = 1 - \beta_i \quad 1 > \alpha_1 > \dots \alpha_m \ge 0 \,, \tag{6.21}$$

and

$$\bar{E}_{\alpha_i}(y_i, s) = \exp\left(-y_i s^{\alpha_i}\right) \,. \tag{6.22}$$

The required functions $\Phi_k(x,t)$ can be obtained as

$$\Phi_k(x,t) = \frac{t^{\lambda_k}}{\Gamma(\lambda_k + 1)} * E_{\alpha_1}(y_1, t) * E_{\alpha_2}(y_2, t) * \dots * E_{\alpha_m}(y_m, t) \,, \tag{6.23}$$

where $*$ denotes the convolution from 0 to t, and the generic function $E_\alpha(y,t)$ is expressed by the formal series

$$E_\alpha(y,t) = \sum_{n=1}^{\infty} \frac{(-y)^n\, t^{-(\alpha n+1)}}{n!\,\Gamma(-\alpha n)}, \qquad t > 0. \tag{6.24}$$

We note that Buchen & Mainardi were unaware that the function $t\, E_\alpha(y,t)$, expressed in the variable

$$z = \frac{y}{t^\alpha} \tag{6.25}$$

as $f_\alpha(z)$, turns out to be an entire function in the complex z-plane, related to the *Wright function*[58], which reads (in our notation)

$$W(z;\lambda,\mu) := \sum_{n=0}^{\infty} \frac{z^n}{n!\,\Gamma(\lambda n+\mu)}, \qquad \lambda > -1,\ \mu > 0. \tag{6.26}$$

In fact, basing on our investigations on the fractional diffusion-wave equation, see[59-64],

$$\frac{\partial^{2\alpha} r}{\partial t^{2\alpha}} = D\,\frac{\partial^2 r}{\partial x^2}, \qquad 0 < \alpha < 1,\quad D > 0, \tag{6.27}$$

we now recognize that

$$f_\alpha(z) = \alpha\, z\, M(z;\alpha), \qquad M(z;\alpha) := W(-z;-\alpha,1-\alpha), \tag{6.28}$$

where $M(z;\alpha)$ is the auxiliary function from which the fundamental solutions of (6.27) can be obtained. However, Buchen & Mainardi, albeit unaware of the Wright function, have provided analytical representations of $E_\alpha(y,t)$ in the following special cases

$$\begin{cases} E_{1/3}(y,t) = \dfrac{y}{3^{1/3}\, t^{4/3}}\, \mathrm{Ai}\left(\dfrac{y}{3^{1/3}\, t^{1/3}}\right), \\[2ex] E_{1/2}(y,t) = \dfrac{y}{2\sqrt{\pi}\, t^{3/2}}\, \exp\left(-\dfrac{y^2}{4t}\right), \end{cases} \tag{6.29}$$

where Ai is the Airy function. More important, they have sought efficient methods of obtaining the inversion formulas. For this purpose they have introduced the functions

$$F_\alpha(z,\lambda_k) = t^{-\lambda_k}\left[\frac{t^{\lambda_k}}{\Gamma(\lambda_k+1)} * E_\alpha(y,t)\right], \tag{6.30}$$

with z given by (6.25), and have provided the following recurrence relation

$$\lambda_k\, F_\alpha(z,\lambda_k) = -\alpha\, z\, F_\alpha(z,\lambda_k-\alpha) + F_\alpha(z,\lambda_k-1), \tag{6.31}$$

which can make easy the determination of the functions $\Phi_k(x,t)$.

Particular examples are the models for which $J(t)$ admits a Taylor expansion about $t = 0^+$ with $J_0 > 0$, without necessarily being an entire function of exponential type; we refer to these models as *simply analytical models*, to be distinguished from the *full analytical models* considered in the previous Section. It is clear that for the *full* or *simply analytical models* the method must provide the solution (5.10) with a wave-front expansion of the type (5.21); of course, this expansion is expected to be convergent or only asymptotic, correspondingly. In fact, in these cases, we obtain

$$\begin{cases} \nu_i = i, \quad \lambda_k = k, \\ p_1 = -1/2, \quad q_1 = \alpha, \quad r_1 = -\beta, \\ p_i = q_i = 0, \quad r_i = \psi_{i+1}/2, \quad i \geq 2, \end{cases} \tag{6.32}$$

and

$$\Phi_k(x, t) = e^{-\alpha x} \frac{t^k}{k!}. \tag{6.33}$$

Thus, the wave-front expansion is originated by the recurrence relation (5.20), but with $A_{k,0} = \delta_{k\,0}$.

To better illustrate the importance of the Buchen-Mainardi method we need to consider *non-analytical models*, for which we may have, for example,

$$\gamma^+(s) = b_0 + b_1 s^{\beta_1}, \quad 0 < \beta_1 < 1. \tag{6.34}$$

Instructive examples which conform to (6.34) are the simple *Voigt model* [see *LV*, Sect 3, (17)] and the *fractional Maxwell model of order 1/2*. [see *LV*, Sect. 5, (59)]. In fact, after a suitable normalization, these models are described as follows.
(*a*) *Voigt model* :

$$\sigma(t) = \epsilon(t) + \frac{d\epsilon}{dt} \implies J(t) = 1 - e^{-t}, \tag{6.35a}$$

so that

$$\gamma(s) = s(s+1)^{-1/2} \implies b_0 = 0, \quad b_1 = 1, \quad \beta_1 = 1/2; \tag{6.36a}$$

(*b*) *fractional Maxwell model of order 1/2* :

$$\sigma(t) + \frac{d^{1/2}\sigma}{dt^{1/2}} = \frac{d^{1/2}\epsilon}{dt^{1/2}} \implies J(t) = 1 + \frac{t^{1/2}}{\Gamma(3/2)}, \tag{6.35b}$$

so that

$$\gamma(s) = s(1 + s^{-1/2})^{1/2} \implies b_0 = -1/8, \quad b_1 = 1/2, \quad \beta_1 = 1/2. \tag{6.36b}$$

As a consequence, for the above models, we respectively obtain

$$\begin{cases} \nu_1 = 1/2, \, \nu_2 = 1, \, \nu_3 = 3/2, \, \nu_i = 0 \quad \text{for} \quad i \geq 4, \\ \lambda_k = k/2, \qquad j = k-1, \, k-2, \, k-3; \end{cases} \tag{6.37a}$$

and

$$\begin{cases} \nu_1 = 1/2, \ \nu_2 = 1, \ \nu_i = 0 \quad \text{for} \quad i \geq 3, \\ \lambda_k = k/2, \qquad j = k-1, k-2. \end{cases} \tag{6.37b}$$

Then the coefficients $A_{k,h}$ of the polynomials $w_k(x)$ in (6.18)-(6.19) turn out to be obtained from the initial data $A_{k,0} = \delta_{k0}$ by the the following recurrence relations

$$A_{k,h} = \frac{1}{2} A_{k-1,h+1} - \frac{1}{2} A_{k-1,h-1} - A_{k-2,h} + \frac{1}{2} A_{k-3,h+1}, \quad 1 \leq h \leq k, \tag{6.38a}$$

and

$$\begin{aligned} A_{k,h} = &-\frac{1}{2} A_{k-1,h} + \frac{1}{16} A_{k-1,h-1} - \frac{1}{2} A_{k-2,h+1} \\ &- \frac{1}{8} A_{k-2,h} - \frac{1}{128} A_{k-2,h-1} \quad 1 \leq h \leq k. \end{aligned} \tag{6.38b}$$

Furthermore, for both models we obtain

$$\Phi_k(x,t) = e^{-b_0 x} t^{k/2} F_{1/2}(z, k/2), \tag{6.39}$$

with

$$z = \frac{b_1 x}{t^{1/2}}, \quad F_{1/2}(z, k/2) = 2^k I^k \text{erfc} \left(\frac{z}{2} \right) := G_k(z). \tag{6.40}$$

These repeated integrals of the error function are easily computed from the following recurrence relations found in Ref[65], which are a particular case of the most general relations (6.31),

$$\begin{cases} G_{-1}(z) = \frac{1}{\sqrt{\pi}} \exp \left(-\frac{z^2}{4} \right), \quad G_0(z = \text{erf} \left(\frac{z}{2} \right), \\ G_k(z) = -\frac{z}{k} G_{k-1}(z) + \frac{2}{k} G_{k-2}(z), \quad k \geq 1. \end{cases} \tag{6.41}$$

In conclusion, we report below the whole asymptotic expansion for the two models
(a) *Voigt model* :

$$r(x,t) \sim \sum_{k=0}^{\infty} \sum_{h=0}^{k} A_{k,h} \frac{x^h}{h!} t^{k/2} G_k \left(\frac{x}{\sqrt{t}} \right) \qquad \text{as} \quad t \to x^+, \tag{6.42a}$$

where the functions $G_k(z)$ are defined by (6.41) and the coefficients $A_{k,h}$ are obtained from the recurrence relation (6.38a).

The *Voigt model* exhibits a response $\forall t > 0$ and has a non-analytical expansion, which we call *diffusion like* response. Fig. 6.1a displays the essential character for times up to twice the retardation time.

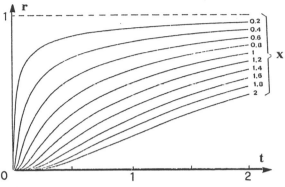

Fig. 6.1a

Pulse shapes for the simple Voigt model

(b) *Fractional Maxwell model of order* $1/2$:

$$r(x,t) \sim e^{x/8} \sum_{k=0}^{\infty} \sum_{h=0}^{k} A_{k,h} \frac{x^h}{h!} (t-x)^{k/2} G_k \left(\frac{x}{2\sqrt{t-x}} \right) \quad \text{as} \quad t \to x^+ , \quad (6.42b)$$

where the function $G_k(z)$ are defined by (6.41) and the coefficients $A_{k,h}$ are obtained from the recurrence relation (6.38b).

The *fractional Maxwell model* displays features in common to both the simple Maxwell and simple Voigt models. There is no motion for $t < x$, but the response at the front $t = x^+$ is zero for $\forall x > 0$ and its expansion is non-analytic, which we call *wave - diffusion like* response. Fig. 6.1b shows the essential characteristics of the pulse in the neighborhood of the onset.

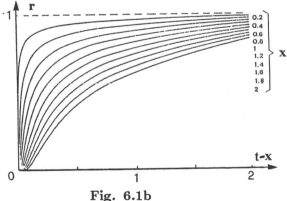

Fig. 6.1b

Pulse shapes for the fractional Maxwell model of order $1/2$

Conclusions

We have presented a review of our work on the topic of transient waves in linear viscoelastic media in the one-dimensional case.

The various concepts of wave speeds, *i.e.* wave-front, phase, group and signal velocities, which are of great relevance in the morphological analysis of dispersive wave phenomena, have been discussed with regard to viscoelastic waves.

The main purpose of this chapter has been to provide efficient methods for computing the evolution of the transient waves in any space-time domain. These methods are essentially based on series recurrence relations which yield (convergent or asymptotic) wave-front expansions which are suitable for numerical computation.

The general topic of transient waves in linear dispersive media has interested many specialists in wave propagation. It is practically impossible to quote all the papers which have attracted the Author's attention with regard to viscoelastic waves, during his activity in different branches of wave propagation. However, as a conclusion, we would like to quote, besides some papers of ours on related topics, see Refs [66-70], a number of additional papers and books by other authors on different aspects of the subject, which were relevant in our formation, without pretending to be exhaustive, see Refs [71-94].

Acknowledgements

This work was supported in part by C.N.R., Gruppo Nazionale per la Fisica Matematica, I.N.F.N., Sezione di Bologna, and the Italian Ministry for Universities (MURST: 60% grants). The Author wishes to thank his colleagues, M. Caputo, P.W. Buchen, G. Turchetti, H. Buggisch, E. Strick, T.B. Moodie, R.J. Tait, P.H. LeBlond, E. van Groesen, F. Tampieri, R. Gorenflo and former students, G. Servizi, R. Nervosi, G. Vitali, E. Grassi, D. Tocci, M. Tomirotti, for their fruitful collaboration on some topics discussed in this review or related to it.

References

1. F. Mainardi (Editor), *Wave Propagation in Viscoelastic Media* (Pitman, London, 1982). [Res. Notes in Maths, Vol. 52]

2. H. Überall, L.R. Dragonette and L. Flax, *J. Acoust. Soc. Am.* **61** (1977) 711-715.

3. L. Flax, L.R. Dragonette and H. Überall, *J. Acoust. Soc. Am.* **63** (1978) 723-731.

4. H. Überall, in *Modern Problems in Elastic Wave Propagation*, eds J. Miklowitz and J.D. Achenbach (Wiley, New York 1978), pp. 239-263.

5. L. Flax, G.C. Gaunaurd, W. Madigosky and H. Überall, in *Physical Acoustics*, eds W.P. Mason and R.N. Thurston (Academic Press, New York, 1981), Vol. 15, pp. 191-294.

6. G.C. Gaunaurd, W. Madigosky, H. Überall and L.R. Dragonette in *ref* [1], pp. 235-257.

7. R. Fiorito, W. Madigosky and H. Überall, *J. Acoust. Soc. Am.* **77** (1985) 489-498.

8. H. Überall (Editor), *Acoustic Resonance Scattering* (Gordon and Breach, Philadelphia, 1992).

9. H. Überall, B.F. Howell and E.L. Diamond, *J. Appl. Phys.* **73** (1993) 3441-3445.

10. F. Mainardi, *Linear Viscoelasticity*, this book (World Scientific, Singapore, 1999).

11. M. Caputo and F. Mainardi, *Riv. Nuovo Cimento* (Ser. II) **1** (1971) 161-198.

12. F. Mainardi, *Pure Appl. Geophys.* [Pageoph] **99** (1972) 72-84.

13. P.W. Buchen and F. Mainardi, *J. Mécanique* **14** (1975) 597-608.

14. F. Mainardi and G. Turchetti, *Mech. Research Comm.* **2** (1975) 107-112.

15. G. Turchetti and F. Mainardi, in *Padè Approximants Method and its Applications to Mechanics*, ed. H. Cabannes (Springer Verlag, Berlin, 1976), pp. 187-207. [Lecture Notes in Physics, Vol. 47]

16. F. Mainardi, G. Servizi and G. Turchetti, *J. Geophys.* **43** (1977) 83-94.

17. F. Mainardi and R. Nervosi, *Lett. Nuovo Cimento* **29** (1980) 443-447.

18. F. Mainardi, *Wave Motion* **5** (1983) 33-41.

19. F. Mainardi, *Il Nuovo Cimento* B **74** (1983) 52-58.

20. F. Mainardi, in *Wave Phenomena: Modern Theory and Applications*, eds C. Rogers and T.B. Moodie (North-Holland, Amsterdam, 1984), pp. 307-317.

21. F. Mainardi, *Wave Motion* **9** (1987) 201-208.

22. F. Mainardi and E. van Groesen, *Il Nuovo Cimento* B **104** (1989) 487-496.

23. E. van Groesen and F. Mainardi, *Wave Motion* **11** (1989) 201-209.

24. F. Mainardi and G. Vitali, in *Asymptotic and Computational Analysis*, ed. R. Wong (Marcel Dekker, New York, 1990), pp. 639-651.

25. F. Mainardi and D. Tocci, in *Nonlinear Hyperbolic Problems: Theoretical, Applied, and Numerical Aspects*, eds A. Donato and F. Oliveri (Vieweg, Braunschweig, 1993), pp. 409-415.

26. F. Mainardi, *Radiofisika* [Radiophys. & Quantum Electr.] **36** (1993) 650-664.

27. D.R. Bland, *The Theory of Linear Viscoelasticity* (Pergamon Press, Oxford, 1960).

28. S.C. Hunter, in *Progress in Solid Mechanics*, eds I.N. Sneddon and R. Hill (North-Holland, Amsterdam, 1960), Vol. 1, pp. 3-60.

29. R.M. Christensen, *Theory of Viscoelasticity* (Academic Press, New York, 1971).

30. A.C. Pipkin, *Lectures on Viscoelastic Theory* (Springer Verlag, New York, 1972).

31. R.C.Y. Chin, in *Physics of the Earth's Interior*, eds A. Dziewonski and E. Boschi (North-Holland, Amsterdam, 1980), pp. 213-246.

32. D. Graffi, in *ref* [1], pp. 1-27.

33. D.S. Berry, *Phil. Mag* (Ser. VIII) **3** (1958) 100-102.

34. S.A. Thau in *Non Linear Waves*, eds S. Leibovich and A.R. Seebass (Cornell Univ. Press, Ithaca, 1974), pp. 44-81.

35. W.I. Futterman, *J. Geophys. Res.* **67** (1962) 5279-5291.
36. E. Strick, *Geophysics* **35** (1970) 387-403.
37. K. Aki and P.G. Richards, *Quantitative Seismology* (Freeman, San Francisco, 1980), Vol. 1.
38. A. Ben-Menahem and S.J. Singh, *Seismic Waves and Sources* (Springer-Verlag, Berlin, 1981).
39. L. Brillouin, *Wave Propagation and Group Velocity* (Academic Press, New York, 1960).
40. M.J. Lighthill, *J. Inst. Maths. Appl.* **1** (1965) 1-28.
41. G.B. Whitham, *Linear and Nonlinear Waves* (Wiley, New York, 1974).
42. G.R. Baldock and T. Bridgeman, *Mathematical Theory of Wave Motion* (Ellis Horwood, Chichester, 1981).
43. L. Brillouin, *Ann. Physik* (Ser. IV) **44** (1914) 203-240 [reported in Ref[39]].
44. H.G. Baerwald, *Ann. Physik* (Ser. V) **7** (1930) 731-760.
45. M. Elices and F. García-Moliner in *Physical Acoustics*, ed. W.P. Mason (Academic Press, New York, 1968), Vol. 5, pp. 163-219.
46. M.J. Rubin, *J. Appl. Phys.* **25** (1954) 528-536.
47. G.F. Carrier, M. Krook and C.E. Pearson, *Functions of a Complex Variable* (McGraw-Hill, New York, 1966), Ch. 6.
48. B.T. Chu, *J. Mécanique* **1** (1962) 439-462.
49. N. Bleistein and R. A. Handelsman, *Asymptotic Expansions of Integrals* (Dover, New York, 1986).
50. E.M. Lee and T. Kanter, *J. Appl. Phys.* **24** (1956) 1115-1122.
51. G. Doetsch, *Introduction to the Theory and Application of the Laplace Transformation* (Springer-Verlag, Berlin, 1974).
52. D.V. Widder, *An Introduction to the Transform Theory* (Academic Press, New York, 1971).
53. J.D. Achenbach and D.P. Reddy, *Z. Angew. Math. Phys.* **18** (1967) 141-144.
54. C.T. Sun, *J. Appl. Mech.* **37** (1970) 1141-1144.
55. P.W. Buchen, *Pure and Appl. Geophys.* **112** (1974) 1011-1023.
56. A. Erdélyi, *Asymptotic Expansions* (Dover, New York, 1956), p. 31.
57. F.G. Friedlander and J.B. Keller, *Comm. Pure and Appl. Math.* **8** (1955) 387-394.
58. A. Erdélyi (Editor), *Higher Transcendental Functions* (McGraw-Hill, New York, 1955), Vol. 3, Ch. 18, pp. 206-227.
59. F. Mainardi, in *Waves and Stability in Continuous Media*, eds S. Rionero and T. Ruggeri (World Scientific, Singapore, 1994), pp. 246-251.
60. F. Mainardi, in *Nonlinear Waves in Solids*, eds J.L. Wegner and F.R. Norwood (ASME book No AMR 137, Fairfield N,J 1995), pp. 93-97. [Abstract in *Appl. Mech. Rev.* **46** (1993) 549]
61. F. Mainardi and M. Tomirotti, in *Transform Methods and Special Functions, Sofia '94*, eds P. Rusev, I. Dimovski and V. Kiryakova (Science Culture Technology, Singapore, 1995), pp. 171-183.
62. F. Mainardi, *Chaos, Solitons & Fractals* **7** (1996) 1461-1477.

63. F. Mainardi, in *Fractals and Fractional Calculus in Continuum Mechanics*, eds A. Carpinteri and F. Mainardi (Springer Verlag, Wien and New York, 1997), pp. 291-348.
64. F. Mainardi and M. Tomirotti, *Annali di Geofisica* **40** (1997) 1311-1328.
65. M. Abramowitz and I.E. Stegun (Editors), *Handbook of Mathematical Functions* (Dover, New York, 1965).
66. F. Mainardi and H. Buggisch, in *Nonlinear deformation waves*, eds U. Nigul and J. Engelbrecht (Springer Verlag, Berlin, 1983), pp. 87-100.
67. T.B. Moodie, F. Mainardi and R.J. Tait, *Meccanica* **20** (1985) 33-37.
68. P.H. LeBlond and F. Mainardi, *Acta Mechanica* **68** (1987) 203-222.
69. F. Mainardi, F. Tampieri and G. Vitali, *Il Nuovo Cimento* **14 C** (1991) 391-399.
70. E. Grassi and F. Mainardi, in *Topics on Biomathematics*, eds I. Barbieri et al. (World Scientific, Singapore, 1993), pp. 265-271.
71. D.S. Berry and S.C. Hunter, *J. Mech. Phys. Solids* **4** (1956) 72-95.
72. E.H. Lee and J.A. Morrison, *J. Polym. Sci.* **19** (1956) 93-110.
73. L. Knopoff and G.J.F. MacDonald, *Rev. Mod. Physics* **30** (1958) 1178-1192.
74. D. Graffi, *Ann. Mat. Pura ed Appl.* (Ser. IV) **60** (1963) 173-194.
75. H. Kolsky, *Stress Waves in Solids* (Dover, New York, 1963).
76. B.D. Coleman, M.E. Gurtin and I.R. Herrera, *Arch. Rat. Mech. Anal.* **19** (1965) 1-19.
77. B.D. Coleman and M.E. Gurtin, *Arch. Rat. Mech. Anal.* **19** (1965) 239-265.
78. H.F. Cooper and E.I. Reiss, *J. Acoust. Soc. Amer.* **38** (1965) 23-34.
79. J.L. Sackman and I. Kaya, *J. Mech. Phys. Solids* **16** (1968) 349-356.
80. M. Caputo, *Elasticità e Dissipazione*, (Zanichelli, Bologna, 1969).
81. E.I. Reiss, *SIAM J. Appl. Math.* **17** (1969) 526-542.
82. B.J. Matkowsky and E.I. Reiss, *Arch. Rat. Mech. Anal.* **42** (1971) 194-212.
83. J.D. Achenbach, *Wave Propagation in Elastic Solids* (North-Holland, Amsterdam, 1973).
84. L. Brun, *J. Mécanique* **13** (1974) 449-498.
85. D. Gamby, *Mech. Research Comm.* **2** (1975) 131-135.
86. L.A. Vainshtein, *Sov. Phys. Usp.* **19** (1976) 189-205.
87. N.H. Ricker, *Transient Waves in Viscoelastic Media* (Elsevier, Amsterdam, 1977).
88. M. Hayes, in *ref* [1], pp. 28-40.
89. L. Brun and A. Molinari, in *ref* [1], pp. 65-94.
90. E. Strick, in *ref* [1], pp. 169-193.
91. P. Renno, *Atti Acc. Lincei, Rend. fis.* [Ser. VIII] **75** (1983) 195-204.
92. A. Narain and D.D. Joseph, *Rheologica Acta* **22** (1983) 528-538.
93. M. Renardy, W.J. Hrusa and J.A. Nohel, *Mathematical Problems in Viscoelasticity* (Longman, Essex, 1987).
94. M. Fabrizio and A. Morro, *Mathematical Problems in Linear Viscoelasticity* (SIAM, Philadelphia, 1992).

Dynamics with Friction: Modeling, Analysis and Experiment, Part II, pp. 191–225
edited by A. Guran, F. Pfeiffer and K. Popp
Series on Stability, Vibration and Control of Systems, Series B, Vol. 7
© World Scientific Publishing Company

DYNAMIC STABILITY AND NONLINEAR PARAMETRIC VIBRATIONS OF RECTANGULAR PLATES

G.L. OSTIGUY

Dept. of Mechanical Engineering, Ecole Polytechnique
P.O.B. 6079, Succ. "Centre-ville"
Montreal (Quebec), H3C 3A7, CANADA

ABSTRACT

The present work reviews the author's recent developments on the dynamic stability and nonlinear parametric vibrations of rectangular plates acted upon by periodic in-plane forces. General rectangular plates are considered, the aspect ratio of the plate being regarded as an additional parameter of the system. The problem is solved for different sets of boundary conditions and the effects of various system parameters are evaluated. Numerical results are compared to ·experimental data to form a qualitative and quantitative verification of the solution.

1. Introduction

The problem of dynamic instability for elastic systems under parametric excitation has been intensively studied by numerous investigators. Such problem is quite interesting from the standpoint of system stability, since the system will respond or vibrate only if certain conditions are met by the excitation and system parameters. Comprehensive analyses of the dynamic stability problem for elastic structures together with many references, can be found in the literature [1-3].

Parametric vibration is a generic term for a class of oscillating motions which can occur in structures or mechanical systems. Parametrically excited systems are governed by differential equations in which the excitation appears as a time-dependent coefficient. This is in contrast with externally excited systems, where the excitation appears as a non-homogeneous term. An interesting property of such systems is that a small excitation can produce a large response when the excitation frequency is not close to the natural frequency of the system. General reviews of problems of parametric instability including instability of rectangular plates were published by Evan-Iwanowski[4], Mettler[5], Ibrahim et al.[6-10], Ostiguy and Nguyen[11], and Ostiguy and Evan-Iwanowski[12]. Books by Evan-Iwanowski[13] and Ibrahim[14] are devoted to the problem of parametric excitations. Other books, by Nayfeh and Mook[15], and Mitropolsky[16], contain one or more chapters on the subject.

A typical example in regard to parametric (dynamic) instability of structures is the case of a rectangular plate acted upon by periodic in-plane forces. When a flat plate sustains an in-plane load of the form $N_y(t) = N_{y0} + N_{yt} \cos \eta t$, the plate will generally experience forced in-plane vibrations, and for certain excitation frequencies, in-plane resonance will take place. However, an entirely different type of resonance will occur when certain relationships exist between the natural frequencies of transverse vibration, the frequency of the periodic in-plane force and the conditions of loading. Thus, apart from in-plane vibrations, transverse vibrations may be induced in the plate, and the plate is said to be dynamically unstable. These resonances occur when the excitation frequency η and a modal frequency Ω_i satisfy approximately the relationship

$$\eta = 2\,\Omega_i / k, \qquad (k = 1, 2, 3, \dots) \tag{1}$$

The case $\eta = 2\Omega_i$ is generally the most important and is called principal parametric resonance.

In contrast with this case of simple parametric resonance, simultaneous and combination resonances may also occur in structures subjected to parametric excitation. These kinds of resonances are characterized by the fact that the system in question resonates simultaneously in more than one normal mode and at different frequencies while only one or all resonant modes are excited directly by the parametric excitation.

The problem of parametric (dynamic) instability of flat elastic plates subjected to periodic in-plane forces has been studied by a number of authors. The first investigation on rectangular plates was done by Einaudi[17]. Subsequent works by Bodner[18] and Chelomei[19] used linear plate theories to establish that a parametric vibration could manifest over multiple continuous regions of the parameter space.

Bolotin[20] was apparently the first to investigate the nonlinear problem of parametric response of a rectangular plate. Somerset and Evan-Iwanowski[21] reinvestigated Bolotin's nonlinear problem and included the effects of the distributed in-plane inertia of the plate. Numerous references pertaining to parametric resonance of plates can be found in the book by Bolotin[2], in the survey article by Evan-Iwanowski[4], and in the doctoral dissertations by Duffield[22], Ostiguy[23], Nguyen[24] and Sassi[25].

A survey of the literature by Ostiguy[23] reveals that work on the subject has been confined mainly to square plates or almost square plates, and mostly limited to the investigation of parametric resonance in the first spatial mode. Using the aspect ratio of the plate as a new parameter of the system, Ostiguy and Evan-Iwanowksi[26] have investigated the effects of various parameters on the dynamic stability and nonlinear parametric response of rectangular plates. More recently, Ostiguy and Nguyen[11,27] have examined the effect of internal resonances and the possibility of simultaneous and combination resonances on the response of parametrically excited rectangular plates. In 1989, Nguyen and Ostiguy[28] have presented the results of an analytical and experimental investigation of the effect of boundary conditions on the dynamic stability and nonlinear parametric response of rectangular plates. Other analytical studies were published by

Berezovskii and Mitropolsky[29], Yamaki and Nagai[30], Kisliakov[31], Takahashi et al.[32], Pierre and Dowell[33], Takahashi and Konishi[34].

The first experimental studies on plates were conducted by Somerset and Evan-Iwanowski[35] and they pertain mainly to the large amplitude, nonlinear parametric response of simply supported square plates. Dixon and Wright[36] investigated experimentally the parametric instability of rectangular plates under various boundary conditions, and subjected to periodic in-plane direct and "shear" type forces. Extensive experiments were carried out by Ostiguy[23] on the parametric instability and resonance of rectangular plates of various aspect ratios. Analytical and experimental results on the stationary and non-stationary parametric responses of nonlinear rectangular plates were presented by Ostiguy and Evan-Iwanowski[37,38]. Recently, Ostiguy, Samson and Nguyen[39,40] performed an analytical and experimental investigation of the occurrence of simultaneous and combination resonances in parametrically excited rectangular plates.

The dynamic instability and parametric resonance of anisotropic and sandwich plates under periodic in-plane loadings was also investigated by a number of authors. For instance, the first known research in the area of parametric instability of anisotropic plates was done by Ambartsumyan and Khachatrian[41]. Ambartsumyan and Gnuni[42] studied the linear and nonlinear problems for a three-layered plate. The problem of parametric instability of orthotropic plates was investigated by Feldman[43]. Schmidt[44] analyzed the possibility of simultaneous longitudinal and lateral parametric resonances of a sandwich plate.

The onset of parametric resonance of stiffened rectangular plates was studied by Duffield and Willems[45]. Yu and Lai[46] considered the influence of transverse shear and edge conditions on the nonlinear dynamic buckling of homogeneous and sandwich plates in plane-strain motion. Parametric instability of unsymmetrically laminated cross-ply rectangular plates was investigated by Birman[47] and by Srinivasan and Chellapandi[48]. Recently, Birman and Bert[49] studied the nonlinear parametric instability of antisymmetrically laminated angle-ply plates. Parametric instability of laminated composite shear-deformable flat panels subjected to in-plane edge loads was examined by Librescu and Thangjitham[50].

It is well known that small initial geometric imperfections are almost inevitable in practice and may affect considerably the static and dynamic behavior of these structures. For instance, Hui and Leissa[51] and Ilanko and Dickinson[52] have shown that initial imperfections may significantly increase the vibrations frequencies of simply supported plates under in-plane compression. Experiments conducted by Somerset[53] and Ostiguy[23] indicated that initial geometric imperfections can significantly change the dynamic stability of the plate and introduce new phenomena, unpredictable through the classical theories established for perfectly flat plates.

One of the first theories concerning the effects of geometric imperfections on the parametric vibration of rectangular plates is due to Silver[54]. Kisliakov[55] studied separately forced and parametrically excited nonlinear vibrations of thin elastic plates with initial imperfections. Experimental evidence of simultaneous and combination resonances, due to the interaction of forced and parametric vibrations, presented by Ostiguy[23] and Ostiguy, Samson and Nguyen[39,40] constitute an original and significant contribution.

The first analytical studies considering the effect of initial geometric imperfections on the interaction between forced and parametric or combination resonances are seemingly due to Ostiguy and Sassi[56], Sassi and Ostiguy[57-61]. Recently, Librescu and Chang[62] have investigated the effects of geometric imperfections on vibration of compressed shear deformable laminated composite curved panels.

The present work review the author's recent developments on the subject and presents a rational analysis of the influence of various system parameters on the dynamic stability and the nonlinear response of parametrically excited rectangular plates.

2. Theoretical Analysis

2.1 Analytical Model

The mechanical system under investigation is a rectangular plate, simply supported or loosely clamped along its edges and acted upon by periodic in-plane forces uniformly distributed along two opposite edges. The two vertical edges are stress-free. The geometry of the plate, the load configuration and the coordinate system are shown in Fig. 1. The x-y plane is selected in the middle plane of the undeformed plate.

The plate is assumed to be thin, of uniform thickness and the plate material is elastic, homogeneous and isotropic. It is also assumed that the loading frequencies over which parametric vibrations occur are considerably below the natural frequencies of in-plane vibrations and, consequently, in-plane inertia forces can be neglected. General rectangular plates are considered, the aspect ratio "r" of the plate being a new parameter of the system.

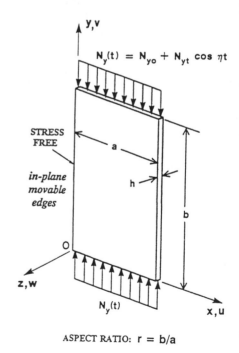

$$N_y(t) = N_{yo} + N_{yt} \cos \eta t$$

ASPECT RATIO: $r = b/a$

Fig. 1. Plate and load configuration.

2.2 Basic Equations

The plate theory used in the analyses may be described as the dynamic analog of the von Kármán large-deflection theory and is derived in terms of the stress function F and the lateral displacement w. The nonlinearity arising in the problem under investigation is due to large amplitudes generating membrane stresses that necessitate the use of nonlinear strain-displacement relations.

The basic equations governing the nonlinear flexural vibrations of perfectly flat rectangular plates are written as:

$$\nabla^4 F = E \left[w_{,xy}^2 - w_{,xx} w_{,yy} \right] \tag{2a}$$

$$\nabla^4 w = \frac{h}{D} \left[F_{,yy} w_{,xx} + F_{,xx} w_{,yy} - 2 F_{,xy} w_{,xy} - \rho w_{,tt} \right] \tag{2b}$$

in which a comma denotes partial differentiation with respect to the corresponding coordinates and where $w = w(x,y,t)$ is the lateral displacement, $F = F(x,y,t)$ is the Airy's stress function defined by:

$$F_{,yy} = \frac{N_x}{h} \; ; \quad F_{,xx} = \frac{N_y}{h} \; ; \quad F_{,xy} = - \frac{N_{xy}}{h} \tag{3}$$

and the operator:

$$\nabla^4 = \frac{\partial^4}{\partial x^4} + \frac{2 \partial^4}{\partial x^2 \partial y^2} + \frac{\partial^4}{\partial y^4} \tag{4}$$

In the foregoing, N_x, N_y and N_{xy} are membranes forces per unit length, h is the plate thickness, ρ the density, t the time, $D = Eh^3/12(1 - \nu^2)$ the flexural rigidity where E and ν are Young's modulus and Poisson's ratio, respectively.

In using Eq. (2), transverse shear deformations and in-plane and rotatory inertia effects are assumed to be negligible. This is a common assumption in the nonlinear analysis of plate vibrations[63], and it restricts the study to an investigation of the lower flexural modes.

2.3 Boundary Conditions

Four different sets of boundary conditions were considered in the analysis (Fig. 2). These boundary conditions are related to both the lateral displacement w and the stress function F.

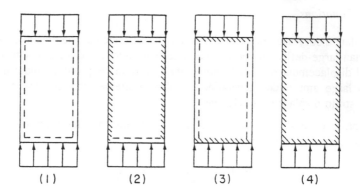

$- - - -$, simply supported edge; $\overline{//////}$, loosely clamped edge.

Fig. 2. The four sets of boundary conditions considered in the analysis.

The boundary stress conditions are expressed as:

$$F_{,yy} = 0 ; \qquad\qquad F_{,xy} = 0 \qquad \text{along } x = 0, a$$
$$F_{,xx} = -N_y(t)/h ; \qquad F_{,xy} = 0 \qquad \text{along } y = 0, b \tag{5}$$

The supporting conditions for the case of a simply supported (S-S) plate on all edges are expressed as:

$$w = w_{,xx} + \nu w_{,yy} = 0 \qquad \text{along } x = 0, a$$
$$w = w_{,yy} + \nu w_{,xx} = 0 \qquad \text{along } y = 0, b \tag{6}$$

and for a loosely clamped (C-C) plate, we have:

$$w = w_{,x} = 0 \qquad \text{along } x = 0, a$$
$$w = w_{,y} = 0 \qquad \text{along } y = 0, b \tag{7}$$

The problem then consists in determining the appropriate functions F and w which will satisfy the governing equations, together with the boundary conditions.

2.4 Method of Solution

An approximate solution of the governing equations (2), in the case of standing flexural waves, is sought in the form:

$$F(x,y,t) = \sum_m \sum_n F_{mn}(t)\, X_m(x)\, Y_n(y) - \frac{x^2}{2h} N_y(t) \tag{8a}$$

$$w(x,y,t) = \sum_p \sum_q W_{pq}(t)\, \Phi_p(x)\, \Psi_q(y) \tag{8b}$$

where $F_{mn}(t)$ are the time-dependent load factors and $W_{pq}(t)$ are the time-dependent generalized coordinates of the system. The indices p and q correspond to the number of half-waves in the direction of axes O_x and O_y, respectively. The spatial forms $X_m(x)$, $\Phi_p(x)$, $Y_n(y)$, $\Psi_q(y)$ are beam eigenfunctions which satisfy the relevant boundary conditions and are orthogonal in their respective intervals.

The particular spatial forms used in the analysis for the stress function are:

$$X_m(x) = (\cosh \alpha_m x - \cos \alpha_m x)$$

$$- \left[\frac{\cosh \alpha_m a - \cos \alpha_m a}{\sinh \alpha_m a - \sin \alpha_m a} \right] (\sinh \alpha_m x - \sin \alpha_m x) \tag{9}$$

$$Y_n(y) = \ldots.$$

The spatial forms for the lateral displacement are:

$$\Psi_p(x) = \sin \frac{p\pi x}{a}, \qquad \Psi_q(y) = \sin \frac{q\pi y}{b} \tag{10}$$

in the case of simply supported plates, and

$$\Phi_p(x) = \left[\frac{\sinh \alpha_p a - \sin \alpha_p a}{\cosh \alpha_p a - \cos \alpha_p a} \right] (\cosh \alpha_p x - \cos \alpha_p x)$$

$$- (\sinh \alpha_p x - \sin \alpha_p x) \tag{11}$$

$$\Psi_q(y) = \ldots.$$

in the case of loosely clamped plates. These beam functions were previously used by Prabhakara and Chia[64] to study the nonlinear vibrations of orthotropic rectangular plates, with excellent results.

Applying the Galerkin's method to the governing equations, using the orthogonality properties of the assumed functions, solving for coefficients F_{mn} in terms of the W_{pq}

coefficients and considering only the first spatial mode in the O_x direction ($p = 1$) lead to the following system of nonlinear ordinary differential equations:

$$\ddot{W}_v + \omega^2 W_v - \frac{\pi^2 v^2}{r^2} N_y W_v + \sum_l \sum_q \sum_s M_v^{lqs} W_l W_q W_s = 0 \qquad (12)$$

in the case of simply supported plates, and

$$\ddot{W}_v + \sum_q K_v^q W_q + \sum_s N_y P_v^s W_s + \sum_l \sum_q \sum_s M^{lqs} W_l W_q W_s = 0 \qquad (13)$$

in the case of loosely clamped plates.

An examination of expressions (12) and (13) reveals that the two sets of differential equations differ significantly from one another. In the case of simply supported plates, we can observe that there is no linear coupling terms and that the in-plane force excites only one mode at a time. This explains why combination resonances are not predictable for simply supported flat plates. For loosely clamped plates, we do have linear coupling terms and the in-plane force excites all the modes simultaneously. Therefore, combination resonances are possible for this particular case of boundary conditions.

Solving the eigenvalue problem of the linearized system, Eq. (13) can be written in terms of the normal coordinates. Introducing the following parameters:

$$
\begin{aligned}
\Omega_v &= \omega_v \sqrt{1 - N_{y0}/N_v} \\
\mu_v &= N_{yt}/2(N_v - N_{y0})
\end{aligned}
\qquad (14)
$$

and adding linear damping, Eqs (12) and (13) take the final form:

$$\ddot{W}_v + 2C_v \dot{W}_v + \Omega_v^2 (1 - 2\mu_v \cos \eta t) W_v + \sum_l \sum_q \sum_s M_v^{lqs} W_l W_q W w_s = 0 \qquad (15)$$

in the case of simply supported plates, and:

$$\ddot{W}_v + 2C_v \dot{W}_v + \Omega_v^2 W_v - \sum_l 2\cos \eta t \, \mu_v^l \Omega_l^2 W_l + \sum_l \sum_q \sum_s M_v^{lqs} W_l W_q W_s = 0 \qquad (16)$$

in the case of loosely clamped plates.

In the foregoing, the dots denote differentiation with respect to time, η is the instantaneous frequency of excitation , C_v is the coefficient of viscous damping, M_v^{lqs} are the coefficients of the nonlinear cubic terms, Ω_v is the free vibration circular frequency of a rectangular plate loaded by the constant component N_{y0} of the in-plane force, ω_v is the free vibration circular frequency of the unloaded plate and N_v represents the static critical load corresponding to the $(1,v)$ buckling mode. Finally, μ_v is the load parameter.

3. Solution of the Temporal Equations of Motion

Equations (15) and (16) represent systems of second-order nonlinear differential equations with periodic coefficients, which may be considered as extensions of the standard Mathieu-Hill equation. Mathematical techniques for solving such nonlinear problems are relatively limited and approximate methods are generally used. The method of asymptotic expansions in powers of a small parameter ϵ, developed by Mitropolsky [16] and generalized by Agrawal and Evan-Iwanowski [65], is an effective tool for studying nonlinear vibrating systems with slowly varying parameters.

If we take three terms in the expansion for the lateral displacement, the continuous system is reduced to a three-dof system. Assuming that the present mechanical system is weakly nonlinear and that the excitation frequency and the load parameter vary slowly with time, the systems of temporal equations of motion (15) and (16) can be rewritten in the following symbolic form:

$$\ddot{W}_v + \Omega_v^2 W_v = \epsilon P(\tau, \theta, W_v, \dot{W}_v), \quad v = 1, 2, 3 \tag{17}$$

where $\tau = \epsilon t$ represents the "slowing" time, $\dot{\theta}(t) = \eta(\tau)$ is the instantaneous frequency of excitation and P_v denotes the perturbation terms in the equation.

Confining ourselves to the first order of approximation in ϵ, we seek a solution of Eq. (17) in the following form:

$$W_m = a_m(\tau) \cos \psi_m(\tau) \tag{18}$$

where a_m and ψ_m are functions of time defined by the system of differential equations:

$$da_m / dt = \dot{a}_m = \epsilon A_1^m(\tau, \theta, a_m, \psi_m) \tag{19a}$$

$$d\psi_m / dt = \dot{\psi}_m = \Omega_m(\tau) + \epsilon B_1^m(\tau, \theta, a_m, \psi_m) \tag{19b}$$

Function $A_1^m(\tau, \theta, a_m, \psi_m)$ and $B_1^m(\tau, \theta, a_m, \psi_m)$ are selected in such a way that Eq. (18) will, after substituting a_m and ψ_m by the functions defined in Eq. (19), represent a solution of Eq. (17).

Following the general scheme of constructing asymptotic solutions and performing numerous transformations we can arrive finally at a system of equations describing the nonstationary response of the discretized system.

4. Stationary Response

The regions of parametric instability and the stationary responses associated with various types of resonances of our system may be calculated as a special case of the nonstationary motions in the resonant regimes described previously. The stationary mode for the system under consideration is obtained when the parameters $\mu(\tau)$ and $\eta(\tau)$ are constant.

The results of the investigation conducted by Nguyen[24] on flat rectangular plates indicate, besides the possibility of principal parametric resonances, the presence of internal resonances and the occurrence, in some cases, of combination resonances. As mentioned earlier, the solution for simply supported flat plates precludes the possibility of combination resonances[27]. An internal resonance is possible when two or more natural frequencies are commensurable or almost commensurable. When an internal resonance coincides with a parametric resonance, the combination of the two types gives rise to simultaneous resonances[39,40]. This kind of resonances is characterized by the fact that the system in question vibrates simultaneously in more than one normal mode and at different frequencies, although only one of the modes is directly excited by the parametric excitation. The results are summarized in Table 1.

Table 1   Effect of boundary conditions on the various types of resonances possible.

TYPES OF RESONANCES	BOUNDARY CONDITIONS	
	Simply-supported	Clamped
• PRINCIPAL PARAMETRIC RESONANCES $$\eta = 2\Omega_i$$	Possible	Possible
• INTERNAL RESONANCES $$\sum_{i=1}^{n} k_i \Omega_i = 0$$	Possible	Possible
• COMBINATION RESONANCES $$k_0 \eta = \sum_{i=1}^{n} k_i \Omega_i$$	Not Possible	Possible
• SIMULTANEOUS RESONANCES $\left\{ \begin{array}{c} \text{PRINCIPAL} \\ + \\ \text{INTERNAL} \end{array} \right.$	Possible	Possible

4.1 Principal Parametric Resonances

In the absence of internal or combination resonances, the parametric excitation can excite only one mode at a time. In this case, principal parametric resonance occurs when the excitation frequency is approximately equal to twice the natural frequency associated with a particular mode of vibration. Ostiguy and Evan-Ewanowski[26] have presented an extensive and rational analysis of the effects of various system parameters on the location and relative importance of the principal regions of parametric instability associated with the lower mode shapes, and the stationary parametric response of rectangular plates within a principal region of instability.

Stationary values for principal parametric response associated with various spatial forms of vibration are given by:

$$a_m = \sqrt{(4\Omega_m/3M_m)(\eta - 2\Omega_m \pm \sqrt{(2\mu_m\Omega_m^2/\eta)^2 - (\Delta_m\Omega_m/\pi)^2})} \,, \qquad (20)$$

where $M_m = M_m^{mmm}$, $\mu_m = \mu_m^m$ and $\Delta_m = 2\pi C_m/\Omega_m$ is the decrement of viscous damping, and where only positive real values for the amplitude are admitted. The "$\pm$" sign upon the inner radical indicates the possibility of two solutions; the larger solution is stable and attainable by the system, while the lower is unstable and not physically realizable. These solutions are represented in Fig. 3. The overhang of vibrations takes place in the direction of higher frequencies. As is evident from Eq. 20, a physically realizable solution exists only if the load parameter μ is sufficiently large to overcome the effect of the damping forces acting on the plate, and the overhang of the vibrations is possible only for a frequency that does not exceed $2\pi\Omega_m\mu_m/\Delta_m$.

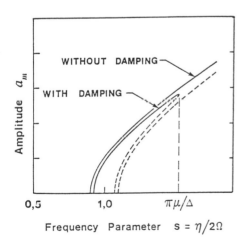

Fig. 3. Stationary parametric response curves.

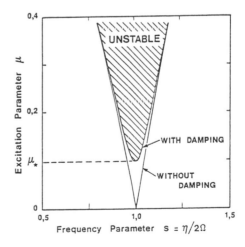

Fig. 4. Principal region of parametric instability.

The base of the resonance curves is the only region in which vibrations may normally initiate. By setting $a_m = 0$ in Eq. 20 and introducing the frequency parameter $s = \eta/2\Omega$, we obtain:

$$s^4 - 2s^3 + \left[1 + \Delta^2/(4\pi^2) \right] s^2 - \mu^2/4 = 0 \qquad (21)$$

This equation makes it possible to locate the boundaries of the principal region of parametric instability associated with any spatial form of vibration in the (μ, s) parameter space (Fig. 4).

4.2 Simultaneous Resonances

When an internal resonance coincides with a parametric resonance, the combination of the two types gives rise to simultaneous resonances. Of all possible internal resonances associated with a flat rectangular plate we will, for convenience, consider in our analysis only an internal resonance of the type $3\Omega_1 \approx \Omega_3$. Consequently, the two following cases of simultaneous resonances will be investigated:

$$\begin{aligned}
&(1): \ \eta \approx 2\Omega_1 \ \text{and} \ 3\Omega_1 \approx \Omega_3 \ ; \\
&(2): \ \eta \approx 2\Omega_3 \ \text{and} \ \Omega_3 \approx 3\Omega_1 \ .
\end{aligned}$$

Case (1): $\eta \approx 2\Omega_1$ and $3\Omega_1 \approx \Omega_3$. It is supposed that the principal parametric resonance $\eta \approx 2\Omega_1$ and the internal resonance $3\Omega_1 \approx \Omega_3$ occur simultaneously. Then, performing numerous transformations and manipulations of the asymptotic solutions, we arrive finally at a system of equations describing the stationary response of the discretized system as follows:

$$-C_1 a_1 + \frac{1}{\eta}\mu_1\Omega_1^2 a_1 \sin\psi + \frac{\Gamma_{12}}{4(\Omega_3 - \Omega_1)} a_1^2 a_3 \sin\psi' = 0; \qquad (22a)$$

$$\eta - 2\Omega_1 - \frac{3\Gamma_{11}}{4\Omega_1} a_1^2 - \frac{\Gamma_{14}}{2\Omega_1} a_3^2 + \frac{2}{\lambda}\mu_1\Omega_1^2 \cos\psi$$

$$- \frac{\Gamma_{12}}{2(\Omega_3 - \Omega_1)} a_1 a_3 \cos\psi' = 0; \qquad (22b)$$

$$- C_3 a_3 - \frac{\Gamma_{31}}{4(3\Omega_1 + \Omega_3)} a_1^3 \sin\psi' = 0; \qquad (22c)$$

$$3\Omega_1 - \Omega_3 + \left[\frac{9\Gamma_{11}}{8\Omega_1} - \frac{\Gamma_{32}}{4\Omega_3} \right] a_1^2 + \left[\frac{3\Gamma_{14}}{4\Omega_1} - \frac{3\Gamma_{36}}{8\Omega_3} \right] a_3^2 - \frac{3}{\lambda}\mu_1\Omega_1^2 \cos\psi$$

$$\qquad (22d)$$

$$+ \left[\frac{3\Gamma_{12}}{4(\Omega_3 - \Omega_1)} a_1 a_3 - \frac{\Gamma_{31}}{4(3\Omega_1 + \Omega_3)} \frac{a_1^3}{a_3} \right] \cos\psi' = 0;$$

where $\psi = \theta - 2\psi_1$ is the phase angle associated with the principal parametric resonance involving the first spatial form, and $\psi' = 3\psi_1 - \psi_3$ represents the phase angle corresponding to the specified internal resonance. The steady-state amplitudes, a_1 and a_3, and the phase angles, ψ and ψ', can be obtained by solving Eq. (22) by a numerical technique.

It follows from Eq. (22) that two possibilities exist : either a_1 is nonzero and a_3 is zero, or neither is zero. The first possibility indicates that the specified internal resonance has no effect on the system response and only the principal parametric resonance involving the first mode may occur. For the latter possibility, as the first mode is the only one excited by the parametric excitation, the presence of the third mode in the response is possible only by the transfer of energy from the first mode to the third mode through an internal mechanism.

Case (2): $\eta \simeq 2\Omega_3$ and $3\Omega_3 \simeq 3\Omega_1$.   As in the previous case, the internal resonance has the same relationship, but this time the third mode is excited parametrically. Using the same analysis as before and after a series of calculations, we obtain a set of four equations defining the stationary solutions [40].

As before, we have two possibilities : either a_1 is zero and a_3 is nonzero or both differ from zero. The first possibility means that only the principal parametric resonance involving the third mode shape may exist. The second possibility indicates the presence of the two spatial forms in the system response and, hence, a transfer of energy from the third mode to the first mode through internal resonance.

4.3 Combination Resonances

This kind of resonance is characterized by the fact that the system in question vibrates simultaneously in more than one normal mode and all resonant modes involved are directly excited by the parametric excitation.

For the combination resonance $\eta = \Omega_1 + \Omega_3$, stationary values for the specified combination response are given by [28]:

$$a_1 = \cfrac{2}{\sqrt{\left[\dfrac{3\Gamma_{11}}{2\Omega_1} + \dfrac{\Gamma_{14}}{\Omega_1} + \dfrac{\Gamma_{32}}{\Omega_3} + \dfrac{3\Gamma_{36}}{2\Omega_3}\left[\dfrac{\mu_3^1\Omega_1^3(\eta+\Omega_1-\Omega_3)}{\mu_1^3\Omega_3^3(\eta-\Omega_1+\Omega_3)}\right]\right]}}$$

$$\times \sqrt{\left[\eta-\Omega_1-\Omega_3\pm\left[\dfrac{\Omega_1+\Omega_3}{\sqrt{\Omega_1\Omega_3}}\right]\sqrt{\dfrac{\mu_1^3\mu_3^1\Omega_1^2\Omega_3^2}{(\eta+\Omega_1-\Omega_3)(\eta-\Omega_1+\Omega_3)}-\dfrac{\mu_1\mu_3\Delta^2}{4\pi^2}}\right]}, \quad (23)$$

$$a_3 = a_1 \sqrt{\frac{\mu_3^1 \Omega_1^3 (\eta + \Omega_1 - \Omega_3)}{\mu_1^3 \Omega_3^3 (\eta - \Omega_1 + \Omega_3)}} , \tag{24}$$

where Δ is the decrement of linear damping, as defined previously. The amplitude of vibration associated with the first mode is determined first by Eq. (23), and then the response amplitude associated with the third mode shape is evaluated by Eq. (24). As in the previous case, only positive real values for the amplitudes are admitted and the "$\pm$" sign indicates the possibility of two solutions associated with each spatial form; the larger of the two solutions is stable, while the lower solution is unstable and physically unrealizable.

From Eq. (24), it turns out that when the amplitude of vibration associated with the first mode is equal to zero, the corresponding steady state amplitude associated with the third mode is also null. Hence, the base width of the stationary response associated with the combination additive resonance is the only region in which vibrations may normally initiate. By setting $a_1 = 0$ in Eq. (23), one obtains:

$$\eta^4 - 2(\Omega_1 + \Omega_3)\eta^3 + \{4\Omega_1\Omega_3 + [(\Delta/2\pi)(\Omega_1 + \Omega_3)]^2\}\eta^2 + 2(\Omega_1 + \Omega_3)(\Omega_1 - \Omega_3)^2\eta$$
$$- (\Omega_1 + \Omega_3)^2(\Omega_1 - \Omega_3)^2 [1 + (\Delta/2\pi)^2] - \Omega_1\Omega_3(\Omega_1 + \Omega_3)^2\mu^2 = 0, \tag{25}$$

where $\mu^2 = \mu_1^3\mu_3^1$. Equation (25) makes it possible to locate again in the (μ, η) parameter space the boundaries of the instability zone associated with the specified combination resonance.

5. Nonstationary Responses

When the excitation parameters N_{y0}, N_{yt} or η, vary with time, we encounter the case of nonstationary response. It is expected that the most pronounced differences in the responses of stationary and nonstationary systems will occur near the resonances or near the resonance zones of these systems.

The nonstationary parametric response of a rectangular plate during a logarithmic sweep of the excitation frequency through a system resonance was studied by Ostiguy[23], Ostiguy and Evan-Iwanowski[38], and by Ostiguy and Lavigne[66] using five different techniques of solution. Two types of transition were considered:

1. Transition through the resonance zone, where the initial frequency is inside the resonance zone and the initial conditions correspond to some non-zero stationary values;

2. Complete passage through the resonance zone, where the initial frequency is outside the resonance zone and the initial conditions correspond to the trivial stable case (W = 0).

In the present nonstationary analysis, the sweep of the excitation frequency is taken to be logarithmic as given by:

$$\eta(t) = \eta_0 \, 2^{mt}$$

where η_0 is the initial frequency (at t = 0), m is the rate of sweep (in octaves per unit of time) and t is the time. The rate of sweep m may be positive or negative; in the former case, the sweep is in the direction of increasing frequencies.

The nonstationary response predicted by the perturbation techniques is obtained by numerically integrating the differential equations describing the rate of change of amplitudes and phases associated with the nonstationary motions involved in the analysis. It may be pointed out that numerical integration of first-order equations governing amplitudes and phases is a much simpler process than integration of the original second-order equations since we have to evaluate the envelopes of oscillatory functions and not the functions themselves.

6. Results and Discussion

In order to get more insight into various aspects of the problem and to highlight the influence of various system parameters on the stability characteristics and the response of rectangular plates, numerical evaluation of the solutions was performed for a wide variety of cases. Extensive experiments were also carried out in order to verify the theoretical predictions and to possibly discover new phenomena not predicted by theory. Results presented in Figs 5-26 are typical of those obtained.

The main parameters governing the stability behaviour and the response of the plate are:
   (a) The conditions of in-plane loading,
   (b) the amount of damping,
   (c) the particular set of boundary conditions,
   (d) the aspect ratio of the plate,
   (e) the type and the degree of initial imperfections present in the plate.

For appropriate values of these parameters, the system is capable of developing various types of resonances involving the lower spatial modes of vibration.

Figure 5 shows the principal regions of parametric instability associated with the lower spatial modes of vibration for a simply supported flat plate. The regions of incipient instability are located in the normalized $(\mu^*, \eta/2\Omega^*)$ plane for given values of P_{cr}, Δ and r. In the foregoing, $P_{cr} = N_{y0} / N^*$ is the ratio of critical loading and N^* is the lowest critical load. The boundary curve corresponding to each spatial mode encloses

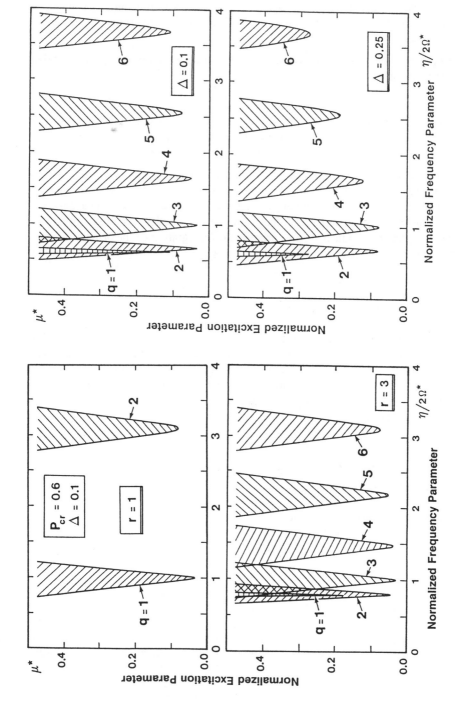

Fig. 6. Effect of varying linear damping on principal regions of parametric instability (S-S plate).

Fig. 5. Principal regions of parametric instability associated with the lower spatial modes for r = 1 and r = 3.

a region of instability. The various spatial modes of vibration are characterized by the number of half-waves q in the direction of compression. The prevalent buckling mode q_c in chosen as the reference mode in the normalization process [26].

The results shown in Fig. 5 indicate that the aspect ratio r plays a crucial role in determining the stability characteristics of rectangular plates. As can be seen, an increase of the aspect ratio does bring the principal instability regions associated with the lower mode shapes closer together; they can even overlap. Consequently, it can be concluded that an increase in r has a significant destabilizing effect on the system.

The relative importance of the principal instability regions associated with the lower mode shapes can be easily established when damping is taken into account. Reference to Fig. 5 shows that the most dangerous region of instability, from the standpoint of width and cutoff value of the load parameter μ^*, is the one associated with the prevalent buckling mode q_c which depends strongly on the aspect ratio of the plate.

Figure 6 shows the effect of varying linear damping on the principal regions of instability associated with the lower spatial modes of vibration. As can be seen, an increase in Δ has the beneficial effect of increasing the amount of withdrawal of the instability regions from the frequency ordinate $\eta/2\Omega^*$ and may preclude the possibility of resonance in certain mode shapes.

Typical stationary parametric response curves associated with the lower mode shapes are shown in Fig. 7. All of the stationary response curves exhibit a right-hand overhang which is typical of a "hard spring" effect, generally due to large deflections. The right-hand overhang is dependent upon the vibratory mode and for increasing q shows an increasing "hard spring" effect.

Numerous experiments were conducted to determine the boundaries of the principal regions of incipient instability and the large amplitude nonlinear parametric res-

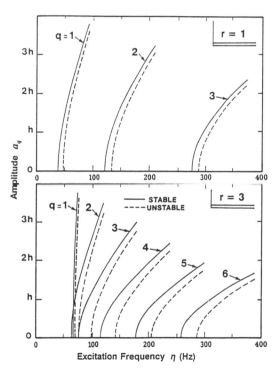

Fig. 7. Parametric response curves associated with the lower spatial modes for r = 1 and r = 3 (S-S).

ponses associated with various mode shapes. Typical results are presented in Figs 8 and 9. As can be seen, the experimental data shown in Fig. 8 exhibit close agreement with the theory. Particularly important is the excellent agreement with the width of the V-shaped instability regions. Reference to Fig. 9 shows that, in general, the experimental curves for the stationary amplitude exhibited good agreement with the corresponding theoretical predictions. It may be seen, however, that the amplitudes measured experimentally are slightly larger than those obtained theoretically at very high amplitudes of vibration. This is partly due to a small relaxation of the constraints at the boundaries and partly due to the fact that the first-order asymptotic approximation tends to exaggerate the effect of non-linearity.

Figures 10 and 11 illustrate the effects of varying the conditions of loading on the stability and parametric response of rectangular plates. As expected, the results indicate that an increase of the static preload N_{y0} or dynamic component N_{yt} of the in-plane force has a destabilizing effect on the system. This manifests itself in two typically contrasting fashions. Firstly, it is seen that an increase in N_{y0} lowers the natural frequencies of the system and shifts the instability zones corresponding to different mode shapes. This means that a variation in N_{y0} may render a stable plate unstable. Secondly, an increase in the static or dynamic com-

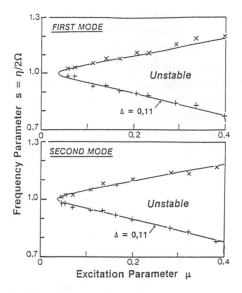

Fig. 8. Comparison of experimental incipient parametric instability regions with theory (S-S plate).

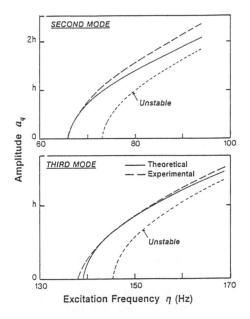

Fig. 9. Comparison of experimental amplitude response curves with theory (S-S plate).

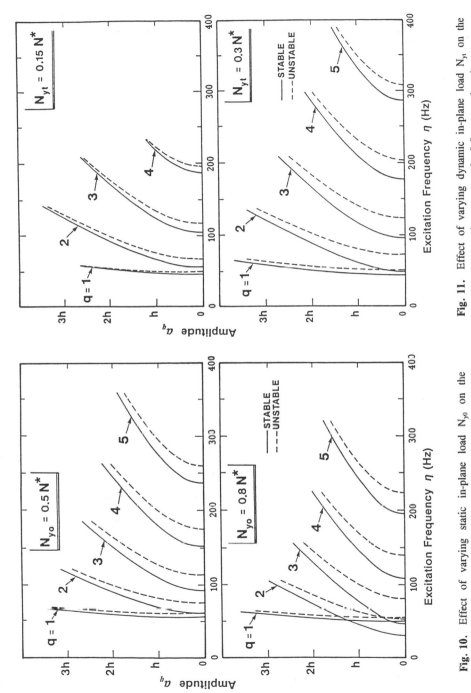

Fig. 11. Effect of varying dynamic in-plane load N_{yt} on the parametric response of a S-S rectangular plate.

Fig. 10. Effect of varying static in-plane load N_{yo} on the parametric response of a S-S rectangular plate.

ponent of the in-plane load widen the instability zones associated with the various mode shapes, augments the maximum value of the corresponding overhangs, and increases the number of possible resonances.

A thorough investigation conducted by Nguyen and Ostiguy[28] indicate that the boundary conditions also play a crucial role in determining the stability and resonances of rectangular plates. The results are summarized in Table 1. Figure 12 shows the regions of parametric instability and the stationary response curves associated with the lower spatial modes of vibration of a rectangular plate for the four different sets of boundary conditions considered in the analysis. An examination of Fig. 12 indicates that the boundary conditions have a strong influence on the relative location of the instability regions, the resonance frequencies of the plate and the degree of overhang of the amplitude response curves.

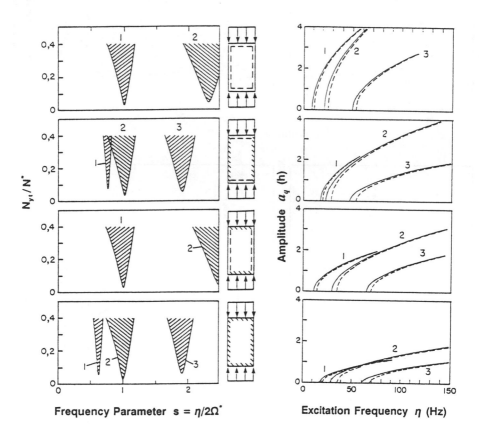

Fig. 12.   Effect of boundary conditions on the dynamic stability and nonlinear parametric response of a rectangular plate (r = 1,25).

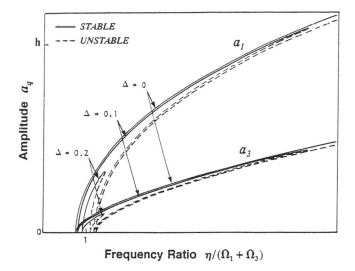

Fig. 13. Stationary response curves associated with the combination additive resonance $\eta = \Omega_1 + \Omega_3$.

Figure 13 shows typical stationary amplitude-frequency response curves associated with the particular case of combination additive resonance analyzed above. All of the response curves exhibit a right-hand overhang which is typical of a hard spring effect. The results also indicate the domination of the lower mode over the stationary response when two mode shapes are involved in a combination additive resonance. The beneficial effect of linear viscous damping is demonstrated by the fact that an increase in damping narrows the instability zone and reduces the maximum values of the overhangs.

The regions of incipient instability associated with various principal parametric and combination resonances of a given plate configuration are illustrated in Fig. 14. The cross-hatched areas represent regions of parametric instability and are characterized by the number of half-waves q in the direction of compression. On the other hand, the point-shaded area represents the region of instability associated with the combination resonance $\eta = \Omega_1 + \Omega_3$, and is identified by the combined number $1 + 3$.

A perusal of Fig. 14 prompts the following observations. The most dangerous region of instability, from the standpoint of length and width of those regions, is the principal region of parametric instability associated with the prevalent buckling mode. Furthermore, the instability zone corresponding to the specified combination resonance is the shortest and the narrowest of all instability regions considered in this particular case. As a consequence, it can be said that a principal parametric resonance is generally more important than a combination resonance.

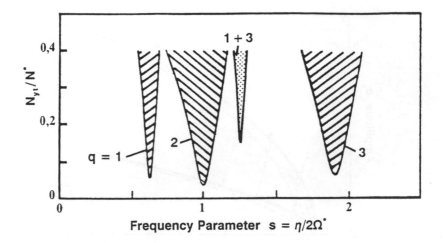

Fig. 14. Regions of instability associated with various principal parametric and combination resonances for $r = 1,25$ (C-C plate).

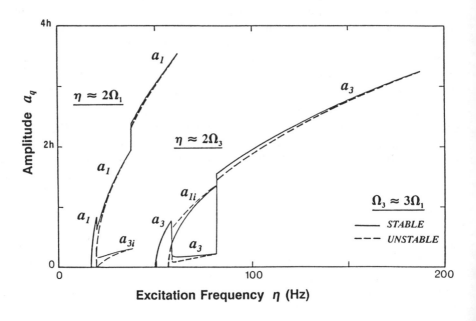

Fig. 15. Effect of internal resonance $\Omega_3 \approx 3\Omega_1$ on the response curves of a S-S rectangular plate.

Typical results associated with the two cases of simultaneous resonances analyzed above are shown in Fig. 15. Nonlinear modal interaction between an internal resonance and a principal parametric resonance on the frequency response curves is also illustrated in this figure. As can be seen, the parametric resonance of the flat plate in the first mode occurs when $\eta \approx 2\Omega_1$. At a certain frequency, however, a small part of energy from the first mode is transferred to the third mode, due to modal coupling between the two modes. Consequently, the amplitude of the first mode decreases slightly but remains larger than the one for the third mode. After a certain range, the energy transfer stops, the amplitude of the third mode disappears and the amplitude of the first mode regains its full strength.

In contrast with the previous case, the response of the system at $\eta \approx 2\Omega_3$ is particularly interesting. We can observe that when the excitation frequency reaches the point where the first mode can be excited through internal resonance, the amplitude of the third mode, which is directly excited by the parametric excitation, drops drastically and becomes less than the amplitude of the first mode which is due to internal resonance. This implies that there is a strong modal interaction and a significant transfer of energy from the third mode to the first one.

All of the results presented above are related to the parametric instability of *perfectly flat* plates. It is well known that small initial geometric imperfections are almost unavoidable in practice and may affect significantly their pre- and post-critical behaviour. Ostiguy[23] observed that the presence of small initial geometric imperfections could modify considerably the dynamic behaviour of parametrically-excited rectangular plates and produce various types of resonances hardly predictable beforehand. Further experiments on the subject were performed by Nguyen[24] and Samson[67] and some of the results were published by Ostiguy et al.[39,40]. The first theoretical works considering the possibility of coexistence of forced and parametric or combination resonances and possible modal interaction are seemingly due to Ostiguy and Sassi[56], Sassi and Ostiguy[57-61].

When the plate is stable, any imperfection, however slight, induces some small lateral vibrations of frequency η. However, when the excitation frequency η is in the vicinity of a resonance frequency Ω_1, the amplitude of the forced vibrations becomes relatively large, as illustrated in Fig. 16. One may note the characteristic gradual entrance into the forced resonance region, whereas entrance into a parametric unstable region is abrupt.

When forced and parametric or combination resonance frequencies are well separated from each other on the frequency scale, the interaction between them is found to be very weak. However, if two or more different types of resonances may occur independently for the same excitation frequency, and interaction between them may be expected. Particularly, when a forced resonance region overlaps a principal region of parametric or combination resonance, the interaction between them manifests itself in different ways,

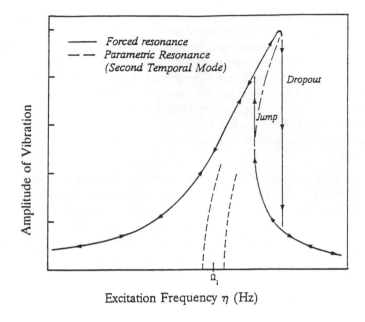

Fig. 16.   Sketch of typical experimental dynamic (forced) response.

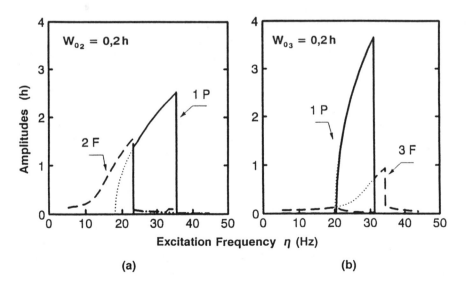

Fig. 17.   Modal interaction between forced and parametric resonances.

depending on the loading conditions, on the relative positions of the resonance frequencies and on the degree of overlapping of their corresponding resonance zones.

When a forced resonance precede a parametric resonance on the frequency scale, parametric vibrations do not appear until the excitation frequency has reached a value large enough to cause a natural drop of the forced oscillations. This is followed by a sudden jump to the upper branch of the stationary parametric response curve, and the plate begins to vibrate laterally at half the excitation frequency, as illustrated in Fig. 17(a). When the forced resonance frequency is higher than the parametric resonance frequency, results indicate that parametric vibrations are generally dominating, reducing the forced oscillations to small amplitudes (Fig. 17b). When the excitation frequency reaches the end of the parametric instability zone, parametric vibrations cease to exist, the forced oscillations regain their normal level and the plate vibration frequency switches from half to one time the excitation frequency.

When a forced resonance completely overlaps a principal region of parametric resonance, the interaction of the two mechanisms is quite evident and manifests itself in two different ways (Fig. 18). Firstly, we observe again the penetration and jump phenomena described previously. Secondly, both parametric and forced vibrations can coexist, and various types of simultaneous resonances can also occur, under suitable conditions.

Figures 19 and 20 illustrate different types of simultaneous or combination resonances observed experimentally. As can be seen, the various possibilities of multiple resonances exhibited by a real system are very difficult to predict beforehand and differ from those generally assumed in analytical

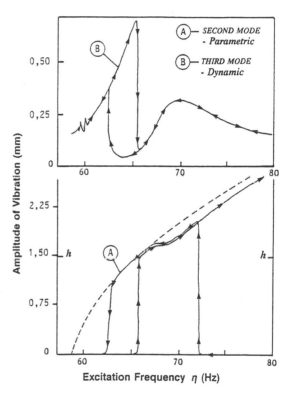

Fig. 18. Modal interaction between forced and parametric resonances.

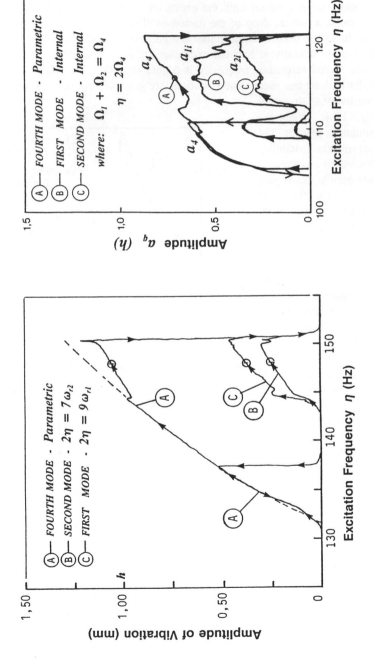

Fig. 20. Experimental response curves associated with a particular case of simultaneous resonances.

Fig. 19. Experimental response curves associated with a particular case of combination resonance.

studies. In general, the experimental results tend to indicate that if different types of resonances may occur independently for approximately the same excitation frequency η, a combination of these resonances is highly probable.

The nonstationary response of the plate during a logarithmic sweep of the excitation frequency $\eta(t)$ through a principal parametric resonance was studied for a wide variety of cases and the results shown in Figs 21-26 are typical of those obtained. The main parameters responsible for the modification of the parametric responses in nonstationary regimes are:

(a) the conditions of in-plane loading,
(b) the amount of damping,
(c) the initial conditions,
(d) the rate as well as the direction of the sweep.

For transitions starting within the resonance zone, the initial conditions selected for the nonstationary processes were those of the stationary stable branches. The results shown in Fig. 21(a) indicate that when a small amount of damping is present, the nonstationary response overshoots and oscillates about the stationary stable branch,

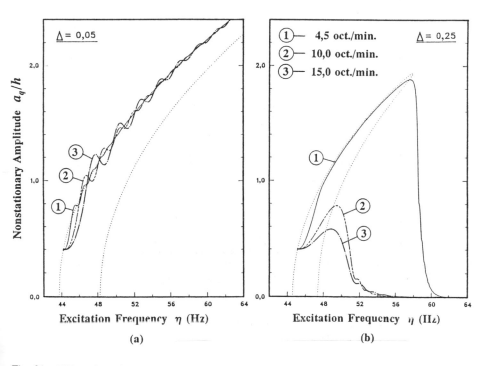

Fig. 21. Effect of varying linear damping and sweep rate on the nonstationary parametric response of a S-S rectangular plate.

eventually becoming asymptotic to the stationary curve as the transition continue. If damping is sufficiently large (Fig. 21b), the amplitude oscillation phenomenon is almost completely eliminated. The stabilizing influence of damping reduces the sweep rate required to drop out of the stationary curve and return to the trivial solution.

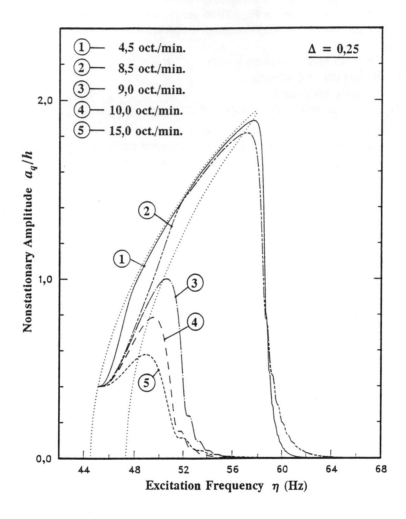

Fig. 22.   Effect of varying sweep rate on the nonstationary parametric response of a S-S rectangular plate.

It is also obvious from the results shown in Figs 21(b) and 22 that the rate of sweep of the excitation frequency η in the direction of the overhang may play a significant role in the modification of the nonstationary response. When the frequency sweep is relatively slow, the nonstationary response of a damped system follows closely the stable branch of the stationary response. As the sweep rate is increased, the lag in amplitude build-up at the start increases but the response eventually catches up with the stationary curve. For rapid increasing transitions, the nonstationary response increases somewhat inside the resonance zone but returns rapidly to the zero-amplitude stable position afterwards. When the sweep is in the opposite direction to the overhang, the effect of the sweep rate is much less pronounced.

Results presented in Figs 23 and 24 indicate that an increase in the initial amplitude of vibration or in-plane loading has a significant destabilizing influence on the system

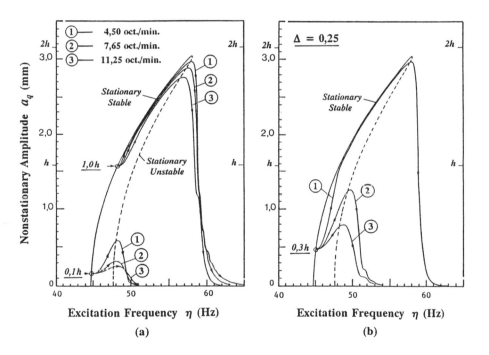

Fig. 23. Effect of varying initial (stationary) amplitude on the nonstationary parametric response of a S-S rectangular plate.

when the sweep is in the direction of the overhang. It was observed, however, that the value of the initial amplitude of vibration or in-plane loading has a negligible effect on the nonstationary response when the sweep is in the opposite direction.

The nonstationary motion during a complete passage through a resonance zone was investigated analytically using different techniques. Starting with trivial (zero) initial conditions to the left or to the right of the incipient instability boundaries and sweeping the excitation frequency either up of down, we found in all cases that the response stays on the trivial stationary solution even inside the boundaries of the instability region where the trivial solution is unstable. In order to get a non trivial response within the resonance zone, we have to perturb the system slightly by starting the integration process with non-zero but small initial conditions. These initial conditions represent small disturbances that are usually encountered in any real physical system.

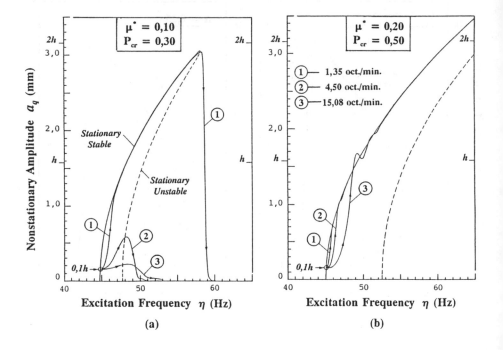

Fig. 24. Effect of varying loading conditions on the nonstationary parametric response of a S-S rectangular plate.

The character of the nonstationary response during a transition of the excitation frequency through a resonance zone was also investigated experimentally, and is illustrated in Figs 25 and 26. For an increasing transition into the resonance zone (Fig. 25), the plate does not respond until the excitation frequency has penetrated the instability zone for some finite distance, thus giving rise to a rapid jump to the stable branch. The penetration effect increases with increasing sweep rates; for rapid transitions, the nonstationary response increases somewhat inside the resonance zone but returns to the trivial solution afterwards. For sufficiently large sweep speeds, the jump phenomenon is not manifest and no parametric vibrations are apparent at all. Similar behaviour is also observed for decreasing transition (Fig. 26). This means that the sweep rate may have a pronounced effect on the lateral stability of the plate.

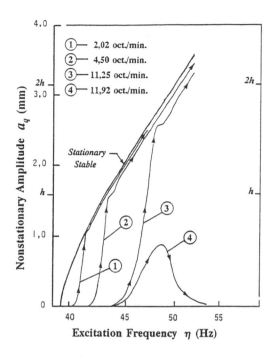

Fig. 25. Experimental nonstationary parametric response during a passage through a resonance zone (increasing frequency).

It is hoped that the results reported here will be useful towards a better understanding of the dynamic stability and nonlinear parametric vibrations of rectangular plates.

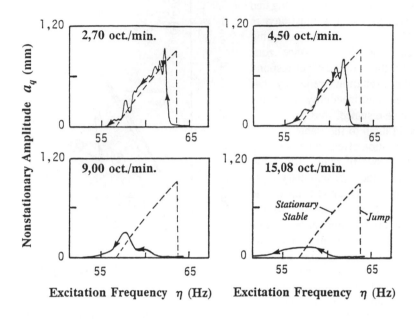

Fig. 26. Experimental nonstationary parametric response during a passage through a resonance zone (decreasing frequency).

Acknowledgments

I would like to express my gratitude to my Ph.D. supervisor and true mentor, Professor R.M. Evan-Iwanowski, for his inspiration and guidance, and acknowledge the contributions of my graduates students, H. Nguyen, L.P. Samson, S. Sassi and P. Lavigne. The long standing support of the National Sciences and Engineering Research Council of Canada (grant OGP0004207) is recognized with gratitude.

References

1. A.S. Vol'mir, *Stability of Elastic Systems* (Fizmatgiz, in Russian, 1963).
2. V.V. Bolotin, *The Dynamic Stability of Elastic Systems* (Holden-Day Inc., San Francisco, 1964).
3. G. Herrmann, *Dynamic Stability of Structures* (Pergamon Press, New York, 1967).
4. R.M. Evan-Iwanowski, *Applied Mechanics Reviews* **18** (9), (1965), pp. 699-702.
5. E. Mettler, in *Dynamic Stability of Structures* (ed. G. Herrmann; Pergamon Press, New York, 1967), pp. 169-188.
6. R.A. Ibrahim and A.D.S. Barr, *Shock and Vibration Digest* **10** (1), (1978), pp. 15-29.
7. R.A. Ibrahim and A.D.S. Barr, *Shock and Vibration Digest* **10** (2), (1978), pp. 9-24.
8. R.A. Ibrahim, *Shock and Vibration Digest* **10** (3), (1978), pp. 41-57.
9. R.A. Ibrahim, *Shock and Vibration Digest* **10** (4), (1978), pp. 19-47.
10. R.A. Ibrahim and J.W. Roberts, *Shock and Vibration Digest* **10** (5), (1978), pp. 17-38.
11. G.L. Ostiguy and H. Nguyen, in *Design and Analysis of Plates and Shells*, PVP-**105** (ASME Book G00356), (eds G.E.O. Widera, H. Chung and D. Hui; ASME, New York, 1986), pp. 127-135.
12. G.L. Ostiguy and R.M. Evan-Iwanowski, in *Dynamics and Vibration of Time-Varying Systems and Structures*, DE-**56** (ASME Book G00817), (eds S.C. Sinha and R.M. Evan-Iwanowski; ASME, Ney York, 1993), pp. 465-474.
13. R.M. Evan-Iwanowski, *Resonance Oscillations in Mechanical Systems* (New York: Elsevier, 1976).
14. R.A. Ibrahim, *Parametric Random Vibrations* (New York: Wiley-Interscience, 1985).
15. A.H. Nayfeh and D.T. Mook, *Nonlinear Oscillations* (New York: Wiley-Interscience, 1979).
16. Yu. A. Mitropolsky, *Problems of the Asymptotic Theory of Nonstationary Vibrations* (New York: D. Davey & Co, 1965).
17. R. Einaudi, *Acta Societatis Gioeniae Catinensis Naturalium Scientiarum, Serie 6*, **1** and **11** (1935-1936). Nota Prima: pp. 1-5; Nota Seconda: pp. 1-20.
18. V.A. Bodner, *Prikl. Mat. i Mekhan.* (Applied Mathematics and Mechanics) **2** (1), (in Russian, 1938), pp. 87-104.
19. V. Chelomei, "The Dynamic Stability of Plates", *Trudy Kiev. Aviats.* **8** (in Russian, 1938).
20. V.V. Bolotin, *Izvestija Akademii Nauk SSSR, Otdelenie Techniceskaja Nauk* **10** (in Russian, 1954), pp. 47-59.
21. J.H. Somerset and R.M. Evan-Iwanowski, *Int. J. Non-Linear Mechanics* **2** (3), (1967), pp. 217-232.
22. R.C. Duffield, *An Investigation of Parametric Stability of Stiffened Rectangular Plates*, (Ph.D. Dissertation, University of Kansas, 1968).
23. G.L. Ostiguy, *Effects of Aspect Ratio on Parametric Response of Nonlinear Rectangular Plates - Analysis and Experiment*, (Ph.D. Thesis, Syracuse University, Syracuse, NY, 1976).
24. H. Nguyen, *Effect of Boundary Conditions on the Dynamic Instability and Responses of Rectangular Plates*, (Ph.D. Thesis, Ecole Polytechnique de Montréal, Québec, 1987).

25. S. Sassi, *Stabilité dynamique et résonances des plaques rectangulaires imparfaites* (Ph. D. Dissertation, École Polytechnique de Montréal, Québec, 1993).

26. G.L. Ostiguy and R.M. Evan-Iwanowski, *ASME J. of Mechanical Design* **104** (1982), pp. 417-425.

27. G.L. Ostiguy and H. Nguyen, *Proc. Third Int. Conf. on Recent Advances in Structural Dynamics* **II** (eds M. Petyt, H.F. Wolfe and C. Mei; ISVR, University of Southampton, England, 1988), pp. 819-829.

28. H. Nguyen and G.L. Ostiguy, *J. Sound and Vibration* **133** (3), (1989), Part I : pp. 381-400; Part II : pp. 401-422.

29. A.A. Berezovskii and Yu. A. Mitropolsky, *Selected Problems of Applied Mechanics* (Akademii Nauk SSSR, Moscow; in Russian, 1974), pp. 119-131.

30. N. Yamaki and K. Nagai, *Report of the Institute of High Speed Mechanics* **32** (288), (1975), pp. 103-127.

31. S.D. Kisliakov, *Int. J. Non-Linear Mechanics* **11** (4), (1976), pp. 219-227.

32. K. Takahashi, M. Tagawa, and T. Ikeda, *Theoretical and Applied Mechanics* **33** (Proc. of the 33rd Japan National Congress for Applied Mechanics, 1983), pp. 311-318.

33. C. Pierre and E.H. Dowell, *ASME J. Applied Mechanics* **52** (4), (1985), pp. 693-697.

34. K. Takahashi and Y. Konishi, *J. Sound and Vibration* **123** (1), (1988), pp. 115-127.

35. J.H. Somerset and R.M. Evan-Iwanowski, in *Developments in Theoretical and Applied Mechanics* **3** (ed. W.A. Shaw; Pergamon Press, Oxford, 1967), pp. 331-355.

36. P.R. Dixon and J.R. Wright, *Symposium on Non-Linear Dynamics* (eds D.J. Johns, P.A.T. Christopher and A. Simpson; Loughborough Univ. of Technology, 1972),Paper D.2.

37. G.L. Ostiguy and R.M. Evan-Iwanowski, *Mécanique Matériaux Electricité* **394-395** (G.A.M.I., 1982), pp. 472-478.

38. G.L. Ostiguy and R.M. Evan-Iwanowski, *Proc. Second Int. Conf. on Recent Advances in Structural Dynamics* **II** (eds M. Petyt and H.F. Wolfe; ISVR, Univ. of Southampton, England, 1984), Part 1 : pp. 535-546; Part 2 : pp. 547-558.

39. G.L. Ostiguy, L.P. Samson and H. Nguyen, in *Dynamics of Plates and Shells*, PVP-**178** (ASME Book H00493), (eds H. Chung, G. Yamada and Y. Narita; ASME, New York, 1989), pp. 23-31.

40. G.L. Ostiguy, L.P. Samson and H. Nguyen, *ASME J. Vibration and Acoustics* **115** (1993), pp.344-352.

41. S.A. Ambartsumyan and A.A. Khachaturian, *Izvestija Akademii Nauk SSSR, Otdelenie Techniceskaja Nauk, Mekhanika i Mashinostroenie* **1** (in Russian, 1960), pp. 113-122.

42. S.A. Ambartsumyan, and V.T. Gnuni, *J. Appl. Math. & Mech.* **25** (4), (1961), pp. 1102-1108.

43. M.P. Feldman, *Theory of Plates and Shells* (Kiev; in Russian, 1962), pp. 244-248.

44. G. Schmidt, *Proc. of Vibration Problems* **2** (6), (Warsaw, 1965), pp. 209-228.

45. R.C. Duffield and N. Willems, *ASME J. Applied Mechanics* **39** (1), (1972), pp. 217-226.

46. Y.Y. Yu and J.-L. Lai, *ASME J. Applied Mechanics* **33** (4) Ser. E, (1966), pp. 934-936 .

47. V. Birman, *Mechanics Research Communications* **12** (1985), pp. 81-86.

48. R.S. Srinivasan and P. Chellapandi, *Computers and Structures* **24** (1986), pp. 233-238.

49. V. Birman and C.W. Bert, *Dynamics and Stability of Systems* **3** (1 & 2), (1988), pp. 57-68.

50. L. Librescu and S. Thangjitham, *Int. J. Non-Linear Mechanics* **25** (2/3), (1990), pp. 263-273.

51. D. Hui and A.W. Leissa, *ASME J. Applied Mechanics* **50** (1983), pp. 751-756.

52. S. Ilanko and S.M. Dickinson, *J. Sound and Vibration* **118** (1987), pp. 313-336.

53. J.H. Somerset, *ASME J. Engineering for Industry* **89** (4) Ser. B, (1967), pp. 619-625.

54. R.L. Silver, *The Effect of Initial Curvature on the Parametric Vibration of Rectangular Plates Subjected to an In-Plane Sinusoidal Load*, (Ph.D. Thesis, Syracuse University, Syracuse, N.Y., 1972).

55. S.D. Kisliakov, *Bulgarian Academy, Theoretical and Applied Mechanics*, **Year VII** (4), (1976), pp. 40-50.

56. G.L. Ostiguy and S. Sassi, *Nonlinear Dynamics* **3** (1992), pp. 165-181.

57. S. Sassi and G.L. Ostiguy, "On the Interaction Between Forced and Combination Resonances", *J. Sound and Vibration* (to appear).

58. S. Sassi and G.L. Ostiguy, "Effects of Initial Geometric Imperfections on the Interaction Between Forced and Parametric Vibrations", *J. Sound and Vibration* (to appear).

59. S. Sassi and G.L. Ostiguy, "Analysis of the Variation of Frequencies for Imperfect Rectangular Plates", *J. Sound and Vibration* (to appear).

60. S. Sassi and G.L. Ostiguy, in *Structural Dynamics: Recent Advances* I (eds N.S. Ferguson, H.F. Wolfe and C. Mei; ISVR, Univ. of Southampton, England, 1994), pp. 340-351.

61. S. Sassi and G.L. Ostiguy, "Nonlinear Modal Interaction in Parametrically-Excited Imperfect Rectangular Plates", *Nonlinear Dynamics* (to appear).

62. L. Librescu and M.-Y. Chang, *Acta Mechanica* **96** (1993), pp. 203-224.

63. H.N Chu and G. Herrmann, *ASME J. Applied Mechanics* **23** (1956), pp.532-540.

64. M.K. Prabhakara and C.Y. Chia, *J. Sound and Vibration* **52** (4), (1977), pp. 511-518.

65. B.N. Agrawal and R.M. Evan-Iwanowski, *J. American Institute of Aeronautics and Astronautics* **1** (7), (1973), pp. 907-912.

66. G.L. Ostiguy and P. Lavigne, in *Dynamics and Vibration of Time-Varying Systems and Structures*, DE-**56** (ASME Book G00817) (eds S.C. Sinha and R.M. Evan-Iwanowski; ASME, Ney York, 1993), pp. 71-79.

67. L.P. Samson, *Étude expérimentale de la stabilité dynamique des plaques rectangulaires excitées paramétriquement*, (M.Sc.A. Thesis, École Polytechnique de Montréal, 1987).

Dynamics with Friction: Modeling, Analysis and Experiment, Part II, pp. 227–252
edited by A. Guran, F. Pfeiffer and K. Popp
Series on Stability, Vibration and Control of Systems, Series B, Vol. 7
© World Scientific Publishing Company

FRICTION MODELLING AND DYNAMIC COMPUTATION

J.P. MEIJAARD

*Laboratory for Engineering Mechanics, Delft University of.
Technology, Mekelweg 2, NL-2628 CD Delft, The
Netherlands*

ABSTRACT

Some phenomological friction models are discussed and their
consequences for the computational problem and instability phenomena
are discussed. A computational scheme for mechanical systems affected
by friction is presented, and two example systems, an arch under
horizontal base excitation and a four-bar linkage under gravity loading,
are discussed.

1   Introduction

Many kinds of mechanical machinery are affected by friction. This friction
may be beneficial if it contributes to the proper working of the machine or
machine part. In brakes and clutches, the friction force makes possible the
transmission of couples and forces, the friction between an automobile tyre
and the road and between a wheel of a track-guided vehicle and the rail are
essential for the acceleration and deceleration, for surmounting slopes and
the curving behaviour. In these instances, and especially for the design of
anti-skidding systems, one needs to have a good estimate of the value of the
friction and the possibility to control it within a specific range. In many
other cases, however, friction is an unwanted phenomenon with detrimental
consequences. The most important adverse effects are the loss of useful energy, wear of surfaces, loss of accuracy of mechanisms such as used in robotic

manipulators and machine-tools, and loss of stability, the most conspicuous being the stick-slip phenomenon. One way to overcome these problems is trying to reduce the friction, for instance by lubrication or by changing the design by a selection of different materials and using rolling elements instead of sliding elements [Kuntz 1995]. Another approach is to try to compensate for the friction by some control, where it is essential that the friction can be predicted, that it is reproducible.

The study of friction phenomena and their attack has resulted in a branch of science that has relatively recently been named *tribology*. It has, however, already a long history, which has been compiled in [Dowson 1979], which book has to be consulted for an overview of the subject and references to the original sources. The phenomenological modelling started with Amontons in 1699 and Coulomb in 1785, who formulated their laws for contact friction. They state that the sliding friction force is proportional to the normal force at the sliding surfaces, independent of the sliding speed and the apparent area of contact. These laws are still widely used in modelling and calculating problems with friction, not because these laws are so accurate, but more because friction depends on a range of mostly unknown circumstances and we have nothing better at our disposal in most cases. In the course of this century, a more scientific approach to the friction phenomena was initiated, with an interplay between mechanical, physical, chemical, and metallurgical theory and advanced experimental techniques. It appeared that friction depends on a wide class of circumstances which are not always known or can be controlled, and has some stochastic character. Some important textbooks on the subject are [Bowden, Tabor 1954, 1964], [Rabinowicz 1965], [Kragelskii 1965]. A good review article is [Tabor, 1981].

Contacts between solids in machines are divided in conformal contacts, where the apparent contact takes place over an extended surface, and counterformal contacts, which can be divided in line contacts with one extended dimension, and point contacts. Mechanical joints with conformal contacts, such as occur in a revolute joint (hinge), a prismatic joint (slider), or a spherical ball-socket joint, are called lower pairs, while joints with line or point contact, such as a pivot bearing, a pair of gears, or cam and follower, are called higher pairs [Reuleaux 1875]. Furthermore, the kind of lubrication between the contacts gives the distinction between dry friction, that is, pure contact friction, boundary lubrication, in which the lubricant is attached to the surfaces of the solids in a thin layer, full film lubrication, in which the solid surfaces are separated by a layer of fluid, and an intermediate between boundary lubrication and full film lubrication, which is called mixed lubrication. The material of the solids can be classified in metals, non-metallic minerals, polymers and rubbers, and composite materials such as fibre reinforced polymers, wood and woven fabrics. All combinations of these classifications, amplified by the multitude of possible material combinations, can have their own peculiarities, so it seems that the hope for one simple friction law is idle.

The purpose of this chapter is to show how different phenomenological models that have been proposed for friction can be incorporated into a computational scheme, without discussing the relative intrinsic merits of the models. Bearing in mind the quite large factor of arbitrariness in the friction models, we pay attention to desirable properties of models from a computational point of view. The study will be restricted to phenomenological models, applied to systems consisting of rigid bodies, with conformal contacts.

2 Phenomenological models

When two solids approach each other and come close, they will mutually exert short-range forces, which can be modelled by surface tractions on their idealized bounding surfaces. A phenomenological model gives a prescription of these tractions **t** based on known macroscopic, readily observable, quantities, the state variables, and their history if there is a memory effect,

$$\mathbf{t} = \mathbf{t}(\text{state variables, history}). \tag{1}$$

The first step in constructing a model is the identification of the relevant state variables. Variables coming into question are the purely mechanical properties such as relative positions and velocities of the surfaces, the deformation and the rate of deformation of the solids at the surface, and other properties, such as the chemical constitution of the surface layers and the temperature.

Some models for the contact between elastic bodies were discussed by Oden and Martins [Oden, Martins 1985], [Martins *et al.* 1990]. Based on a review of the relevant literature, they presented a fairly simple model without memory effects and showed how it could be implemented in finite element models and calculations. If it is assumed that the solids are rigid, their deformations and deformation rates cannot be taken as state variables. Although the stiffness of the interface is generally smaller than the stiffness of the contacting bodies, it is consistent with the rigid-body modelling to assume that the interface between the solid has no stiffness and no damping properties in the normal direction, so the contact is modelled as a one-sided constraint, without tangential stiffness. Pure sliding of the bodies over each other is possible. If we introduce these assumptions of rigidity, the contact forces appear in the constitutive relation at the interface, which are not state variables in a strict sense. Phenomenological friction models for contacting rigid bodies tend to be more complicated than corresponding models for deformable bodies, because they have to incorporate somehow the effects of the deformations and small vibrations at the surfaces.

2.1   Models without memory effects

The simplest form of a law of friction is the classical Amontons-Coulomb law in which the static and kinetic coefficients of friction are the same. For a two-dimensional contact with normal force $F_n \leq 0$ (positive if it is a traction force and negative if it is a compressive force) and relative sliding velocity v, this law reads

$$\begin{cases} F_f = -\mu F_n & (v > 0) \\ \mu F_n \leq F_f \leq -\mu F_n & (v = 0) \\ F_f = \mu F_n & (v < 0) \end{cases} , \qquad (2)$$

where F_f is the friction force, positive in the direction of negative velocities. It was noted by Painlevé [Painlevé 1895] that this law may lead to contradictory results if it is combined with the assumption of rigidity of bodies. Several remedies were proposed to overcome these difficulties. Hamel [Hamel 1949] draw attention to the infeasibility of initial conditions which lead to the paradoxes, that is, these initial conditions cannot be obtained in the course of some motion. Another remedy is the introduction of impulsive forces, which cause the immediate reduction to zero of some sliding velocities, a process called jamming or jambing.

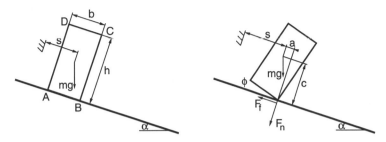

Figure 1: Block sliding on an inclined plane. Sliding without tilting (left), and sliding with tilting (right).

To make these paradoxes and their resolution clearer, we consider an example, similar to one discussed by Painlevé. The system consists of a block ABCD with breadth b, height h and mass m, sliding on an inclined plain under an angle α with the horizontal, under the influence of gravity, see Figure 1 (left). If we assume that the block slides downwards without tilting around B, its equation of motion is $m\ddot{s} = mg(\sin \alpha - \mu \cos \alpha)$, where s is the distance travelled of the mass centre of the block along the slope. This equation is valid as long as $\mu \leq b/h$. For tilting around the lower base edge B of the block, where the tilting angle is denoted by ϕ, while the edge B still slides along the slope, we have the equations of motion

$$
\begin{aligned}
m\ddot{s} &= mg \sin \alpha - F_f, \\
\tfrac{1}{12}(b^2 + h^2)m\ddot{\phi} &= aF_n + cF_f, \\
m[a\ddot{\phi} - c\dot{\phi}^2] &= -F_n - mg \cos \alpha, \\
F_n &\leq 0, \quad F_f = -\mu F_n.
\end{aligned}
\tag{3}
$$

Here, $a = \frac{b}{2} \cos \phi - \frac{h}{2} \sin \phi$ and $c = \frac{b}{2} \sin \phi + \frac{h}{2} \cos \phi$ are the two distances as given in Figure 1 (right). These equations have a meaningful solution if

$\mu < \mu_{cr} = [b^2 + h^2 + 12a^2]/(12ac)$. If $\mu \geq \mu_{cr}$, no solution that is compatible with the initial conditions and the law of friction is possible. To see what happens, we imagine the situation that the coefficient of friction depends on the distance travelled along the slope, and that it increases monotonically, starting from a zero value. We let the block start from zero velocity. Initially, the block will have a purely translational accelerated motion. If we assume that the parameters α, b, h, and the dependence of μ on s are such that the block will not come to permanent rest (*e.g.* h is sufficiently large), pure translation continues until $\mu = b/h$, at which point tilting starts. Near the point at which μ approaches μ_{cr}, the values of the rotational accelerations reach large values until edge B stops slipping and the block purely rotates around the edge B. This point of view supports the resolution of Hamel of disallowed initial conditions: the initial values which lead to the paradox cannot be obtained in a natural way in the course of the motion. Considering now the extreme case in which μ is zero over some distance and then suddenly reaches a large value, we see that the effect is that edge B immediately halts by an impulsive process, as if it hits an obstacle, for which we can write down the impulse equations as

$$
\begin{aligned}
m\Delta\dot{s} &= -S_f, \\
\tfrac{1}{12}(b^2 + h^2)m\Delta\dot{\phi} &= \tfrac{b}{2}S_n + \tfrac{h}{2}S_f, \\
m\tfrac{b}{2}\Delta\dot{\phi} &= -S_n, \\
\dot{s} + \Delta\dot{s} &= \tfrac{h}{2}\Delta\dot{\phi}, \\
S_n \leq 0, \quad \mu S_n &\leq S_f \leq -\mu S_n,
\end{aligned}
\tag{4}
$$

where Δ denotes an instantaneous increment and S_n and S_f are the normal and friction impulses. These equations have the solution

$$
\begin{aligned}
\Delta\dot{\phi} &= \tfrac{3}{2}\tfrac{h}{b^2+h^2}\dot{s}, \\
\Delta\dot{s} &= \tfrac{-4b^2-h^2}{4(b^2+h^2)}\dot{s}, \\
S_n &= \tfrac{-3bh}{4(b^2+h^2)}m\dot{s}, \\
S_f &= \tfrac{4b^2+h^2}{4(b^2+h^2)}m\dot{s}, \\
\left|\tfrac{S_f}{S_n}\right| &= \tfrac{4b^2+h^2}{3bh} \leq \mu.
\end{aligned}
\tag{5}
$$

These results can also be obtained by noting that the moment of momentum with respect to the edge B remains constant during the impact. These results support the view that in case that the initial conditions do occur, which is made possible in this instance by the fact that the friction coefficient changes discontinuously with the travelled distance, the relative motion comes to rest immediately by impulsive jamming.

An extension to the classical model of Amontons-Coulomb is to make the coefficient a function of the sliding velocity. It is assumed that the coefficient decreases rapidly with increasing speed, while for large speeds the coefficient may increase again. Curves of this form are called Stribeck curves after this man's investigations on journal bearings [Stribeck 1902]. The dependence on the sliding velocity appears to be continuous, but in practical cases, the decrease in friction with increasing speeds may be so steep that it is better to model the curve by a discontinuous fall at zero velocity, that is, to make a distinction between static and kinetic friction at vanishing speeds. This sudden fall has the theoretical and computational disadvantage that the solution of the equations of motion of a system affected by friction does not depend in a continuous way on parameters and initial values. We can show this by an example of the harmonically forced oscillations of a mass on which acts a friction force, see Figure 2. After some scaling, we can write the equation of

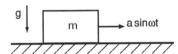

Figure 2: Harmonically forced mass affected by friction.

motion as

$$\ddot{x} = a \sin t \quad F_f(\dot{x}), \qquad (6)$$

where a is a parameter measuring the amplitude of the forcing, and F_f is the following function of the sliding velocity,

$$\begin{cases} F_f = \mu_k + (1 - \mu_k) \exp(\frac{-\dot{x}}{v_0}) & (\dot{x} > 0) \\ -1 \leq F_f \leq 1 & (\dot{x} = 0) \\ F_f = -\mu_k - (1 - \mu_k) \exp(\frac{\dot{x}}{v_0}) & (\dot{x} < 0) \end{cases}.$$

Here, it is assumed that $0 < \mu_k < 1$, while v_0 is a positive parameter. For the limiting case that v_0 approaches zero, we have a model with constant static and kinetic coefficients of friction which differ. We consider the initial conditions $x = 0$, $\dot{x} = 0$ at $t = 0$, and consider a range of the parameter a. As long as $a \leq 1$, the mass remains at rest for all times. For the continuous model, a periodic oscillation starts if a passes the value of one, whose amplitude is asymptotically given by $\hat{x} = 9(a-1)^2/4$, independent of the shape of the function of the friction force. This result can be derived by considering an expansion of the solution around $t = \pi/2$. On the other hand, for the discontinuous system, the amplitude reaches a finite value immediately. For $\mu_k = 0.8$ and the values $v_0 = 0.001, 0.01, 0.1, 1$, the amplitude as a function of a is shown in Figure 3.

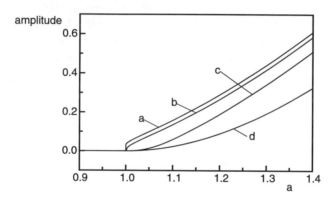

Figure 3: Amplitudes of the harmonically forced mass for the parameter values $\mu_k = 0.8$, $v_0 = 0.0001$ (a), $v_0 = 0.01$ (b), $v_0 = 0.1$ (c), and $v_0 = 1$ (d), for variable forcing amplitude a.

A cause for the difference between static and kinetic friction which has received little attention in the past is the statistical nature of the process of friction [Rabinowicz 1965]: the simple laws only represent some average of the friction. Now kinetic friction can be regarded as an average value of the friction, while static friction is determined by the instantaneous value of the friction. If we assume that the friction is some rapidly changing function of the relative positions of the contacting bodies, it is more likely that the sliding bodies will come to rest at positions with a high friction, and the static friction is higher on the average. Furthermore, if a tangential force increases slowly, some microslip might occur to a near-by position with a higher friction, and gross sliding starts at a higher value of the friction. This last phenomenon might be a partial explanation of the observed dependence of the static friction on the rate of application of the tangential force [Richardson, Nolle 1976].

A further extension of the law is that the coefficient of friction may depend on the normal load, which dependence is often weak for hard materials, but may be appreciable for soft material like rubbers and polymers, and also in the cases in which the coefficient of friction depends on the velocity.

2.2 Models with memory effects

More complicated friction models include memory effects. With the most general formulation, the number of state variables is infinite, for instance in the case of a pure time delay τ, in which the observable state variables over the complete time interval $[t - \tau, t]$ form the state at the time t. Because systems with an infinite number of state variables cannot be easily handled numerically, we have either to approximate these states by a finite number of state variables, or, better, to forgo the most general formulation and to recourse to models with a finite number of degrees of freedom. As an example of the first approach, we can approximate a pure time delay, which has been proposed in some friction models [Hess, Soom 1990], by the differential equation

$$(1 + \frac{\tau}{n} \frac{d}{dt})^n q_s = s, \tag{8}$$

with n auxiliary state variables $d^i q_s / dt^i$, $i = 0, \ldots, n - 1$. Here, q_s is the approximate delayed value, $q_s(t) \approx s(t - \tau)$. The second approach would directly use these state variables with a fixed value of n. The reason for this is that the pure time delay and the approximation (8) are experimentally indistinguishable for large n and slowly varying s.

A model in which the sliding velocity is the only state variable was used in [Rice, Ruina 1983] and [Gu *et al.* 1984]. Their finite-dimensional models have some draw-backs, such as the possibility that the friction force may have the wrong sign for high velocities. Another simple model, which has already six parameters, is

$$F_f = -\mu(1 + c_\theta \theta) F_n \mathrm{sgn}(v),$$
$$\dot{\mu} = (\mu_s - \mu)/\tau_s + (\mu_k - \mu)|v|/d_k, \qquad (9)$$
$$\dot{\theta} = v F_f - \beta \theta.$$

The coefficient of friction μ itself is determined by an evolution equation, while the variable $\theta \geq 0$ represents the temperature rise due to the dissipation of energy at the sliding surfaces. The parameter c_θ represents the dependence of the coefficient of friction on the temperature, μ_s is the ultimate static friction coefficient, τ_s is a time constant for the increase of friction at rest, μ_k is the kinetic coefficient of friction at high sliding velocities, d_k is a characteristic length for the onset of kinetic friction, and β is a cooling coefficient according to Newton's law.

2.3   Stability of stationary sliding

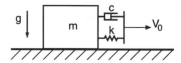

Figure 4: Block drawn over a surface at a constant velocity.

For the system shown in Figure 4, the stability of stationary sliding for some friction models will be investigated. The system consists of a block with mass m which is drawn over a surface at a constant speed V_0, where the drawing point and the block are connected by a linear spring with stiffness k and damper with damping coefficient c. The acceleration due to gravity is g. The time-independent solution is given by a constant speed, V_0, and a constant elongation of the spring corresponding to the constant friction force at this sliding velocity. If the influence of disturbances on this steady state is investigated, the equation of motion is linearized around the stationary motion, and also the equations describing the friction process are linearized around their stationary values.

For models without memory, the stability is determined by $dF_f/dv + c$: if this effective damping is positive, small disturbances are damped out, while if it is negative, small perturbations are amplified and stationary sliding is unstable. A necessary condition for instability is that the magnitude of the friction force decreases with increasing sliding speeds.

The more complicated friction model given by equation (9) has a stationary solution with

$$\mu_0 = \mu_s - \frac{(\mu_s - \mu_k)V_0/d_k}{1/\tau_s + V_0/d_k} = \mu_k + \frac{(\mu_s - \mu_k)/\tau_s}{1/\tau_s + V_0/d_k},$$
$$F_{f0} = \mu_0 mg/(1 - \mu_0 mgc_\theta V_0/\beta),$$
$$\theta_0 = V_0 F_{f0}/\beta, \tag{10}$$

where we assume that $\beta - \mu_0 mgc_\theta V_0 > 0$. The stability of this solution is determined by the linearized equations,

$$\begin{pmatrix} \dot{s} - V_0 \\ \dot{v} \\ \dot{\mu} \\ \dot{\theta} \end{pmatrix} = \mathbf{A} \begin{pmatrix} s - V_0 t + F_{f0}/k \\ v - V_0 \\ \mu - \mu_0 \\ \theta - \theta_0 \end{pmatrix}, \tag{11}$$

$$\mathbf{A} = \begin{pmatrix} 0 & 1 & 0 & 0 \\ -k/m & -c/m & -(1 + c_\theta\theta_0)g & -\mu_0 c_\theta g \\ 0 & (\mu_k - \mu_0)/d_k & -(1/\tau_s + V_0/d_k) & 0 \\ 0 & \mu_0 mg(1 + c_\theta\theta_0) & mg(1 + c_\theta\theta_0)V_0 & -\beta + \mu_0 mgc_\theta V_0 \end{pmatrix}.$$

For two special cases, the stability criterion will be derived. In the first case, the friction is independent of the temperature, and the corresponding characteristic equation is

$$\lambda^3 + \lambda^2[c/m + 1/\tau_s + V_0/d_k]$$
$$+ \lambda[(c/m)(1/\tau_s + V_0/d_k) - (1 + c_\theta\theta_0)g(\mu_0 - \mu_k)/d_k + k/m] \tag{12}$$
$$+ (k/m)(1/\tau_s + V_0/d_k) = 0.$$

The condition for stability according to the Routh-Hurwitz criterion [Hurwitz 1895] is

$$(c/m)^2(1/\tau_s + V_0/d_k)$$
$$+(c/m)[(1/\tau_s + V_0/d_k)^2 - (1 + c_\theta\theta_0)g(\mu_0 - \mu_k)/d_k + k/m] \qquad (13)$$
$$-(1/\tau_s + V_0/d_k)(1 + c_\theta\theta_0)g(\mu_0 - \mu_k)/d_k > 0.$$

From this we conclude that stationary sliding is unstable if the external damping is small. If, for a given combination of parameters, stationary sliding is unstable, stability can be regained by several means: one can increase the damping, or increase the stiffness, or increase the sliding velocity.

In the second case, we assume that the coefficient of friction is independent of the sliding velocity, $\mu = \mu_0 = \mu_k = \mu_s$. In this case the characteristic equation becomes

$$\lambda^3 + \lambda^2[c/m + \beta - \mu_0 mgc_\theta V_0]$$
$$+\lambda[(c/m)(\beta - \mu_0 mgc_\theta V_0) + \mu_0^2 g^2 mc_\theta(1 + c_\theta\theta_0) + k/m] \qquad (14)$$
$$+(k/m)(\beta - \mu_0 mgc_\theta V_0) = 0.$$

The condition for stability is now

$$(c/m)^2[\beta - \mu_0 mgc_\theta V_0]$$
$$+(c/m)[(\beta - \mu_0 mgc_\theta V_0)^2 + k/m + \mu_0^2 g^2 mc_\theta(1 + c_\theta\theta_0)] \qquad (15)$$
$$+(\beta - \mu_0 mgc_\theta V_0)[\mu_0^2 g^2 mc_\theta(1 + c_\theta\theta_0)] > 0.$$

From this we conclude that stationary sliding is always stable if c_θ is positive, while it can be unstable when this parameter is negative. The influence of varying the parameters is not so univocal in this case.

The instability is of the Hopf type, which means that perturbations from the stationary solution lead to oscillatory responses with increasing amplitudes, while the motion ultimately approaches a limit cycle.

From these examples one could draw the erroneous conclusion that it is necessary that the coefficient of friction depends on the sliding velocity or some internal state variables in order to obtain instability. This conclusion only holds if the normal force acting at the sliding surfaces is constant and the friction is the only non-linearity in the system. Examples of instabilities

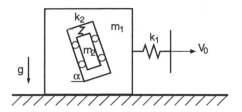

Figure 5: Block with an internal degree of freedom drawn over a surface at a constant velocity.

for systems in which the normal force is not constant were already given in [Oden, Martins 1985]. Another example is described now.

Figure 5 shows a system similar to the system of Figure 4, but the block with mass m_1 now contains a slider with mass m_2 under an angle α, which is connected to the block by a spring with stiffness k_2, while no viscous damping is modelled. The friction between the block and the ground is modelled by the law (2). This system may be regarded as a simple model for an elastic block. The equations of motion of the linearized system read

$$m \begin{pmatrix} 1 & \epsilon \cos\alpha(1 - \mu\tan\alpha) \\ \epsilon\cos\alpha & \epsilon \end{pmatrix} \begin{pmatrix} \ddot{s}_1 \\ \ddot{s}_2 \end{pmatrix} + \begin{pmatrix} k_1 & 0 \\ 0 & k_2 \end{pmatrix} \begin{pmatrix} s_1 \\ s_2 \end{pmatrix} = \begin{pmatrix} 0 \\ 0 \end{pmatrix}.$$
(16)

Here, $m = m_1 + m_2$ is the total mass of the block, $\epsilon = m_2/m$, and s_1 and s_2 are the displacement deviations from the stationary motion of the two bodies. The characteristic equation reads

$$[\epsilon - \epsilon^2 \cos^2\alpha(1 - \mu\tan\alpha)]\lambda^4 + [\epsilon\omega_1^2 + \omega_2^2]\lambda^2 + \omega_1^2\omega_2^2 = 0, \qquad (17)$$

where $\omega_1^2 = k_1/m$, and $\omega_2^2 = k_2/m$. The solutions become unstable if the two eigenfrequencies merge at

$$[\epsilon\omega_1^2 - \omega_2^2]^2 + 4\omega_1^2\omega_2^2\epsilon^2 \cos^2\alpha(1 - \mu\tan\alpha) = 0. \qquad (18)$$

This can happen if $\omega_2^2 \approx \epsilon\omega_1^2$ and $\tan\alpha > 1/\mu$. The instability stems from a so-called Hamiltonian Hopf bifurcation [Van der Meer 1985]. When some small amount of viscous damping is added, the instability remains.

2.4   Two-dimensional sliding

Until now, we have only considered the case in which sliding at contacting surfaces took place in a well-defined direction, for instance due to the two-dimensional character of the problem. In general, the direction of sliding is determined by the dynamics. For isotropic friction at a point contact, the generalization of the friction laws is that the friction force is in the direction of the relative sliding velocity and determined by the absolute value of this velocity,

$$\mathbf{F}_f = \mu F_n \mathbf{v}/|\mathbf{v}|. \tag{19}$$

This law has the disadvantage that it makes the resultant differential equations stiff for low sliding velocities, that is, small perturbations in the velocity may give rise to large variations of the friction force. A more serious problem is that the criterion of sticking will never be met if numerical errors are present. A way to overcome this problem is to consider both conditions $v_x = 0$ and $v_y = 0$ and to consider a threshold value for the other if one condition is satisfied. This threshold value must be of the order of the accuracy of the integration method. For contacting surfaces, the distribution of the normal force is generally unknown and the friction moment counteracting relative rotations is not uniquely determined. Some assumptions have to be made in order to obtain a unique solution. The same difficulty for the condition of sticking arises as three sliding velocities become zero at the transition to sticking.

3   Analysis of systems of several rigid bodies

3.1   Analysis of mechanical systems

We present here a general framework for the analysis of mechanical systems as it was proposed by Besseling [Besseling 1979], [Besseling *et al.* 1985]. Although the original formulation is in terms of finite elements, it can be applied to any kind of mechanical system. We shall not try to give its most general formulation, but content ourselves in applying it to the class of problems addressed in the present chapter. In the present case we assume that the system is two dimensional and we have a number of rigid bodies whose configuration can be described by the positions and orientations of their centres of mass. For two-dimensional systems, these are the two Cartesian coordinates of the centre of mass and an angle of rotation. The assembled coordinates for all bodies are denoted by the vector **x**, whose dimension will be denoted by n.

The contacting surfaces impose some constraints on these coordinates. In two-dimensional systems they impose two constraints, while there is one sliding degree of freedom. If there is sticking, all three relative degrees of freedom are constrained. The contacting surface can be considered as a finite element having the centres of mass of the contacting bodies as its two nodes and the configuration coordinates as its nodal variables. Violations of the constraints of contact and sliding distances can be regarded as generalized deformations, which are functions of the coordinates of the two contacting bodies. (Or of one body if it is a contact between this body and the ground; the ground can be regarded as the second body) These deformations can be assembled for all contacts and written symbolically as

$$\epsilon_i = D_i(x_j), \quad i = i, \ldots, m, j = 1, \ldots, n. \tag{20}$$

Here, m is the total number of deformations. Note that if elastic elements are present, the deformations of these may be added to the total number of deformations. The deformations can be split in prescribed deformations, ϵ^0, which have a given value and impose constraints on the system, and free or calculable deformations, ϵ^c, $\epsilon^T = [\epsilon^{0T}, \epsilon^{cT}]$. In our case, normal displacements are zero if two surfaces have contact, and their sliding distance has a prescribed value if sticking occurs: these are prescribed deformations. The normal displacements for non-contacting surfaces and sliding distances if slipping occurs or if there is no contact, are calculable deformations, that is, they are directly calculable from the configuration coordinates of the system. The constraint equations can be written as

$$\epsilon_i^0 = D_i^0(x_j), \quad i = i, \ldots, n - s + r, j = 1, \ldots, n, \tag{21}$$

where $n - s + r$ is the number of active constraints; the meaning of the numbers s and r will be explained next. We start from a given configuration of the system, in which all constraints are satisfied. Differentiating equation (21) with respect to time gives

$$\dot{\epsilon}_i^0 = D_{i,j}^0 \dot{x}_j, \tag{22}$$

where the comma denotes a partial derivative, the index giving the component with respect to which is differentiated, and the summation convention applies to repeated indices. In matrix-vector notation we may write equation (22) as $\dot{\epsilon}^0 = \mathbf{D}^0 \dot{\mathbf{x}}$. Applying the Gauss-Jordan algorithm to this equation, it can be solved for $\dot{\mathbf{x}}$ as

$$\dot{\mathbf{x}} = \boldsymbol{\Sigma}^0 \dot{\epsilon}^0 + \mathbf{Z}^* \dot{\mathbf{x}}^m, \tag{23}$$

Here, $\mathbf{x}^m$, having dimension s, are the degrees of freedom, consisting of a subset of the configuration coordinates, $\boldsymbol{\Sigma}^0$ is an $n \times m$ matrix, a kind of pseudo-inverse of the matrix $\mathbf{D}^0$, and $\mathbf{Z}^*$ is an $n \times s$ transfer matrix for the velocities, whose columns form a basis for the right null space of the matrix $\mathbf{D}^0$, $\mathbf{D}^0 \mathbf{Z}^* = \mathbf{0}$. The number of redundant constraints is given by r. The compatibility equations for the redundant constraints are written as $\mathbf{Z}\dot{\epsilon}^0 = \mathbf{0}$, where the rows of the $r \times m$ matrix $\mathbf{Z}$ form a basis for the left null space of $\mathbf{D}^0$, $\mathbf{Z}\mathbf{D}^0 = \mathbf{0}$. In most practical cases, as here, the prescribed velocities of deformations are taken to be zero, which simplifies the expressions.

An expression for the accelerations is obtained by differentiating equation (22) again with respect to time,

$$\ddot{\epsilon}^0 = \mathbf{D}^0 \ddot{\mathbf{x}} + D_{i,jk}^0 \dot{x}_j \dot{x}_k, \tag{24}$$

which can be solved as

$$\ddot{\mathbf{x}} = \boldsymbol{\Sigma}^0 (\ddot{\epsilon}^0 - D_{i,jk}^0 \dot{x}_j \dot{x}_k) + \mathbf{Z}^* \ddot{\mathbf{x}}^m. \tag{25}$$

The equations of motion can now be found by introducing generalized stresses dual to the generalized deformations, namely σ_i^0 dual to the prescribed deformations ϵ_i^0 as Lagrangian multipliers, and σ_l^c dual to calculable deformations ϵ_l^c, which are determined by constitutive equations. The unreduced equations of motion read

$$\mathbf{D}^{0T} \sigma^0 + \mathbf{D}^{cT} \sigma^c = \mathbf{f} - \mathbf{M}\ddot{\mathbf{x}} - \mathbf{h}, \tag{26}$$

where $\mathbf{M}$ is the mass matrix for the system, assembled from the masses and moments of inertia of the individual bodies, and $\mathbf{h}$ are inertia terms quadratic in velocities, which are zero in the two-dimensional case considered here. The reduced equations of motion are obtained by pre-multiplying equation (26) by $\mathbf{Z}^{*T}$, which eliminates σ^0,

$$(\mathbf{Z}^{*T}\mathbf{MZ}^*)\ddot{\mathbf{x}}^m = -\mathbf{g} + \mathbf{Z}^{*T}\mathbf{f} - \mathbf{Z}^{*T}\mathbf{D}^{cT}\sigma^c, \tag{27}$$

where all inertia terms that are quadratic in the velocities have been collected in $\mathbf{g} = \mathbf{Z}^{*T}\mathbf{h} - \mathbf{Z}^{*T}\mathbf{M}\mathbf{\Sigma}^0 D^0_{i,jk}\dot{x}_j\dot{x}_k$. The unknown Lagrangian multipliers, which can be physically interpreted as the constraint forces, are obtained from the equation

$$\sigma^0 = \mathbf{\Sigma}^{0T}\mathbf{f} - \mathbf{\Sigma}^{0T}\mathbf{D}^{cT}\sigma^c_l - \mathbf{\Sigma}^{0T}(\mathbf{M}\ddot{\mathbf{x}} + \mathbf{h}) + \mathbf{Z}^T\sigma^r, \tag{28}$$

where σ^r are the redundant stresses. In systems without dry friction, first, the equation of motion (27) can be solved, and then the reaction forces can be calculated with the help of equation (28). For systems with dry friction, however, the two systems are coupled, because the calculable stresses σ^c depend on the reaction forces. Generally, the dependence will have a non-linear character, and the accelerations and reaction forces have to be determined iteratively. If we start from some current approximation, the equation for the corrections according to the Newton-Raphson scheme, which are denoted by a prefixed Δ, are

$$\begin{pmatrix} \mathbf{Z}^{*T}\mathbf{MZ}^* & \mathbf{Z}^{*T}\mathbf{D}^{cT}\mathbf{F} \\ \mathbf{\Sigma}^{0T}\mathbf{MZ}^* & \mathbf{I} + \mathbf{\Sigma}^{0T}\mathbf{D}^{cT}\mathbf{F} \end{pmatrix} \begin{pmatrix} \Delta\ddot{\mathbf{x}}^m \\ \Delta\sigma^0 \end{pmatrix} = \\ \begin{pmatrix} -(\mathbf{Z}^{*T}\mathbf{MZ}^*)\ddot{\mathbf{x}}^m - \mathbf{g} + \mathbf{Z}^{*T}\mathbf{f} - \mathbf{Z}^{*T}\mathbf{D}^{cT}\sigma^c \\ -\sigma^0 + \mathbf{\Sigma}^{0T}\mathbf{f} - \mathbf{\Sigma}^{0T}\mathbf{D}^{cT}\sigma^c_l - \mathbf{\Sigma}^{0T}(\mathbf{M}\ddot{\mathbf{x}} + \mathbf{h}) + \mathbf{Z}^T\sigma^r, \end{pmatrix}. \tag{29}$$

Here, $\mathbf{F}$ is the derivative of the calculable stresses σ^c with respect to the reaction forces σ^0. In order to make the problem solvable, we have to assume that there are either no degrees of freedom, or no redundant constraints. The iteration has to be repeated until convergence is obtained. If the friction forces depend linearly on the reaction forces, the solution is found in one iteration. The paradoxes of Painlevé appear here by the fact that the iteration matrix on the left-hand side of equation (29) goes through a singularity if the coefficient of friction is increased.

If one starts from the equations (24) and (26), an alternative iteration scheme is found as

$$
\begin{pmatrix} \mathbf{M} & \mathbf{D}^{0T} + \mathbf{D}^{cT}\mathbf{F} \\ \mathbf{D}^0 & 0 \end{pmatrix} \begin{pmatrix} \Delta\ddot{\mathbf{x}} \\ \Delta\sigma^0 \end{pmatrix} = \begin{pmatrix} -\mathbf{D}^{0T}\sigma^0 - \mathbf{D}^{cT}\sigma^c + \mathbf{f} - \mathbf{M}\ddot{\mathbf{x}} - \mathbf{h} \\ -\mathbf{D}^0\ddot{\mathbf{x}} - D^0_{i,jk}\dot{x}_j\dot{x}_k \end{pmatrix}.
$$
$$(30)$$

If we compare the formulation of equation (29) with equation (30), we see that the former is more involved, but a smaller number of equations have to be solved, only the equations for the degrees of freedom and the reaction forces that affect sliding friction have to be included, whereas in the latter a system of equations involving all coordinates and reaction forces has to be solved, which is, however, simpler.

In a numerical integration process, only the independent coordinates are advanced. In a new configuration, one first has to determine the dependent coordinates in a Newton-Raphson iteration on equation (21), with the independent coordinates held fixed.

3.2   Arch loaded by a horizontal base motion

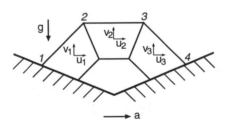

Figure 6: Arch loaded by a horizontal base motion.

In Figure 6, an arch consisting of three wedged blocks with a top angle of 45 degrees and equal mass and dimensions is shown. We shall assume that the blocks always remain in contact and hence no rotation of the blocks can occur. The masses and the acceleration due to gravity are scaled to unity. The position coordinates for this system are then the horizontal and vertical displacements of the mass centres of the blocks. The eight generalized deformations are the normal displacements ϵ_{in} and tangential displacements ϵ_{it}, $i = 1, 2, 3, 4$, at the four contact surfaces. The matrix $\mathbf{D}$ is constant in the present case, and given as

$$\epsilon = \mathbf{D}x,$$

$$
\begin{pmatrix}
\epsilon_{1n} \\
\epsilon_{1t} \\
\epsilon_{2n} \\
\epsilon_{2t} \\
\epsilon_{3n} \\
\epsilon_{3t} \\
\epsilon_{4n} \\
\epsilon_{4t}
\end{pmatrix}
=
\begin{pmatrix}
\sin\phi_1 & \cos\phi_1 & 0 & 0 & 0 & 0 \\
\cos\phi_1 & -\sin\phi_1 & 0 & 0 & 0 & 0 \\
-\sin\phi_2 & -\cos\phi_2 & \sin\phi_2 & \cos\phi_2 & 0 & 0 \\
-\cos\phi_2 & \sin\phi_2 & \cos\phi_2 & -\sin\phi_2 & 0 & 0 \\
0 & 0 & -\sin\phi_3 & -\cos\phi_3 & \sin\phi_3 & \cos\phi_3 \\
0 & 0 & -\cos\phi_3 & \sin\phi_3 & \cos\phi_3 & -\sin\phi_3 \\
0 & 0 & 0 & 0 & -\sin\phi_4 & -\cos\phi_4 \\
0 & 0 & 0 & 0 & \cos\phi_4 & \sin\phi_4
\end{pmatrix}
\begin{pmatrix}
u_1 \\
v_1 \\
u_2 \\
v_2 \\
u_3 \\
v_3
\end{pmatrix},
$$

$$(31)$$

where $\phi_1 = \pi/8$, $\phi_2 = 3\pi/8$, $\phi_3 = 5\pi/8$, and $\phi_4 = 7\pi/8$.

If no motion occurs, there are two redundant constraints, which we choose as ϵ_{4n} and ϵ_{4t}. With this choice, the constant matrices $\mathbf{\Sigma}^0$ and $\mathbf{Z}$ are given by

$$
\mathbf{\Sigma}^0 =
\begin{pmatrix}
\sin\phi_1 & \cos\phi_1 & 0 & 0 & 0 & 0 & 0 & 0 \\
\cos\phi_1 & -\sin\phi_1 & 0 & 0 & 0 & 0 & 0 & 0 \\
\sin\phi_1 & \cos\phi_1 & \sin\phi_2 & \cos\phi_2 & 0 & 0 & 0 & 0 \\
\cos\phi_1 & -\sin\phi_1 & \cos\phi_2 & -\sin\phi_2 & 0 & 0 & 0 & 0 \\
\sin\phi_1 & \cos\phi_1 & \sin\phi_2 & \cos\phi_2 & \sin\phi_3 & \cos\phi_3 & 0 & 0 \\
\cos\phi_1 & -\sin\phi_1 & \cos\phi_2 & -\sin\phi_2 & \cos\phi_3 & -\sin\phi_3 & 0 & 0
\end{pmatrix}, \quad (32)
$$

$$
\mathbf{Z} =
\begin{pmatrix}
-\tfrac{1}{2}\sqrt{2} & -\tfrac{1}{2}\sqrt{2} & 0 & 1 & \tfrac{1}{2}\sqrt{2} & \tfrac{1}{2}\sqrt{2} & 1 & 0 \\
\tfrac{1}{2}\sqrt{2} & -\tfrac{1}{2}\sqrt{2} & -1 & 0 & -\tfrac{1}{2}\sqrt{2} & \tfrac{1}{2}\sqrt{2} & 0 & 1
\end{pmatrix}. \quad (33)
$$

The equation (28) now reduces to the equilibrium equation

$$\sigma^0 = \mathbf{\Sigma}^{0T}\mathbf{f} + \mathbf{Z}^T\sigma^r. \quad (34)$$

If no base motion is present, equilibrium can be maintained as long as $\mu \geq$ 0.131652. For a friction coefficient of $\mu = 0.3$, the stationary solution can be maintained as long as the horizontal acceleration satisfies $|a| \leq 0.465080\,g$, where the redundant stresses take different values as a is positive or negative, so in the non-ideal real situation, some microslip will occur if the structure is loaded by a harmonic excitation with an amplitude just below this critical value. This microslip may accumulate to large values if the structure is loaded by many cycles.

There are four possible mechanisms with one degree of freedom, depending on which surface sticks. There are also four basically different mechanism with two degrees of freedom. The possibility of two sliding directions doubles the number of states. If we take u_2 as the independent coordinate and in addition v_2 if there are two degrees of freedom, the regions with different kinds of motion can be depicted as in Figure 7. On the axes are the two independent velocities $\dot{u}_2$ and $\dot{v}_2$. In total, 17 different states are possible, one state without motion in the origin, eight states with one degree of freedom depicted as the lines with slopes $\pm\sqrt{2}\pm 1$, and eight states corresponding to regions between these lines with two degrees of freedom. The conditions for

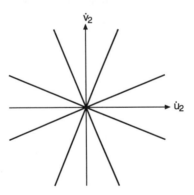

Figure 7: Regions in the space of independent velocities with different states.

transitions are in most cases clear, except for the case in which the speeds become zero. An analysis shows that the velocity will stay zero as long as

$|a| \leq 0.465080\,g$. If the acceleration is just larger than this value, a sliding mechanism with one degree of freedom will set in, with $\epsilon_{1t} < 0$, $\epsilon_{2t} > 0$, $\epsilon_{3t} = 0$, and $\epsilon_{4t} < 0$ if a is positive, and a mechanism with $\epsilon_{1t} > 0$, $\epsilon_{2t} = 0$, $\epsilon_{3t} < 0$, and $\epsilon_{4t} > 0$ if a is negative. If a is still larger, $|a| > 0.574923\,g$, a mechanism with two degrees of freedom sets in, with sliding at all four surfaces.

For this example, the equations of motion are linear for each state, and the only difficulty is in the multitude of possible states. We will not pursue this example further, but refer to the work of Lötstedt [Lötstedt 1981] and Pfeiffer and his collaborators, *e.g.* [Glocker, Pfeiffer 1993], where a convenient method for determining the transitions is presented.

3.3   Four-bar linkage under gravity loading

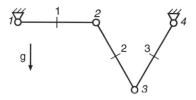

Figure 8: Four-bar linkage under gravity loading.

As a second example, we consider a four-bar linkage as shown in Figure 8, which moves under gravity loading and whose joints are affected by dry friction. All three links are one unit long and of unit mass, which is uniformly distributed over their lengths. The ground is two units long, and the acceleration due to gravity is scaled to one unit. The friction moment in the joints opposes the motion and is proportional to the magnitude of the reaction forces at the joints with a coefficient of proportionality μ, the reduced friction coefficient. We shall choose $\mu = 0.01$ in the unit system used. The coordinates that describe the positions of the links are the coordinates of their centres and a rotation angle in a positive direction, measured from the horizontal position. There are eight constraints in normal directions at the joints described by the generalized deformations ϵ_{ix} and ϵ_{iy}, $i = 1, 2, 3, 4$, so the system has maximally one degree of freedom. The rotation angle of the middle link ϕ_2 will be chosen as the independent coordinate. The eight constraint violations and the four relative rotation angles at the hinges, ϵ_{ir}, are the generalized strains. The expressions for these strains are

$$
\begin{aligned}
\epsilon_{1x} &= x_1 - \tfrac{1}{2}\cos\phi_1 + 1, & \epsilon_{1y} &= y_1 - \tfrac{1}{2}\sin\phi_1, \\
\epsilon_{2x} &= -x_1 - \tfrac{1}{2}\cos\phi_1 + x_2 - \tfrac{1}{2}\cos\phi_2, & \epsilon_{2y} &= -y_1 - \tfrac{1}{2}\sin\phi_1 + y_2 - \tfrac{1}{2}\sin\phi_2, \\
\epsilon_{3x} &= -x_2 - \tfrac{1}{2}\cos\phi_2 + x_3 - \tfrac{1}{2}\cos\phi_3, & \epsilon_{3y} &= -y_2 - \tfrac{1}{2}\sin\phi_2 + y_3 - \tfrac{1}{2}\sin\phi_3, \\
\epsilon_{4x} &= -x_3 - \tfrac{1}{2}\cos\phi_3, & \epsilon_{4y} &= -y_3 - \tfrac{1}{2}\sin\phi_3,
\end{aligned}
$$

$$
\begin{aligned}
\epsilon_{1r} &= \phi_1, \\
\epsilon_{2r} &= -\phi_1 + \phi_2, \\
\epsilon_{3r} &= -\phi_2 + \phi_3, \\
\epsilon_{4r} &= -\phi_3.
\end{aligned}
\tag{35}
$$

If motion at a hinge occurs, the constitutive relation for the friction moment is given by

$$
\sigma_{ir} = \mu\,\mathrm{sgn}(\epsilon_{ir})\sqrt{\sigma_{ix}^2 + \sigma_{iy}^2}, \quad i = 1, 2, 3, 4,
\tag{36}
$$

which is a non-linear relation. There are only three possible states of the system, either $\phi_2 > 0$, in which case $\epsilon_{1r} < 0$, $\epsilon_{2r} > 0$, $\epsilon_{3r} < 0$, and $\epsilon_{4r} > 0$, or $\phi_2 < 0$, in which case $\epsilon_{1r} > 0$, $\epsilon_{2r} < 0$, $\epsilon_{3r} > 0$, and $\epsilon_{4r} < 0$; or $\phi_2 = 0$. The sticking region, in which all subsequent motion ceases, is determined numerically as $|\phi_2| < 0.0217282$. For the solution of the accelerations, the iteration scheme of equation (30) is used, because the system (29) is more complicated and does not give rise to a much smaller number of equations in the present case.

First, the resulting motion as we start from the initial conditions as shown in Figure 8 with $\phi_2 = -\pi/3$ and $\dot{\phi}_2 = 0$ is determined. Figure 9 shows the results for ϕ_2 and for the vertical coordinate y_2, which has a double frequency. Note that at the end, ϕ_2 is constant, but unequal to zero.

In order to show that the mass matrix of the unconstrained system may be singular in the present formulation, we have also considered the case in which the first and third link are massless and only the middle link carries

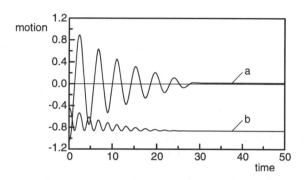

Figure 9: Motion of the four-bar linkage; (a) angle of the middle link; (b) vertical coordinate of the centre of the middle link.

mass. The same initial values and coefficients of friction were used. The results are shown in Figure 10.

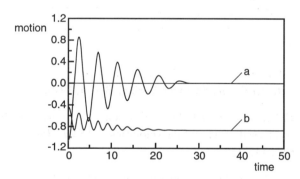

Figure 10: Motion of the four-bar linkage with two massless links; (a) angle of the middle link; (b) vertical coordinate of the centre of the middle link.

References

[1] BESSELING, J.F., "Finite element methods," In: BESSELING, J.F., AND HEIJDEN, A.M.A. VAN DER, *Trends in Solid Mechanics*, Delft University Press, Delft, and Sijthoff & Noordhoff, Alphen aan den Rijn, 1979, pp.53–78.

[2] BESSELING, J.F., JONKER, J.B., AND SCHWAB, A.L., "Kinematics and dynamics of mechanisms," *Delft Progress Report* **10** (1985), 160–172.

[3] BOWDEN, F.P., AND TABOR, D., *The Friction and Lubrication of Solids*, revised edition, Clarendon, Oxford, 1954.

[4] BOWDEN, F.P., AND TABOR, D., *The Friction and Lubrication of Solids, Part II*, Clarendon, Oxford, 1964.

[5] DOWSON, D., *History of Tribology*, Longman, London, 1979.

[6] GLOCKER, C., AND PFEIFFER, F., "Complementarity problems in multibody systems with planar friction," *Archive of Applied Mechanics* **63** (1993), 452–463.

[7] GU, J.-C., RICE, J.R., RUINA, A.L., AND TSE, S.T., "Slip motion and stability of a single degree of freedom elastic system with rate and state dependent friction," *Journal of the Mechanics and Physics of Solids* **32** (1984), 167–196.

[8] HAMEL, G., *Theoretische Mechanik, Eine einheitliche Einführung in die gesamte Mechanik*, Springer-Verlag, Berlin, 1949.

[9] HESS, D.P., AND SOOM, A., "Friction at a lubricated line contact operating at oscillating sliding velocities," *ASME Journal of Tribology* **112** (1990), 147–152.

[10] HURWITZ, A., "Ueber die Bedingungen, unter welchen eine Gleichung nur Wurzeln mit negativen reellen Theilen besitzt," *Mathematische Annalen* **46** (1895), 273–284.

[11] KRAGELSKII, I.V., *Friction and Wear*, Butterworths, London, 1965.

[12] KUNTZ, J.P., *Rolling Link Mechanisms*, Dissertation, Delft University of Technology, Delft, 1995.

[13] MEER, J.-C. VAN DER, *The Hamiltonian Hopf Bifurcation*, Springer-Verlag, Berlin, 1985.

[14] LÖTSTEDT, P., "Coulomb friction in two-dimensional rigid body systems," *Zeitschrift für angewandte Mathematik und Mechanik* **61** (1981), 605–615.

[15] MARTINS, J.A.C., ODEN, J.T., AND SIMÕES, F.M.F., "A study of static and kinetic friction," *International Journal of Engineering Science*, **28** (1990), 29–92.

[16] ODEN, J.T., AND MARTINS, J.A.C., "Models and computational methods for dynamic friction phenomena," *Computer Methods in Applied Mechanics and Engineering*, **52** (1985), 527–634.

[17] PAINLEVÉ, P., "Sur les lois du frottement de glissement," *Comptes Rendus de l'Académie des Sciences* **121** (1895), 112–115.

[18] RABINOWICZ, E., "The nature of the static and kinetic coefficients of friction," *Journal of Applied Physics*, **22** (1951), 1373–1379.

[19] RABINOWICZ, E., "The intrinsic variables affecting the stick-slip process," *Proceedings of The Physical Society*, London, **71** (1958), 668–675.

[20] RABINOWICZ, E., *Friction and Wear of Materials*, Wiley, New York, 1965.

[21] REULEAUX, F., *Theoretische Kinematik; Grundzüge einer Theorie des Maschinenwesens*, 2 Vols., Vieweg, Braunschweig, 1875.

[22] RICE, J.R., AND RUINA, A.L., "Stability of steady frictional slipping," *ASME Journal of Applied Mechanics* **50** (1983), 343–349.

[23] RICHARDSON, R.S.H., AND NOLLE, H., "Surface friction under time-dependent loads," *Wear* **37** (1976), 87–101.

[24] STRIBECK, R., "Die wesentlichen Eigenschaften der Gleit- und Rollen-lager," *Zeitschrift des Vereines Deutscher Ingenieure* **46** (1902), 1341–1348, 1432–1438, 1463–1470.

[25] TABOR, D., "Friction - the present state of our understanding," *ASME Journal of Lubrication Technology* **103** (1981), 169–179.

Dynamics with Friction: Modeling, Analysis and Experiment, Part II, pp. 253–308
edited by A. Guran, F. Pfeiffer and K. Popp
Series on Stability, Vibration and Control of Systems, Series B, Vol. 7
© World Scientific Publishing Company

DAMPING THROUGH USE OF PASSIVE AND SEMI-ACTIVE DRY FRICTION FORCES

ALDO A. FERRI

G.W.Woodruff School of Mechanical Engineering
Georgia Institute of Technology
Atlanta, GA 30332-0404, USA

ABSTRACT

A common simplification used in studies of dry or Coulombic friction damping is that the friction force is of constant magnitude. This chapter discusses the damping characteristics exhibited by systems having motion-dependent friction forces. The friction force may vary as a result of passive or active means, and both cases are treated herein. In the passive case, the friction force is made to depend on the relative slip displacement through the geometry of the frictional connection; in the semi-active case, the friction force is controlled through active means in response to sensory feedback. It is seen that the vibratory behavior of systems with motion-dependent friction forces is considerably different than that exhibited by dry friction damped systems with constant-magnitude friction forces.

1. Introduction

Dry friction plays a vital role in many mechanical systems. In lightly damped built-up structures such as truss frames, dry friction at the structural junctures can be responsible for as much as 90% of the passive energy dissipation capacity[1,2]. Unlike viscous and/or viscoelastic damping treatments which add weight and cost to a structure, dry friction is often present and readily available in structural joints.

The analysis of damping in built-up structures is impeded by two related aspects of dry friction. The first is that dry friction is inherently nonlinear. The second is that dry friction damped systems are prone to sticking, hence a dry friction damped system can exhibit dramatically different dynamic behavior depending on whether a particular friction interface is slipping or sticking. This type of phenomenon greatly complicates the numerical and analytical investigation of friction damped structures. Due to the difficulties associated with the analysis of dry friction damped systems, most studies have used very simple models to describe the interfacial friction forces. The simplest and by far the most popular friction model states that during slipping, the friction force is constant in magnitude and acts to oppose the relative sliding motion. This simple friction model still results in numerical difficulties, and simulation results from systems that incorporate the simple friction model often exhibit very complicated behavior. It is for these reasons that very few studies have ventured into the use of more complicated friction models.

Dry friction damping has been reviewed recently by several authors.   A comprehensive review by Guran et al.[3] appears in this volume. A recent two-part article by Ibrahim[4] also surveyed the modeling and analysis of systems with dry friction. Armstrong-Hélouvry et al.[5] reviewed control compensation strategies for systems having dry friction. Ferri[6] focused primarily on the *dissipative properties* of dry friction and on its role in vibration isolation systems in his recent survey article. Finally, the series of review articles by Beards[2] and Jones[7] provide extensive bibliographies of recent papers in the friction damping area.

This chapter discusses the analysis and design of systems for which the normal force across the friction interfaces is not constant. The variation in normal force can occur due to passive means or due to active means. In the next section, background information is presented along with applications that have motivated the study of friction damped structures. Section 2 discusses the behavior of passive systems whose friction interface is parallel to the direction of primary motion. Section 3 contains a description of semi-active friction dampers for which the normal load across the friction interface is controlled using active forces. Section 4 contains concluding remarks.

2. Passive Mechanisms

As stated in the introduction, most of the prior work in dry friction damped systems has invoked two simplifying assumptions: (1) the kinetic friction coefficient is constant, μ=constant, and (2) the force normal to the sliding interface is constant, N=constant. In light of the fact that the magnitude of the friction force is equal to the product f=μN, this results in the magnitude of the resistive force being constant during sliding. Even with these two assumptions, the analysis of dry friction damped systems is challenging and many questions remain unanswered. In reality, however, both assumptions are known to be false in most, if not all, frictionally-damped systems. The friction coefficient is known to depend on a host of system and environmental factors[3-5]. Furthermore, the normal force across the sliding interface (hereafter simply referred to as the *normal force*) generally depends on the sliding motion itself in all but the most careful laboratory experiments.

This chapter presents an overview of research on systems for which the second assumption above is relaxed. The next subsection presents some of the most significant prior work in this area. The remaining subsections detail some of the important findings contained in the author's previous work in this area. It must be noted that research into systems for which the friction coefficient is variable is also important and has received very little attention to date from the perspective of *friction damping* [8]. This topic, however, will be left to another paper.

2.1 Background

In the area of friction damping, one of the earliest studies involving non-constant normal forces is that of Bielawa[9]. In this work, Bielawa models a turbomachinery rotor disk having a part-span segmented shroud. When interblade slip occurs, vibratory energy is dissipated by dry friction damping. What sets Bielawa's work apart from earlier efforts is the ability of his model to account for changes in the normal forces at the shroud caused by elastic deformation of the blades and shrouds. One of the key findings of his work was that beyond some threshold displacement amplitude, the equivalent viscous damping ratio approaches a *constant value* as if it were damped by *purely linear structural damping*. This is in stark contrast to the classical situation of systems with constant μN, where the equivalent viscous damping ratio approaches zero as the slip displacement amplitude grows[10,11]. While Earles and Williams[12] used an *amplitude-dependent* friction force in their study of shrouded turbine blade vibration, it seems to have been used only as a correction factor. They observed that smaller force/displacement levels required smaller friction coefficients than did higher force/displacement levels in order to obtain good agreement with experimental results.

The analysis of dry friction damped turbomachinery blades with displacement-dependent normal forces was also considered by Menq and co-workers[13,14]. An important feature of their analysis was the ability of the model to approximate *microslip* or partial slip in the frictional interface. The analysis confirmed that the damping behavior was fundamentally different than that expected of a dry friction damped system. In particular, they confirmed that unbounded response at resonance (which is related to equivalent viscous damping ratio decreasing with amplitude of response) was prevented from occurring.

The fact that friction can provide a "linear damping" characteristic has very important ramifications with regard to flutter suppression in turbomachinery. As noted by Den Hartog[15], a dry friction interface will dissipate energy at a rate linearly proportional to the slip displacement amplitude. In contrast, aerodynamic forces can feed energy into a structural system *at a rate proportional to the square of the displacement amplitude*. Thus, when dry friction is used to suppress flutter in a bladed rotor disk, it appears that friction can at best attenuate aerodynamically-excited vibration below some initial amplitude. If the vibration amplitude grows too large for any reason, dry friction may be unable to dissipate energy fast enough to maintain bounded oscillations[16-20]. It is also important to note that in high-temperature applications, such as jet engine turbines, damping treatments other than dry friction are problematic[7]. Therefore, if dry friction can be modified in such a way that it produces a damping characteristic similar to that of linear structural damping, it can have profound implications for the elimination of turbomachinery flutter problems.

Variable normal forces occur regularly in the vibration of built-up structures such as truss-frames. In an analysis of various types of connecting joints, Hertz and Crawley[21-23] found that if the friction forces were allowed to increase with system slip displacement, the overall passive damping of the structure would be similar to that provided by linear viscous or linear structural damping. This type of behavior was verified experimentally by Folkman and Redd[24] in their study of damping in a truss structure with pin-type connecting joints. Other work that has found viscous-like damping characteristics in dry friction damped systems is that of Beucke and Kelly[25], Ferri[26], Anderson and Ferri[27], Makris and

Constantinou[28], Ferri and Heck[29], Whiteman and Ferri[30], and Inaudi, et al.[31] These studies involve systems for which the normal force grows *linearly* with displacement magnitude; the resistive forces generated by such an arrangement have been termed *linear-Coulomb damping* or *linear-friction damping* by some authors.

A second way in which dry friction can give rise to a viscous-like damping effect is in the case of in-plane slip induced by foreshortening of a beam or plate. In this case, the slipping plane is oriented perpendicular to the direction of dominant vibrational motion. Beards[32] investigated the optimum clamping force to maximize the damping of a rectangular plate which experienced in-plane slip along one boundary. He found experimentally that damping could be increased if the clamping force was reduced. The first analytical treatment of this problem is that of Jézéquel[33]. For an axisymmetric circular plate, he showed that the equivalent viscous damping ratio was *invariant with respect to amplitude* just as in the case of linear structural damping. Dowell[34] extended the work of Jézéquel to the case of beams and rectangular plates and examined how the equivalent viscous damping ratio varied with beam/plate parameters and boundary conditions. A number of experimental studies have validated these predictions[35-37]. In all of the aforementioned work, the normal force was assumed to be constant. Research which considers the case of variable normal force and in-plane slip will be described in Section 2.4 below.

Several studies of dry friction systems with variable normal forces should also be mentioned although they are not explicitly concerned with *passive damping* characteristics. Chaos in a system having dry friction with variable normal force has been studied by Feeny and his co-workers[38-40]. Although several researchers have found chaos in systems with friction-induced oscillations and advanced friction laws[3,4], Feeny's work appears to be the first to uncover chaotic behavior in a system using a constant friction coefficient. The stability of steady sliding in the presence of variable normal force has also been investigated[41,42]. Finally, the variations in normal force induced by vibration normal to the sliding plane have also been considered[43-45].

2.2 Linear-Coulomb Damping

To illustrate the viscous-like damping attribute of dry friction damped systems with displacement-dependent normal forces, consider the single-degree-of-freedom (SDOF) model shown in Figure 1. This system was studied in detail by Anderson and Ferri[27]; similar systems were considered by Beucke and Kelly[25] and by Makris and Constantinou[28]. The normal force increases linearly with slip displacement magnitude due to the ramp profile angle γ; hence the name "linear-Coulomb damping." The friction force for this system can be written concisely using a signum or "sgn" function. (The sgn function is equal to +1, -1, or 0 depending on whether the argument is positive, negative, or zero, respectively). Assuming -x to be the direction of a positive friction force:

$$F_f = \mu\left(N_0 + K\tan\gamma|x|\right)\text{sgn}(\dot{x}) \tag{1}$$

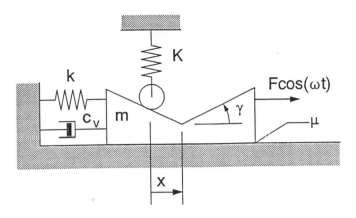

Figure 1: Single-degree-of-freedom dry-friction damped system with displacement-dependent normal load.

Note that the term in parenthesis multiplying the sgn function in (1) is the normal force, which grows linearly with $|x|$ depending on both the normal spring constant K and the ramp angle γ. The quantity N_0 accounts for any preload in the normal spring. (N_0 can also account for the weight of the object if the normal direction is vertical.) The equation of motion for this system for the case of a harmonic external force is given by:

$$m\ddot{x} + c_v\dot{x} + kx + F_f = F\cos(\omega t) \qquad (2)$$

The system described by (2) exhibits qualitatively different behavior than the "classic" dry friction damped SDOF studied by Den Hartog[15]. Note that the classic dry friction damped oscillator is recovered simply by setting K or γ to zero.

First consider the free response (F=0) of this system to nonzero initial conditions. To clarify the role of dry friction energy dissipation, the viscous damping contribution is excluded, c_v=0. Figure 2a shows the free response for the case of γ=30°, N_0=0 while Figure 2b is for γ=0, N_0=30 N. (Both curves use m=1kg, c_v=0, k=100N/m, μ=0.5, K=50N/m.) Note that the envelopes of decay are very different for the system with displacement-dependent friction forces. In particular, while the envelopes of decay are linear for the "classic" case, they have more of an exponential shape for the displacement-dependent case. It can be shown, in fact, that the envelopes are exactly exponential for the case of N_0=0 .

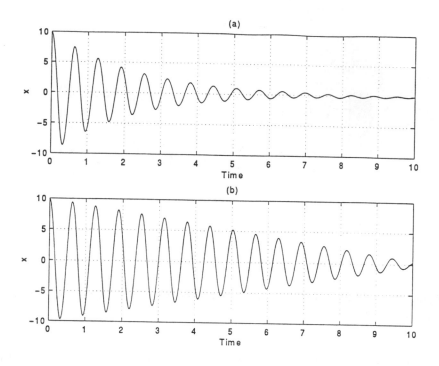

Figure 2: Free response of SDOF system. (a) with displacement-dependent friction force, (b) without displacement-dependent friction force.

This was shown by Caughey and Vijayaraghavan[46] in the analysis of a SDOF system with "linear hysteretic damping" which has a similar mathematical form. The damping of the nonlinear system (2) can be analyzed qualitatively by a number of techniques. One of the most common approaches is the harmonic balance method[6]. The fundamental assumption in the harmonic balance (HB) method is that the response can be expressed as a sum of frequency components or harmonics. For simplicity, only the first-order technique will be pursued here; several authors have performed multi-harmonic analyses of dry friction damped systems with constant normal forces including Pierre et al.[47] and Ferri and Dowell[48]. In the first-order HB method, it is assumed that the displacement is harmonic with unknown amplitude and phase. Without loss of generality, the unknown phase can be eliminated from the response through a time-shift operation. Thus we can assume that the response and forcing have the form:

$$x = X \cos \omega t \tag{3}$$

$$F = F^c \cos \omega t + F^s \sin \omega t \tag{4}$$

Expressions (3) and (4) are first substituted into (1) and (2). Next, either of two equivalent operations can be performed: the nonlinear terms can be expanded in a Fourier series and coefficients of $\sin(\omega t)$ and $\cos(\omega t)$ equated or "balanced". Alternatively, equation (2) can be multiplied by $\cos(\omega t)$ and integrated over a forcing period yielding one algebraic equation; the process is then repeated for $\sin(\omega t)$. In either case two algebraic equations are rendered:

$$(k - m\omega^2)X = F^c \tag{5}$$

$$-c_v \omega X - \tfrac{4}{\pi} \mu N_0 - \tfrac{2}{\pi} \mu K X \tan \gamma = F^s \tag{6}$$

A final equation stems from the relationship that

$$F^2 = (F^c)^2 + (F^s)^2 \tag{7}$$

Inserting (5) and (6) into (7) yields a quadratic equation in X:

$$aX^2 + bX + c = 0 \tag{8}$$

where

$$a = \left[\left(k - m\omega^2\right)^2 + \left(c_v \omega + \tfrac{2}{\pi} \mu K \tan \gamma\right)^2 \right] \tag{9a}$$

$$b = 2\left(\tfrac{4}{\pi} \mu N_0\right)\left(c_v \omega + \tfrac{2}{\pi} \mu K \tan \gamma\right) \tag{9b}$$

$$c = \left(\tfrac{4}{\pi} \mu N_0\right)^2 - F^2 \tag{9c}$$

Due to the approximations invoked in the derivation of (6), only positive solutions to (8) are valid. An examination of (8) and (9) reveals that:

- No positive solutions of X exist if $F < \tfrac{4}{\pi} \mu N_0$. This is the so-called threshold force relationship[20]. Below this critical value of F, the harmonic balance solution suggests that the mass does not move. However, based on physical arguments, it is easy to establish that the true threshold force is only μN_0; thus the HB technique predicts a threshold force value that is approximately 27% too high.

- Provided that $F > \frac{4}{\pi}\mu N_0$, there will be only one positive solution to (8); i.e., the steady-state response is always single-valued.

- For the case of no viscous damping, $c=0$, the response amplitude is always bounded so long as $\mu, K, \gamma > 0$. This should be contrasted with the classic dry friction damped system which experiences unbounded response at resonance (for the case $F > \frac{4}{\pi}\mu N_0$) unless viscous damping is present[15].

Further insight can be gained by examining the equivalent viscous damping coefficient for the system, c_{eq}. This is most easily obtained in this case by comparing the viscous and non-viscous terms in (6):

$$c_{eq} = c_v + \frac{4\mu N_0}{\pi \omega X} + \frac{2}{\pi \omega}\mu K \tan \gamma \qquad (10)$$

Note that this expression can also be obtained as a sinusoidal describing function[49] for the viscous and friction terms in (2). The first term on the right hand side of (10) is simply the viscous damping coefficient; the second term is the contribution from the spring pre-load and/or system weight and matches that contained in standard texts; e.g., Timoshenko et al.[50]. (Typically, the expression is obtained by equating the energy loss per cycle of the dry friction damped system with that of a system damped only by viscous forces. Since that technique also assumes a harmonic response, it is not surprising that the equivalent viscous damping expression is the same.) The third term on the right hand side of (10) is a constant, except for the presence of the excitation frequency in the denominator. In this way, it is very similar to *linear structural damping*[51]. For an investigation of a variety of different techniques for the determination of c_{eq} (besides the HB or sinusoidal describing function techniques), see Beucke and Kelly[25].

It is also interesting to compute the equivalent viscous damping ratio for the system using (10):

$$\zeta_{eq} \equiv \frac{c_{eq}}{2\sqrt{k_{eq}m_{eq}}} = \zeta + \frac{2\mu N_0}{\pi m \omega \omega_n X} + \frac{\mu K \tan \gamma}{\pi m \omega \omega_n} \qquad (11)$$

where $k_{eq} = k$, $m_{eq} = m$, ζ is the original viscous damping ratio ($c_v/2m\omega_n$) and ω_n is the original natural frequency of the system ($\sqrt{k/m}$). It can be seen that the contribution to the system damping from the spring preload N_0 diminishes with increasing displacement amplitude, while that of $K\tan\gamma$ is invariant with respect to amplitude.

The accuracy of the first-order harmonic balance technique can be demonstrated through a comparison of steady-state amplitudes of response with those predicted by a time-domain technique. The time-domain technique is based on splicing together the exact

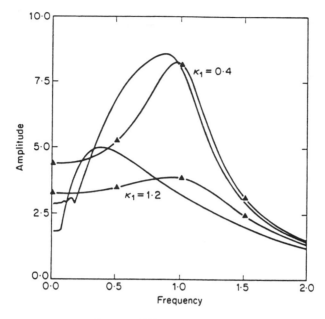

Figure 3: Frequency response for two different values of $\kappa_1 = (\mu K \tan\gamma)/k$. Comparison of harmonic-balance Δ and time-domain technique (unmarked lines). From Ref. 27.

solutions to (2) in each of four piecewise-linear domains of the state-space[27]. Figure 3 shows the frequency response curves for 2 different values of $\kappa_1 = \mu K \tan\gamma/k$. Both harmonic balance and time-domain solution amplitudes are plotted. Three phenomena are evident in this plot. First, the accuracy of the HB solution technique degrades as $K\tan\gamma$ gets larger. Second, the HB solution accuracy degrades as the frequency of excitation gets smaller. (This occurs due to the prevalence of stick-slip motion when the forcing frequency is much less than the system natural frequency.) Third, an interesting "cross-over" effect occurs in certain frequency regimes. In particular, in the nondimensional frequency range 0.2 to 0.4, increasing $K\tan\gamma$ actually *increases* the steady-state amplitude of response, as judged by the time-domain solution curves.

One of the important characteristics of dry friction damped systems which is not present in other linear or nonlinear mechanical systems is that of sticking. In the case of free vibration, a response trajectory can terminate in a condition where it is permanently stuck away from the origin of the phase plane. In essence, there is a *zone of equilibria* rather than an isolated equilibrium at the origin of the state space. One way of characterizing the sticking behavior of a system is through its *sticking regions*. Although the sticking conditions can be derived from physical force balance relations, it is easier in many cases to examine the vector fields on either side of the sticking boundary, $\dot{x} = 0$. The two conditions that must be met in order for sticking to occur are:

$$\dot{x} = \varepsilon \Rightarrow \ddot{x} < 0 \tag{12}$$

$$\dot{x} = -\varepsilon \Rightarrow \ddot{x} > 0 \tag{13}$$

where ε is an infinitesimal positive quantity. If both (12) and (13) are satisfied for a particular value of displacement and forcing, then the system sticks until one relation or the other becomes violated. Substitution of $\dot{x} = \pm\varepsilon$ into (2) results in the following conditions for sticking:

$$\dot{x} = +\varepsilon \Rightarrow \quad m\ddot{x} = \left(F(t) - kx - \mu(N_0 + K\tan\gamma|x|)\right) < 0 \tag{14a}$$

$$\dot{x} = -\varepsilon \Rightarrow \quad m\ddot{x} = \left(F(t) - kx + \mu(N_0 + K\tan\gamma|x|)\right) > 0 \tag{14b}$$

Equations (14a) and (14b) are both satisfied so long as:

$$|F(t) - kx| < \mu(N_0 + K\tan\gamma|x|) \tag{15}$$

The way to interpret (15) is that if $\dot{x}=0$ and if (15) is satisfied at any time in this system's free or forced response, the mass will stick and x(t)=constant. In the case of free response, the mass will remain stuck indefinitely; in the case of forced response, the mass will remain fixed unless and until the force is such that (15) is violated.

Equation (15) can also be used to generate sticking regions in the x-t plane. Figure 4 shows the relationship between the sticking region in the x-t plane (shown shaded in the figure) and the response trajectory in the extended phase space (x, $\dot{x}$,t). Starting from an initial condition $(x_0, \dot{x}_0)$, the trajectory "unwinds" forward in time. Portions of the trajectory above the x-t plane in Figure 4 are shown with solid lines while portions below the x-t plane are shown dashed. If a trajectory lands in a sticking region (as, for example, the first time that the trajectory hits the x-t plane in Figure 4) it will stick with constant displacement until it reaches the boundary of the shaded sticking region. The trajectory continues off the x-t plane after the boundary is reached.

Figures 5-7 show some representative examples of sticking regions for three different values of $\kappa_1=\mu K\tan\gamma/k$. Note that for small values of $K\tan\gamma$ such as shown in Figure 5, the sticking regions are only slightly distorted from their shapes in the "classic" friction damped case[52,53]. As $K\tan\gamma$ grows, the sticking regions grow. At the critical value shown in Figure 6, the sticking region extends to infinite displacement, but note that permanent lock-up is not possible. As t increases, any trajectory trapped in a sticking region on the x-t plane will eventually reach a region boundary, beyond which the system resumes slipping. In contrast, the sticking region shown in Figure 7 extends to infinity in both directions. If any trajectory hits the x-t plane with a nondimensional displacement greater than approximately 20, the mass will remain locked up permanently.

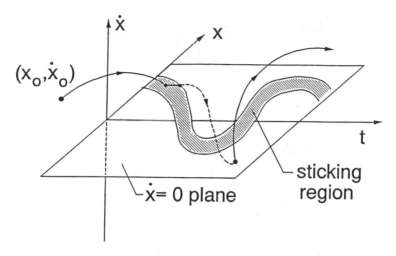

Figure 4: System trajectory in $(x, \dot{x}, t)$ space. Sticking region in x-t plane is shown as shaded. (Trajectories below the x-t plane are shown as dashed.)

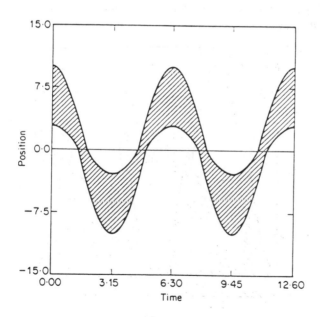

Figure 5: Sticking region for $\kappa_1 = (\mu K \tan\gamma)/k = 0.4$ Combinations of nondimensional x and t for which sticking occurs when $\dot{x} = 0$ is shaded. From Ref. 27.

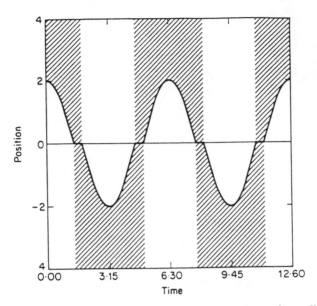

Figure 6: Sticking region for $\kappa_1 = (\mu K \tan\gamma)/k = 1.0$ Combinations of nondimensional x and t for which sticking occurs when $\dot{x} = 0$ is shaded. From Ref. 27.

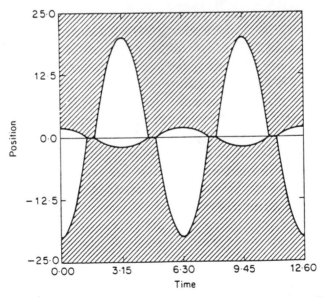

Figure 7: Sticking region for $\kappa_1 = (\mu K \tan\gamma)/k = 1.2$ Combinations of nondimensional x and t for which sticking occurs when $\dot{x} = 0$ is shaded. From Ref. 27.

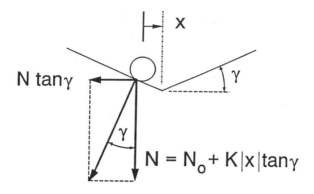

Figure 8: Components of roller force parallel to and perpendicular to the sliding plane.

2.3 Profiled Block

One effect omitted in the analysis above is the component of the normal spring force that is tangent to the slip plane. Figure 8 shows that when x is displaced from the origin, an in-plane component of the force between the roller and the mass is present. This force produces a qualitatively different type of dynamic behavior from that discussed in the previous subsection (Whiteman and Ferri, 1995a, 1995b). The equation of motion for this system has one extra term when compared with (2):

$$m\ddot{x} + c_v\dot{x} + kx + F_f + (N_0 + K|x|\tan\gamma)\tan\gamma\,\text{sgn}(x) = F\cos(\omega t) \tag{16}$$

where the expression in parenthesis, $N_0 + K|x|$ tanγ, is recognized as the normal force in the spring assuming that N_0 is due only to spring preload and not to the weight of the system. It can be shown that the added term has a significant effect on the stiffness of the system and, thus, on its tendency to return to the origin.

Again, this system can be examined for its qualitative traits through the application of the HB method. Assuming that the forcing and the response have the harmonic forms given in (3) and (4), and balancing harmonics yields:

$$(k - m\omega^2)X + (KX\tan^2\gamma + \tfrac{4}{\pi}N_0\tan\gamma) = F^c \tag{17}$$

$$-c_v\omega X - \tfrac{4}{\pi}\mu N_0 - \tfrac{2}{\pi}\mu KX\tan\gamma = F^s \tag{18}$$

The left-hand side of (17) can be equated with $\left(k_{eq} - m\omega^2\right)X$ in order to obtain the combined "equivalent in-plane stiffness" of the system, k_{eq}:

$$k_{eq} = k + K \tan^2 \gamma + \frac{4N_0 \tan \gamma}{\pi X} \tag{19}$$

Note that the contribution of the spring preload decreases with displacement amplitude, while the stiffness influence from $K \tan^2 \gamma$ is *purely linear*. This results from the fact that absolute value and sgn nonlinearities "cancel" each other: $|x| \, \mathrm{sgn}(x)=x$. Defining the equivalent natural frequency of the system to be $\Omega_n = \sqrt{k_{eq}/m}$, the equivalent viscous damping ratio for this system is:

$$\zeta_{eq} = \frac{c_v}{2m\Omega_n} + \frac{2\mu N_0}{\pi m\omega\Omega_n X} + \frac{\mu K \tan \gamma}{\pi m\omega\Omega_n} \tag{20}$$

If one evaluates the expression above at resonance, $\omega = \Omega_n$, and makes use of (19), the equivalent viscous damping ratio at resonance, ζ_{res} is obtained:

$$\zeta_{res} = \frac{c_v}{2m\sqrt{\dfrac{k}{m} + \dfrac{K \tan^2 \gamma}{m} + \dfrac{4N_0 \tan \gamma}{\pi m X}}} + \frac{2\mu N_0 + \mu K X \tan \gamma}{\pi \left[\left(k + K \tan^2 \gamma\right)X + \dfrac{4}{\pi} N_0 \tan \gamma \right]} \tag{21}$$

The first term on the right hand side of (21) is simply the effect of the viscous term applied to a system that is stiffened by the in-plane force. The second term shows a complicated dependence on displacement amplitude, X. Two limits are of interest:

$$\lim_{X \to 0} \quad \zeta_{res} = \frac{\mu}{2 \tan \gamma} \tag{22}$$

$$\lim_{X \to \infty} \quad \zeta_{res} = \frac{c_v}{2m\sqrt{\dfrac{k}{m} + \dfrac{K \tan^2 \gamma}{m}}} + \frac{\mu K \tan \gamma}{\pi \left(k + K \tan^2 \gamma\right)} \tag{23}$$

The analysis shows that even in the absence of viscous damping, the equivalent viscous damping of the system is nonzero and is bounded. The limit (23) is especially important as it suggests that the phenomena of unbounded response at resonance, which is an undesirable feature of the classic dry friction damped situation, is not present in systems with displacement-dependent normal forces.

Returning to equations (17) and (18), a single quadratic equation in X can be obtained following the procedure described earlier:

$$a' X^2 + b' X + c' = 0 \tag{24}$$

where

$$a' = \left[(k - m\omega^2 + K \tan^2 \gamma)^2 + (c\omega + \tfrac{2}{\pi}\mu K \tan \gamma)^2 \right] \tag{25a}$$

$$b' = 2(\tfrac{4}{\pi}N_0)\left[\tan \gamma \left(k - m\omega^2 + K \tan^2 \gamma \right) + \mu(c\omega + \tfrac{2}{\pi}\mu K \tan \gamma) \right] \tag{25b}$$

$$c' = (\tfrac{4}{\pi}N_0)^2 (\tan^2 \gamma + \mu^2) - F^2 \tag{25c}$$

As before, only positive solutions for X are permitted due to the assumptions implicit in the derivation of equations (17) and (18). One positive solution can be shown to exist as long as $c' < 0$:

$$F > F_{\text{threshold}} = (\tfrac{4}{\pi}N_0)\sqrt{\tan^2 \gamma + \mu^2} \tag{26}$$

Equation (26) is a sufficient condition for there to be only one positive solution for X; however, it is not necessary for positive solutions to exist. In fact, two positive solutions for X can exist provided that $F < F_{\text{threshold}}$, $b' < 0$, and $(b')^2 - 4a'c' > 0$. Note that b in (9b) is always positive, hence, in the absence of the in-plane stiffening force, at most one solution can be positive for any combination of system and excitation parameters. A necessary condition for $b' < 0$ is that $N_0 \neq 0$ and:

$$\omega > \omega_1 = \frac{\mu c_v + \sqrt{(\mu c_v)^2 + 4m \tan \gamma \left\{ (k + K \tan^2 \gamma) \tan \gamma + (2/\pi)\mu^2 K \tan \gamma \right\}}}{2m \tan \gamma} \tag{27}$$

In fact, a careful time-domain analysis reveals that the occurrence of double-valued steady-state solutions in this system is much more prevalent than suggested by (26) and (27). Figure 9 shows a comparison of the HB (solid line) and time-integration solutions(denoted with + marks) for a particular set of parameters. Note that over certain frequency ranges, there are two sets of time-integration solution amplitudes. The steady-state displacement amplitude that develops depends on the system's starting conditions. Figure 10 shows two time solutions corresponding to another set of system and excitation parameters. By virtue of the fact that they are obtained by forward integration in time, they are seen to be stable. This is significant as it means that two different amplitudes of response are possible depending on the system's initial disturbance. The conclusions drawn for SDOF systems were found to be valid for the case of Multi-DOF systems[30]. As might be expected, good quantitative agreement existed in the low frequency ranges but deteriorated for excitation frequencies beyond the first resonance. Additionally, several frequency ranges were noted over which internal resonances occurred; these pertained to

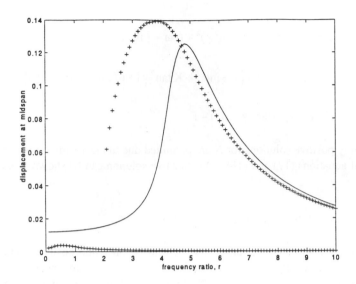

Figure 9: Frequency response for SDOF system including in-plane restoring force. Normalized displacement versus $r=\omega/\omega_n$. Comparison of harmonic balance (solid line) and time integration (+). From Ref. 30.

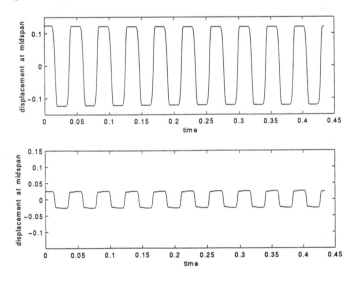

Figure 10: Steady-state response resulting from two different initial conditions. Both curves pertain to the same system and excitation parameters. From Ref. 30.

cases where higher harmonics of the excitation frequency coincided with higher system modes.

The ability of dry friction with variable normal forces to suppress flutter in aeroelastic systems was investigated by Whiteman and Ferri[54,55]. These studies confirmed that so long as the equivalent viscous damping ratio of the dry-friction damped system is greater than the effective *negative viscous damping ratio* of the aerodynamic forces, the system remains stable for all initial conditions and disturbance amplitudes.

2.4 Shock and Vibration Isolation

This section addresses shock and vibration isolation through the aid of dry friction. In many respects, friction is an ideal mechanism with which to provide shock isolation. The reason is that friction elements can be made to saturate their transmitted force. This is especially true in the case of constant normal force, which limits the magnitude of the transmitted force to μN. This property of dry friction has been used extensively in the areas of rail-road vehicles and in aseismic isolation of buildings[6,56].

Several investigators have considered the use of friction elements to isolate SDOF systems. The earliest such investigation appears to be that of Levitan[57] who used a Fourier series to develop an "exact solution" for a system subjected to harmonic base excitation. Other examples include Yeh[58], Hundal[59], Schlesinger[60], Westermo and Udwadia[61], Marui and Kato[53], Parnes[62], Schwarz et al.[63], and Ravindra and Mallik[64]. The response of friction isolation systems to random excitation has also been a topic of considerable interest in the earthquake engineering community[56,65-67].

In all of the aforementioned references, the normal force is assumed to be constant. In fact, it can be shown that for the special case of a SDOF system subjected to a step increase in base velocity, an ideal friction element with constant normal force is the optimal mount which gives the smallest peak acceleration level for a given maximum relative displacement. The optimality, however, does not apply universally to all types of input and/or to other metrics of performance[68]. The earliest mention of non constant normal forces in a vibration isolation system is by Mercer and Rees[69]. This work presents a novel concept of a friction mount where the normal forces are varied hydraulically depending on the relative displacement. Later, Tadjbakhsh and Lin[70] considered the performance of a seismic isolation device in which the normal force varied as a function of the relative displacement and relative velocity between the structure and the foundation.

As an example of how displacement-dependent normal forces effect the isolation performance of friction devices, consider the system shown in Figure 11. The system is similar to the externally forced system considered by Whiteman and Ferri[30] and discussed in Section 2.3. Changes in relative displacement generate changes in the normal force across the friction interface and also produce a force component tangent to the friction interface because of the profile angle γ. The equation of motion for this system can be written in terms of the relative displacement z as:

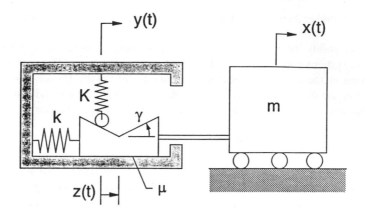

Figure 11: Base-isolation system using dry friction with displacement-dependent normal force.

$$m\ddot{z} + kz + \mu\left[N_0 + K|z|\tan\gamma\right]\text{sgn}(\dot{z}) + \left[N_0 + K|z|\tan\gamma\right]\tan\gamma\,\text{sgn}(z) = -m\ddot{y} \qquad (28)$$

The bracketed term in (28) is seen to be the normal force which consists of a pre-load portion plus a displacement-dependent portion. Note the resemblance of (28) to (16). This system can be analyzed qualitatively using first-order harmonic balance if one first assumes harmonic base excitation and response:

$$y(t) = Y\cos(\omega t + \phi) = Y_c\cos(\omega t) + Y_s\sin(\omega t) \qquad (29)$$

$$z(t) = Z\cos(\omega t) \qquad (30)$$

Substitution of (29) and (30) into (28) and balancing the first harmonics of all terms yields the algebraic system of equations:

$$m\omega^2 Y_c = (k - \omega^2)Z + \frac{4}{\pi}N_0\tan\gamma + KZ\tan^2\gamma \qquad (31)$$

$$m\omega^2 Y_s = -\frac{4}{\pi}\mu N_0 - \frac{2}{\pi}\mu KZ\tan\gamma \qquad (32)$$

Squaring and adding (31) and (32) yields a quadratic equation in Z:

$$a''Z^2 + b''Z + c'' = 0 \qquad (33)$$

where

$$a'' = \left[\left(k + K\tan^2\gamma - \omega^2 m\right)^2 + \left(\frac{2}{\pi}\mu K\tan\gamma\right)^2\right]$$ (34a)

$$b'' = 2\left[\left(k + K\tan^2\gamma - \omega^2 m\right)\left(\frac{4}{\pi}N_0\tan\gamma\right) + \left(\frac{2}{\pi}\mu K\tan\gamma\right)\left(\frac{4}{\pi}\mu N_0\right)\right]$$ (34b)

$$c'' = \left[\left(\frac{4}{\pi}N_0\tan\gamma\right)^2 + \left(\frac{4}{\pi}\mu N_0\right)^2 - m^2\omega^4 Y^2\right]$$ (34c)

Again, only positive solutions for Z are permitted due to the assumptions made in deriving equations (31) and (32). Examination of this quadratic reveals the following interesting facts. First, Z can become unbounded at resonance if and only if $\mu K\tan\gamma=0$. Only one positive solution is possible as long as $c''<0$; i.e., if Y is sufficiently large. However, two positive solutions are possible if $c''>0$, $b''<0$, and $(b'')^2-4a''c''\geq0$. As noted earlier, this is significant in that if this behavior is evident in the actual system (28), it means that two different amplitudes of response are possible depending on the system's initial disturbance. Proceeding as before, an equivalent natural frequency Ω_n and an equivalent viscous damping ratio ζ_n can be obtained:

$$\Omega_n = \sqrt{\frac{k}{m} + \frac{K\tan^2\gamma}{m} + \frac{4N_0\tan\gamma}{\pi m Z}}$$ (35)

$$\zeta_n = \frac{2\mu N_0 + \mu KZ\tan\gamma}{\pi m Z\omega\Omega_n}$$ (36)

It is seen that Ω_n decreases with Z (if $N_0\tan\gamma>0$) in the manner of a system with a "softening spring." The displacement-dependent normal force also gives rise to an equivalent viscous damping that does not approach zero as $Z\to\infty$.

The merits of a vibration isolation system are judged by its relative displacement transmissibility, $T_{rd}=Z/Y$, and its acceleration transmissibility, $T_a=(\omega^2 X)/(\omega^2 Y)=X/Y$, where X is the amplitude of the absolute displacement of the mass, $x(t)=y(t)+z(t)$. The relative displacement transmissibility describes the stroke or *rattlespace* requirements for the isolation mount. The acceleration transmissibility describes the isolation capability of the mount. Figures 12 and 13 show the effect on T_{rd} and T_a of changes in the profile angle γ. Note that for sufficiently low frequencies there are no positive solutions to (33), consequently, there is no HB prediction for T_{rd} and T_a. (The plots show zero values solely for plotting purposes.) The absence of positive solutions indicates that the relative motion

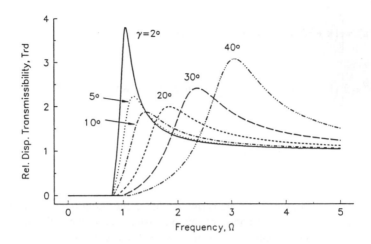

Figure 12: Relative displacement transmissibility, T_{rd}, versus nondimensional frequency. $\Omega = \dfrac{\omega}{\omega_n}$, $\omega_n^2 = \dfrac{k}{m}$, $\dfrac{N_0}{Yk} = 1$, $\dfrac{K}{k} = 10$, $\mu = 0.5$, various ramp angles γ.

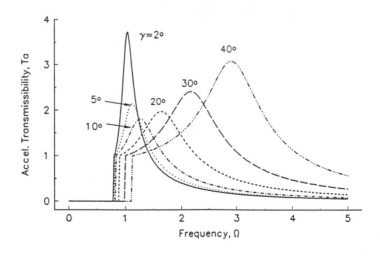

Figure 13: Absolute acceleration transmissibility, T_a, versus nondimensional frequency; same parameter values as for Figure 11.

is most probably of a stick-slip nature. It is seen that the peak transmissibilities decrease with increasing γ up to a point and then begin increasing. Fortunately, both transmissibilities are minimized at approximately the same value of ramp angle, $\gamma=12°$. Figure 14 shows the equivalent viscous damping ratio evaluated at resonance. This value is obtained using (36) with $\omega= \Omega_n$ and setting Z to the maximum value of displacement amplitude over all frequencies. It is seen that the damping ratio is also maximized at $\gamma=12°$. Finally, Figure 15 shows the factor by which the equivalent natural frequency increases (over its value at $\gamma=0$) as a function of γ. It is seen that increases in γ increase the effective stiffness because of the component of the roller force tangent to the friction interface. In light of the trends in damping and stiffness, the high value of transmissibility for small values of γ are most likely due to inadequate damping, while the high value of transmissibility at high values of γ is most likely due to the increase in stiffness.

The influence of the preload force N_0 on the relative displacement and acceleration transmissibilities is seen in Figures 16 and 17, respectively. It is seen that N_0 reduces the peak values of both T_{rd} and T_a but does so at the expense of high-frequency isolation (T_a).

2.5 In-Plane Slip

As mentioned earlier, several researchers have considered the damping contribution from in-plane slip at frictional boundaries of beams and plates. This subsection describes the analysis of such systems when the normal force is dependent on the in-plane slip displacement. Two such systems shown in Figures 18 and 19 were considered by Ferri and Bindemann[71]. Figure 18 shows a long slender beam which is hinged ideally at its left support and allowed to slip against a stationary support on its right end. The profiled nature of the beam's end together with the spring/roller system oriented normal to the slipping plane causes the normal force to increase as a function of the in-plane slip displacement. The system in Figure 19 also has a variable normal load on its right end, but in this case, the normal force increases with *transverse* beam tip displacement.

The governing equation for the transverse motion of both of these systems has the form[72]:

$$EI\frac{\partial^4 w(x,t)}{\partial x^4} - N_x\frac{\partial^2 w(x,t)}{\partial x^2} + m\frac{\partial^2 w(x,t)}{\partial t^2} = F(x,t) \tag{37}$$

where EI is the beam flexural rigidity, $w(x,t)$ is the beam's transverse displacement at location x and time t, N_x is the axial tensile force in the beam, m is the beam mass per unit length, and $F(x,t)$ is the distributed force per unit length. If one neglects the axial inertia, it can be shown N_x is a function of t only:

$$N_x = -\mu N\,\mathrm{sgn}(\dot{u}) - N\tan\gamma\,\mathrm{sgn}(u) \tag{38}$$

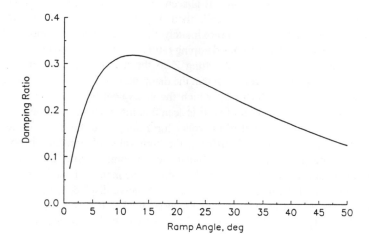

Figure 14: Equivalent viscous damping ratio at resonance versus ramp angle. Same parameter values as for Figure 11.

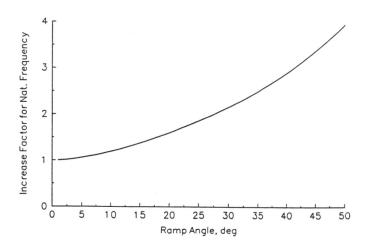

Figure 15: Ratio of Ω_n to ω_n versus ramp angle. Same parameter values as for Figure 11.

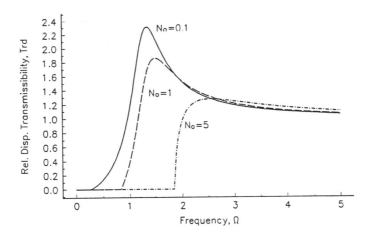

Figure 16: Relative displacement transmissibility, T_{rd}, versus nondimensional frequency. $\Omega = \dfrac{\omega}{\omega_n}$, $\omega_n^2 = \dfrac{k}{m}$, $\dfrac{K}{k} = 10$, $\mu = 0.5$, $\gamma = 12^\circ$, various levels of preload.

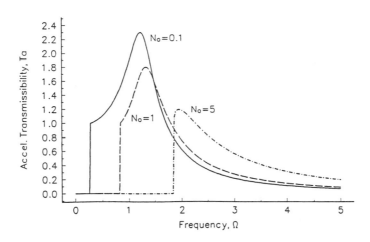

Figure 17: Absolute acceleration transmissibility, T_a, versus nondimensional frequency; same parameter values as for Figure 16.

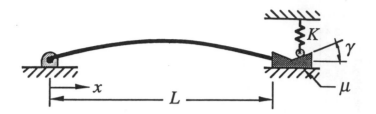

Figure 18: Flexible beam system with in-plane slip and friction on right support. Normal force dependent on in-plane slip displacement. From Ref. 71.

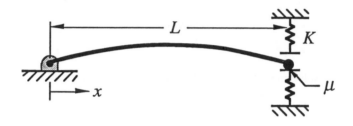

Figure 19: Flexible beam system with in-plane slip and friction on right support. Normal force dependent on transverse beam displacement. From Ref. 71.

where μ is the coefficient of friction, γ is the profile angle, u is the in-plane or axial slip displacement of the beam's right end, and N is the normal force. For the system shown in Figure 18, N is given by:

$$N = N_0 + K|u|\tan\gamma \qquad (39)$$

where N_0 is the spring preload and K is the spring constant of the normal spring. (Note that the contribution to the normal load from the shear contribution on the beam at x=L has been neglected.) A simplifying assumption that is justified for moderate in-plane force is that the slip displacement is due entirely to beam foreshortening, i.e., the beam is inextensible. In this case, the expression for u simplifies to:

$$u = -\frac{1}{2}\int_0^L \left(\frac{\partial w(x,t)}{\partial x}\right)^2 dx \qquad (40)$$

A first approximation to this system's vibrational characteristics can be obtained by assuming that the transverse beam displacement is dominated by a single spatial mode:

$$w(x,t) = z(t)\phi(x) \tag{41}$$

where $z(t)$ is the time-varying "modal amplitude" and $\phi(x)$ is an assumed mode for the transverse beam displacement. Substitution of (41) into equations (37)-(40) and applying the Galerkin Procedure yields the following one-mode approximation to the beam vibration problem[72]:

$$m_n(\ddot{z} + \omega_n^2 z) + N_x \beta z = f(t) \tag{42}$$

where

$$m_n = \int_0^L m\phi^2 dx \tag{43a}$$

$$m_n \omega_n^2 = \int_0^L EI(\phi'')^2 dx \tag{43b}$$

$$\beta = \int_0^L (\phi')^2 dx \tag{43c}$$

$$f(t) = \int_0^L F(x,t)\phi(x) dx \tag{43d}$$

The subscript n is used to designate a particular "mode" or resonance number. A one mode approximation for the axial force has the same form as (38) with N as given in (39) and with u and $\dot{u}$ given by:

$$u = -\frac{1}{2}\beta z^2 \tag{44a}$$

$$\dot{u} = -\beta z\dot{z} \tag{44b}$$

Combining the above yields:

$$m_n(\ddot{z} + \omega_n^2 z) + \mu\left(N_0 + \frac{1}{2}\beta Kz^2 \tan\gamma\right)\beta z\,\text{sgn}(z\dot{z}) + \left(N_0 + \frac{1}{2}\beta Kz^2 \tan\gamma\right)\beta z\tan\gamma = f(t) \tag{45}$$

Equation (45) can be time integrated to determine the transient or forced response of the system depicted in Figure 18. Qualitative information can be obtained by applying the first-order HB procedure. As before, the response and forcing are assumed to have the following harmonic form:

$$z(t) = A \cos(\omega t) \tag{46}$$

$$f(t) = f \cos(\omega t + \alpha) = f^c \cos(\omega t) + f^s \sin(\omega t) \tag{47}$$

Substitution of (46) and (47) into (45), expanding nonlinear terms using only the first harmonic and balancing harmonics yields the following two equations:

$$m_n(\omega_n^2 - \omega^2)A + c_0 A + c_1 A^2 = f^c \tag{48}$$

$$-c_2 A - c_3 A^3 = f^s \tag{49}$$

where:

$$c_0 = \beta N_0 \tan \gamma \tag{50a}$$

$$c_1 = \frac{3}{8} K \beta^2 \tan^2 \gamma \tag{50b}$$

$$c_2 = \frac{2}{\pi} \mu \beta N_0 \tag{50c}$$

$$c_3 = \frac{\mu}{2\pi} K \beta^2 \tan \gamma \tag{50d}$$

Equations (48) and (49) can be combined using the relation $f^2 = (f^c)^2 + (f^s)^2$:

$$
\begin{aligned}
&(c_1^2 + c_3^2)A^6 + 2(c_0 c_1 + c_1 m_n(\omega_n^2 - \omega^2) + c_2 c_3)A^4 \\
&+ (c_0 + 2m_n c_0(\omega_n^2 - \omega^2) + m_n^2(\omega_n^2 - \omega^2)^2 + c_2^2)A^2 - f^2 = 0
\end{aligned} \tag{51}
$$

As before, only positive real solutions for the amplitude A are valid; at any value of excitation frequency, there are at most 3 positive solutions.

Following a similar procedure to that described in Section 2.2, expressions can be obtained for the equivalent natural frequency Ω_n and the equivalent viscous damping ratio ζ_n:

$$\Omega_n = \sqrt{\omega_n^2 + \frac{c_0}{m_n} + \frac{c_1 A^2}{m_n}} \qquad (52)$$

$$\zeta_n = \frac{c_2 + c_3 A^2}{2m_n \omega \Omega_n} \qquad (53)$$

When evaluated at the resonant frequency $\omega = \Omega_n$, ζ_n takes the form:

$$\left(\zeta_n\right)_{res} = \frac{c_2 + c_3 A^2}{2\left(m_n \omega_n^2 + c_0 + c_1 A^2\right)} \qquad (54)$$

Qualitatively, it is clear from (52) that the system exhibits hardening spring behavior provided that $c_1 \neq 0$ (K,γ>0). From (54), it is seen that the system's damping is significantly different than that traditionally associated with systems that are dry friction damped. It should also be noted that if the profile angle is set to zero, one recovers the results given by Jézéquel[33] and Dowell[34], namely that $(\zeta_n)_{res}$ is invariant with respect to amplitude of response just as in the case of linear viscous or structural damping.

Returning to the system of Figure 19, a one-mode approximation followed by an HB analysis can be applied to uncover the qualitative damping features of this type of system. There are three differences from the preceding analysis:

1. The transverse spring K adds stiffness directly to the system; equation (42) becomes:

$$m_n(\ddot{z} + \omega_n^2 z) + N_x \beta z + K\phi_L^2 z = f(t) \qquad (55)$$

2. The expression for normal force must be replaced by:

$$N = K|w(L,t)| \approx K\phi_L |z(t)| \qquad (56)$$

where ϕ_L is the value of the assumed mode function $\phi(x)$ evaluated at $x=L$; it is presumed that ϕ is chosen such that $\phi_L > 0$.

3. The axial force N_x is due entirely to the friction force at the beam end:

$$N_x = -\mu N \, sgn(\dot{u}) \approx \mu K\phi_L |z| sgn(z\dot{z}) \qquad (57)$$

Substitution of (57) into (55) yields the one-mode approximation for this system:

$$m_n(\ddot{z} + \omega_n^2 z) + K\phi_L^2 z + \mu\beta K\phi_L z|z|\text{sgn}(z\dot{z}) = f(t) = f\cos(\omega t) \tag{58}$$

From the HB analysis, two algebraic equations can be derived:

$$m_n(\omega_n^2 - \omega^2)A + K\phi_L^2 A = f^c \tag{59}$$

$$-c_4 A^2 = f^s \tag{60}$$

Finally, the equivalent natural frequency and viscous damping ratio at resonance for the system of Figure 19 are given by:

$$\Omega_n = \sqrt{\omega_n^2 + \frac{K\phi_L^2}{m_n}} \tag{61}$$

$$(\zeta_n)_{res} = \frac{c_4 A}{2(m_n\omega_n^2 + K\phi_L^2)} \tag{62}$$

where

$$c_4 = \frac{4\mu\beta K\phi_L^2}{3\pi} \tag{63}$$

The important thing to note about (62) is that the damping increases with transverse displacement amplitude in a manner similar to aerodynamic or *air* (drag) damping. This air-damping characteristic has been observed in a number of other systems that have been studied analytically. One such system consisted of a rigid beam inserted into a nonlinear sleeve joint model[26]. Another example is a recent study of a flexible beam that was pin-supported on one end and restrained by a nonlinear sleeve joint on the other end[73]. The flexible beam model included the effects of axial inertia and extensibility. Furthermore, the beam model used several assumed modes for the beam rather than a single one. A system similar to the one shown in Figure 18 was used as a "testbed" with which to examine alternative friction models[8]. The beam model allowed for beam extensibility and several beam modes were used to describe the transverse beam displacement. The study considered five different models for dry friction and determined their effect on the system's damping and vibratory properties.

3. Semi-Active Friction

As described in the previous section, the behavior of dry friction damped systems with displacement-dependent friction forces is fundamentally different than that of systems with constant normal forces. It is natural to ask if better damping properties can be obtained if one *actively controls* the normal force in real time based on feedback from sensory outputs.

This section focuses on the design and behavior of semi-active frictionally damped systems. The concept of semi-active damping is first reviewed in Section 3.1. The use of semi-active friction damping in a SDOF system is described in Section 3.2 while Section 3.3 discusses its application to vibration control for flexible structures. Section 3.4 describes the application of semi-active friction damping to an automobile suspension system.

3.1 Semi-Active Damping

The concept of semi-active damping was formally proposed by Karnopp et al. in 1974[74]. The concept involves the use of active control theory to augment the damping properties of a passive element in real time. Sometimes referred to as active-passive damping, the technique offers considerable advantages in performance over completely passive damping elements, at the expense of a slight increase in system cost/complexity. On the other hand, semi-active damping cannot deliver the level of performance of a *fully active* system. The tradeoff is that semi-active damping requires much less energy (since one is only changing a passive damping level) and consequently can usually be implemented with less weight and cost than a fully active system. It is interesting to note that a semi-active damping cannot add energy to a system. Because of this, it cannot destabilize the closed-loop system, which is an important consideration in flexible structure control.

There are several means of realizing semi-active damping; see for example those suggested by Karnopp[75]. The most common way is by means of a viscous dashpot with a variable (controlled) orifice. This technique has been explored extensively in the field of semi-active automotive suspension[76-79]. It has also been suggested for use in flexible structure control; see for example, Davis et al.[80]. Another way of achieving semi-active damping is through the use of electrorheological fluids whose viscosity can be controlled through application of an electric field. This technology has also been explored for use in both semi-active suspension and flexible structure control[81-84]. Other ways in which semi-active damping can be accomplished include impact damping[85,86], and piezoelectric networks[87,88].

The concept of semi-active friction damping has been studied recently for use in both flexible structure control and in semi-active automotive suspension systems. The concept uses active control theory to vary the normal force and, thus, the friction force in response to sensory feedback. The earliest use of semi-active friction damping is in the paper by Anderson and Ferri[27] which is described in Section 3.2. The concept was also mentioned by Karnopp[75] as one of several ways of developing semi-active damping forces. Karnopp's idea was based on an antilock braking system (ABS), which of course is a closely related technology. In the case of ABS braking, one is concerned with the avoidance of sticking in a frictional interface while in the case of semi-active damping, one

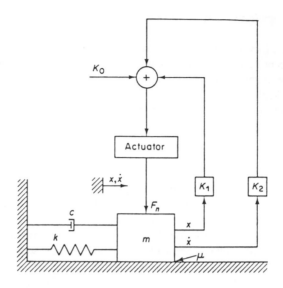

Figure 20: SDOF system with semi-active friction force. From Ref. 27.

is chiefly concerned with dissipating energy as quickly as possible. While the two objectives are related, especially since a sticking interface cannot dissipate energy, the control systems that maximize these two objectives are different.

The use of semi-active friction damping in flexible structure control was first studied by Ferri and Heck[29,89] and extended with the help of Lane[90-93]. This research is described in Section 3.3 below. Other researchers that have considered semi-active friction damping for vibration control are Onoda and Minesugi[94] and Dupont and Stoker[95]. It should also be mentioned that Wang, Kim, and Shea[84] found that a bi-product of controlling the electrorheological fluid was a semi-active friction force.

The use of semi-active friction damping in the development of semi-active automotive suspension was first proposed and analyzed by Ferri and Heck[96]. Further work, including experimental testing of the concept is described in Lane et al.[97] and Huynh[98]. An overview of this work is presented in Section 3.4. The idea has also been the subject of recent papers by Ryba[99] and by Rezeka and Saafan[100].

3.2 Semi-Active Friction Damping in a SDOF System

The use of feedback control to modify the normal force across a frictional interface was studied by Anderson and Ferri[27]. In this paper, both displacement-dependent normal forces and velocity-dependent normal forces were examined. Consider the SDOF system shown in Figure 20. One can envision a situation in which the slip displacement and slip velocity are "fed back" to a normal force actuator. A control law of the following form was considered:

$$N = K_0 + K_1|x| + K_2|\dot{x}| \tag{64}$$

where $K_0 > 0$ is a constant bias, $K_1 > 0$ is a scalar "gain" for displacement magnitude, and $K_2 > 0$ is a gain for slip velocity magnitude. The control law modifies the normal force according to the magnitudes of slip displacement and velocity. The first two terms were analyzed in detail above in Section 2.2; note that $K_1 = K\tan\gamma$. The friction force can still be written in the form given by (1). However, an interesting simplification occurs when (64) and (1) are substituted into (2):

$$m\ddot{x} + c\dot{x} + kx + m\left(K_0 + K_1|x| + K_2|\dot{x}|\right)\mathrm{sgn}(\dot{x}) = F(t)$$
$$m\ddot{x} + \left(c + mK_2\right)\dot{x} + kx + m\left(K_0 + K_1|x|\right)\mathrm{sgn}(\dot{x}) = F(t) \tag{65}$$

Mathematically, the effect of the velocity feedback term is equivalent to that of a purely linear viscous damper, even though the physical mechanism of energy removal is quite different. In fact, if K_0 and K_1 are set to zero, the resulting differential equation is entirely linear. In essence, *the absolute value nonlinearity and the sgn nonlinearity cancel, leaving a purely linear term.*

A convenient property of systems having only a slip velocity feedback term (setting K_0 and K_1 equal to zero in equation (64)) is that it is impossible for them to stick away from the origin. When sticking occurs, $\dot{x} = 0 \Rightarrow N = 0$, which precludes the occurrence of sticking.

3.3 Structural Vibration Control

The semi-active frictional interface was further explored in the context of damping augmentation of flexible structures[29,89]. The system selected for study consisted of two linear elastic beams connected by a semi-active revolute joint as shown in Figure 21. Each beam is simply supported, but resistive moments are transmitted from one beam to the other through the intermediate semi-active joint. The flexural displacements of each beam can be expressed in terms of admissible bases functions:

$$w_1(x_1,t) = \sum_{i=1}^{N} a_i(t)\phi_i(x_1); \qquad w_2(x_2,t) = \sum_{i=1}^{N} b_i(t)\psi_i(x_2) \tag{66}$$

The basis functions ϕ_i and ψ_i are chosen to be the normalized eigenfunctions for simply-supported beams of length L_1 and L_2, respectively:

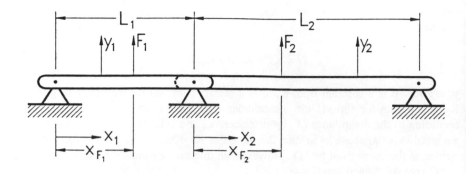

Figure 21: Flexural system with semi-active frictional joint placed between two pinned-pinned flexible beams. From Ref. 29.

$$\phi_n(x_1) = \sqrt{\frac{2}{m_1 L_1}} \sin\left(\frac{n\pi x_1}{L_1}\right) \tag{67a}$$

$$\psi_n(x_2) = \sqrt{\frac{2}{m_2 L_2}} \sin\left(\frac{n\pi x_2}{L_2}\right) \tag{67b}$$

where m_1 and m_2 are the mass per unit lengths of beams 1 and 2. Using these basis functions, the following equations of motion can be obtained:

$$\ddot{a}_i + 2\zeta_{1i}\omega_{1i}\dot{a}_i + \omega_{1i}^2 a_i = F_1(t)\phi_i(x_{F_1}) + M(t)\phi_i'(L_1), \qquad i = 1, N \tag{68}$$

$$\ddot{b}_i + 2\zeta_{2i}\omega_{2i}\dot{b}_i + \omega_{2i}^2 b_i = F_2(t)\psi_i(x_{F_2}) - M(t)\psi_i'(0), \qquad i = 1, N \tag{69}$$

where $M(t)$ is the controlled resistive moment between the two beams acting through the semi-active joint:

$$M(t) = \mu r N \, \text{sgn}(\dot{\theta}) \tag{70}$$

Here, μ is the coefficient of friction, r is a length scale for the joint (effective radius at which the friction force acts), N is the normal force which clamps the two beams into contact, and θ is the relative angle of rotation between the two contacting beam surfaces:

$$\theta(t) = \sum_{i=1}^{N} b_i(t)\psi_i'(0) - \sum_{i=1}^{N} a_i(t)\phi_i'(L_1) \tag{71}$$

Finally, ζ_{1i} and ζ_{2i} are the linear damping ratios for beams 1 and 2, respectively, and ω_{1i} and ω_{2i} are the original natural frequencies of beams 1 and 2, respectively, when no interactive moment is present, $M(t)=0$:

$$\omega_{ij} = j^2\pi^2\sqrt{\frac{EI_i}{m_i L_i^4}}, \qquad i=1,2 \qquad j=1,N \tag{72}$$

where EI_i is the flexural rigidity of beam i.

Equations (68) and (69) can be placed in state-space form by defining the state vector to be

$$x = \left[a_1, a_2, \cdots a_N, b_1, \cdots b_N, \dot{a}_1, \cdots \dot{a}_N, \dot{b}_1, \cdots \dot{b}_N\right]^T \tag{73}$$

and defining

$$F = \left[F_1(t) \quad F_2(t)\right]^T \tag{74}$$

Note that if N assumed modes are used for each beam, x is a 4Nx1 state vector. With these definitions, the system dynamics take the form:

$$\dot{x} = Ax + B_F F + B_M M \tag{75}$$

where the matrices A, B_F and B_M are easily identified from the indicial form (68) and (69). Note that although the closed-loop system is nonlinear, the *open-loop dynamics as represented by (75) are linear*.

The system outputs are chosen to be the transverse displacements of beam 1 at $x_1=0.25L_1$ and of beam 2 at $x_2=0.75L_2$:

$$y_1 = \sum_{i=1}^{N} a_i(t)\phi_i(0.25L_1); \qquad y_1 = \sum_{i=1}^{N} b_i(t)\psi_i(0.75L_2) \tag{76}$$

In vector form, the outputs can be expressed as:

$$y = \begin{bmatrix} y_1 \\ y_2 \end{bmatrix} = Cx \tag{77}$$

where C is a time-invariant, 2x4N output matrix. It is convenient also to express θ and $\dot{\theta}$ in terms of x:

$$\theta = \left[-\phi_1'(L_1), -\phi_2'(L_1), \cdots, -\phi_N'(L_1), \psi_1'(0), \cdots, \psi_N'(0), \underbrace{0, 0, \cdots, 0}_{2N \text{ zeros}} \right] x = D_1 x \qquad (78)$$

$$\dot{\theta} = \left[\underbrace{0, 0, \cdots, 0}_{2N \text{ zeros}}, -\phi_1'(L_1), -\phi_2'(L_1), \cdots, -\phi_N'(L_1), \psi_1'(0), \cdots, \psi_N'(0) \right] x = D_2 x \qquad (79)$$

Note that D_1 and D_2 are 1x4N row matrices, and that θ and $\dot{\theta}$ are linear in the state x.

Whereas the sticking conditions for a SDOF system are relatively easy to obtain, a more systematic approach is necessary when dealing with MDOF flexible systems. Such an approach is afforded by the theory of variable structure systems (VSS)[101]. To the author's knowledge the first application of VSS to dry friction damped systems was that of Ferri and Heck[29,89]. In the terminology of VSS, "sticking" in a dry friction damped systems is referred to as "sliding" because system trajectories are forced to "slide" along a switching surface. In this chapter, sticking and sliding will refer to the physical condition of the frictional interface.

An important feature in VSS systems is the presence of a switching surface, s(x)=0. In many cases, the dynamics of a VSS are discontinuous across this switching surface. When the vector fields on either side of the switching surface point towards the switching surface, the trajectory is forced to remain on this surface. In the dry friction damped system under consideration, the switching surface is simply $s(x)=\dot{\theta}=D_2 x=0$; the closed-loop dynamics as described by (70) and (75) are in general discontinuous across this surface due to the sgn nonlinearity in (70). Using VSS, the sticking condition can be written as:

$$\dot{\theta}\ddot{\theta} = x^T D_2^T D_2 \dot{x} = x^T D_2^T D_2 \left[Ax + B_F F + B_M M \right] < 0 \quad \forall \text{ x such that } |\dot{\theta}| < \varepsilon \qquad (80)$$

where ε is a small positive parameter. The method of equivalent control[101] can now be used to obtain the dynamics of the system during sticking. The equivalent control is found by setting the time derivative of $\dot{\theta}$ in (79) equal to zero and, substituting $\dot{x}$ from (75), solving for M. This yields the following expression:

$$\dot{x} = \left[A - B_M \left(D_2 B_M \right)^{-1} D_2 A \right] x + \left[B_F - B_M \left(D_2 B_M \right)^{-1} D_2 B_F \right] F \qquad (81)$$

As expected, the sticking dynamics are linear, as they would be if the two beams were welded together at some angle θ_0. Less obvious is the fact that the linear dynamics must contain two zero eigenvalues because of the constraint that $\dot{\theta}=0$.

Collocated Control

The simplest type of control system architecture is to make the normal force across the semi-active joint a function of joint variables alone; i.e., $N=N(\theta,\dot{\theta})$. As in the SDOF case, it is possible to fully linearize the equations of motion during slipping by choosing the control structure:

$$N = K|\dot{\theta}| \tag{82}$$

Substitution of (82) into (70) yields:

$$M(t) = \mu r K|\dot{\theta}|sgn(\dot{\theta}) = \mu r K\dot{\theta} = \mu r K D_2 x \tag{83}$$

Thus, the resistive moment is mathematically equivalent to that supplied by a rotary viscous dashpot. For this reason, this control design is referred to as the *viscous joint* controller. The linearity of the closed-loop dynamics are evident if one substitutes (83) into (75):

$$\dot{x} = [A + \mu r K B_M D_2]x + B_F F \tag{84}$$

The control design amounts to the selection of the scalar gain K. Because the closed-loop system is linear, stability is easily determined by examining the eigenvalues of the matrix $[\Lambda+\mu r K B_M D_2]$. Based on physical arguments, a sufficient condition for stability is K>0 as this represents the addition of a passive viscous damper to a system that was already stable and passive.

A viscous joint control system was designed for a two-beam system having the following parameter values:

Table 1. Parameter values for two-beam system

Parameter	Numerical Value
m_1, m_2	0.4 kg/m
EI_1, EI_2	20 N·m^2
L_1	1 m
L_2	1.8 m
μ	0.5
r	2.5 cm
ζ_{1i}, ζ_{2i}, i=1,N	0

Based on a 1-mode per beam model and a 3-mode per beam model, it was determined that selecting K=150 produced nearly maximum damping in all system modes.

State-Feedback Control: Clipped Optimal

The performance of the semi-active joint can be enhanced greatly if one allows the normal force to depend on the entire motion of the system N=N(x). An effective design procedure is that of clipped optimal control[90,91]. The technique makes use of the fact that the system dynamics (66) are linear in the resistive moment. Therefore, one could treat the moment M(t) as the control input and use standard linear optimal control design techniques. Due to the restriction that M(t) be passive, however, the following constraint must be satisfied:

$$M(t)\dot{\theta}(t) \geq 0. \tag{85}$$

This inequality basically says that the control moment must only do negative work on the system. Whenever the moment requested by the control law is consistent with (85), it is applied by the semi-active joint; when it is inconsistent with (85), the control is turned off or *clipped*. The inequality constraint (85) makes the control design problem of semi-active systems fundamentally different from that of fully-active systems. The fact that the moment can only do negative work is both a benefit and detriment of semi-active controllers: The passive nature of the control ensures that the system cannot be de-stabilized by the semi-active element, which is an important consideration in flexible structure control[102]. Furthermore, since one is only controlling a dissipative element, the power and weight requirements of the control actuator become smaller. On the one hand, the system cannot be completely controllable although it may be stable to the origin. Thus, there is a penalty in terms of performance.

A controller for the unforced two-beam system having parameter values specified in Table 1 was designed using the linear quadratic regulator (LQR) theory plus clipping per constraint (85). This control design is termed the *clipped LQR* design. The following performance index is minimized:

$$J = \frac{1}{2} \int_0^{\infty} x^T Q x + 2 x^T R M + \rho_M M^2 \ dt \tag{86}$$

where Q and R are the state and cross-state weighting matrices, and ρ_M is the (scalar) control weighting. It can be shown[102] that the optimal joint moment is given by a linear, time-invariant state feedback with 1x4N gain matrix K_O:

$$M(t) = -K_o x \tag{87}$$

where

$$K_o = \left[B_M^T P + R^T\right]/\rho_M \tag{88}$$

with P the positive-definite solution to the following algebraic matrix Riccati equation:

$$\left[A - B_M R^T / \rho_M\right]P + P\left[A - B_M R^T / \rho_M\right] + Q - RR^T / \rho_M - PB_M B_M^T P / \rho_M = 0 \tag{89}$$

Note that while M was used as the input to the system, it is actually the normal force N that is the control variable. Substitution of (70) into (87) yields the following expression for N:

$$N = \begin{cases} -\dfrac{1}{\mu r}K_o x\, \mathrm{sgn}(\dot\theta) \\ 0 \quad \text{if } N < 0 \end{cases} \tag{90}$$

By appropriate choices of Q, R, and ρ_M, various state-feedback control designs can be obtained to accomplish certain objectives. Several choices are contained in the thesis by Lane[92]. One good choice of quadratic performance index was found to be:

$$J = \int_0^\infty (T + V) + 0.1 M^2 \, dt \tag{91}$$

where T and V are the total system kinetic and potential energies, respectively. The weighting parameters that produce this performance index are easily seen to be:

$$Q = \mathrm{diag}\left(\omega_{11}^2, \omega_{12}^2, \cdots, \omega_{1N}^2, \omega_{21}^2, \omega_{22}^2, \cdots, \omega_{2N}^2, \underbrace{1, 1, \cdots, 1}_{2N \text{ ones}} \right); \qquad R = 0; \qquad \rho_M = 0.2 \tag{92}$$

The advantage of clipped LQR control over the viscous joint design is apparent if one compares the energy dissipation of the two designs. Figure 22 shows the total energy (T+V) as a function of time in response to a pulse disturbance applied to beam 1 at $x_F = 0.25 L_1$; the pulse response is computed numerically using the following functional form for $F_1(t)$:

$$F_1 = \begin{cases} 25 \text{ Newtons} & 0 \le t \le 0.01\text{s} \\ 0 & t > 0.01\text{s} \end{cases}$$

The figure shows clearly that a substantial improvement in energy dissipation is accomplished if the normal force is allowed to depend on the entire state rather than on the single joint rate quantity. The degree to which control clipping occurred is seen in Figure 23 which shows the normal force vs time in the clipped LQR control design. Clipping is

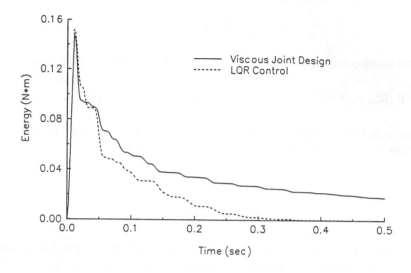

Figure 22: Total energy (T+V) for pulse response of two-beam system. Comparison of clipped LQR and collocated viscous joint designs. From Ref. 90.

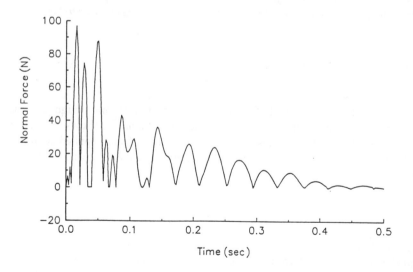

Figure 23: Normal force at semi-active friction joint for pulse response of two-beam system. Clipped LQR control design. From Ref. 90.

evidenced by the flat N=0 segments of the control history which is where the control had to be turned off. It is seen that the amount of time over which the clipped LQR controller requests an energy input to the system, thereby resulting in control force clipping, is very small indeed. This is not surprising when one considers that the performance index (91) is dominated by the total system energy. Obviously, it is unlikely that the controller would call for positive work to be performed on the system under these circumstances.

Nonlinear Optimal Control

It is interesting to compare the performance of the clipped LQR control, which is suboptimal because of the ad-hoc way that the constraint is applied, to the *true optimal performance*. The true optimal control solution to the present problem was developed by Lane and Ferri[90] using constrained optimal control. Due to the complexity of the solution procedure, only a brief overview is presented here.

It is convenient to re-write the unforced state equations (75) in the form:

$$\dot{x} = Ax + \mu r \, sgn(\dot{\theta}) B_M N \tag{93}$$

Since (93) is linear in the normal force N, this form is termed the bilinear form. In this form, the constraint is a simple inequality:

$$N \geq 0 \tag{94}$$

Note that this problem is nonlinear as well as constrained. Thus numerical solutions will be sought, and only fixed final time will be considered. The fixed-time performance criterion used is similar to the infinite time criterion (86):

$$J = \frac{1}{2} x_f^T P_f x_f + \frac{1}{2} \int_0^{t_f} x^T Q x + \rho_N N^2 \, dt \tag{95}$$

where t_f is the fixed final time, x_f denotes the state evaluated at the final time, $x_f = x(t_f)$, and the matrix P_f is used to apply a final-time state weighting. The optimal control problem, then, is to minimize (95) subject to (93), (94) and the additional constraints:

$$x(0), t_f \quad \text{specified} \tag{96}$$

$$x(t_f) \quad \text{free} \tag{97}$$

As a specific example, consider a two-beam system with parameter values given in Table 1 and with Q as specified in (84), with $P_f = Q$, and with $\rho_N = 3.125 \times 10^{-5}$ (this value of

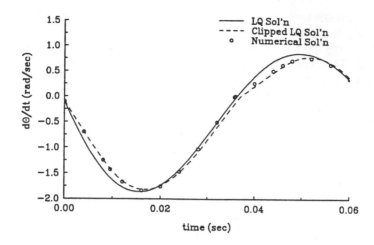

Figure 24: $\dot{\theta}$ vs time for two-beam system with semi-active frictional joint. Comparison of LQ solution, clipped LQ, and numerical optimization control designs. From Ref. 90.

ρ_N gives the same relative weighting between the resistive moment M and the state as used previously.) Thus the performance criterion is:

$$J = (T + V)\Big|_{t=t_f} + \int_0^{t_f} (T + V) + 1.5625 \times 10^{-5} N^2 \, dt$$

$$= (T + V)\Big|_{t=t_f} + \int_0^{t_f} (T + V) + 0.1 M^2 \, dt \tag{98}$$

Due to the difficulty of solving the nonlinear optimization problem, only one beam mode per beam was used to model the system. The numerical optimization computer package BNDSCO[103] was used to solve the two-point boundary value problem.

The first case considered is that of free response from non-zero initial conditions over a time interval of 0.06 seconds; i.e., $t_f = 0.06$s. The initial state corresponds to an initial midspan displacement for beam 1 of 1cm and an initial beam 1 velocity such that $\dot{\theta} = -0.1$ rad/s. Figure 24 compares the response of the joint rate $\dot{\theta}$ using 3 different control strategies. The first is the totally unconstrained LQR solution using (98) and assuming that the semi-active joint is replaced with an active hinge which can perform both negative *and* positive work on the system. Since the problem is linear, unforced and unconstrained, this must be the absolute minimum-cost solution. The second solution plotted in Figure 24 corresponds to clipped LQ control. The control is obtained by taking the time-varying optimal state-variable feedback gain and setting the normal force to zero whenever the solution calls for positive work to be performed. This should be a suboptimal solution

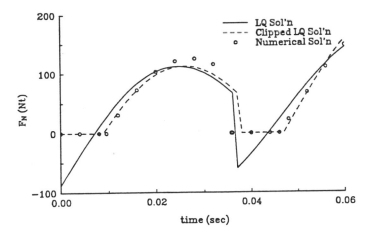

Figure 25: Normal force vs time for two-beam system with semi-active frictional joint. Comparison of LQ solution, clipped LQ, and numerical optimization control designs. From Ref. 90.

because of the ad-hoc way in which the constraint (94) is imposed. The third solution displayed is the true optimal control response as calculated by BNDSCO. Note that similarity between the clipped LQ solution and the true optimal. The closeness is further evidenced if one compares the value of the cost function for the three control techniques:

$$J = 0.010449 \quad \text{unconstrained LQ}$$
$$J = 0.011202 \quad \text{clipped LQ}$$
$$J = 0.011237 \quad \text{BNDSCO numerical solution}$$

Note that unconstrained LQ solution yields the smallest value of J as expected. However, the clipped LQ control actually outperforms the numerically-obtained optimal solution of BNDSCO. The reason for this outcome is most likely due to the numerical limitations of the BNDSCO routine. A similar phenomenon was observed by Hrovat et al.[76] for a semi-active suspension problem. Figure 25 compares the normal forces for the three control techniques. The time periods for which the normal force is negative in the unconstrained LQ response denotes those occasions where the optimal control calls for energy input to the two-beam system. The excellent performance of the clipped LQ control was observed for other combinations of t_f and initial conditions. Figures 26 and 27 show comparisons of the three control designs for the same initial condition as before but with $t_f=0.2s$. Again, the clipped LQ response compares favorably with the constrained optimal solution. The performance indices for these three controls are as follows:

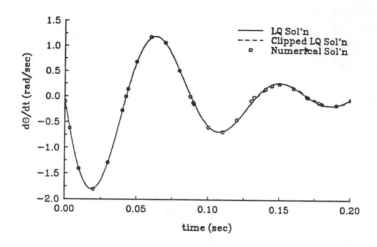

Figure 26: $\dot{\theta}$ vs time for two-beam system with semi-active frictional joint. Comparison of LQ solution, clipped LQ, and numerical optimization control designs. From Ref. 90.

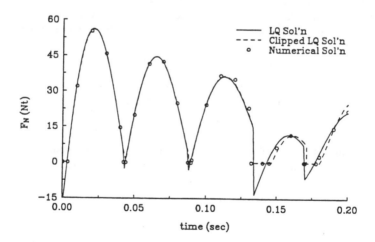

Figure 27: Normal force vs time for two-beam system with semi-active frictional joint. Comparison of LQ solution, clipped LQ, and numerical optimization control designs. From Ref. 90.

$$J = 0.004883 \text{ unconstrained LQ}$$
$$J = 0.004908 \text{ clipped LQ}$$
$$J = 0.004911 \text{ BNDSCO numerical solution}$$

Again, it is seen that the clipped LQ control actually outperforms the constrained optimal solution obtained using BNDSCO.

The good performance of the clipped LQ control design is significant since it suggests that *linear* control design techniques may be used to develop semi-active control strategies with acceptable results. Several other issues have been addressed in the context of semi-active vibration control. Lane and Ferri[93] investigated the combination of active and semi-active control elements into one vibration control system. Two control design techniques were compared: a hierarchical strategy where the semi-active element control design is performed first and one in which the three elements are designed together. Another issue is the tuning of the semi-active joint that is necessary depending on whether disturbances are applied to one beam or the other. Since the beams have different lengths and therefore different natural frequencies, different viscous joint control gains are optimal depending on whether the disturbance enters the slow subsystem or the fast subsystem.

3.4 Semi-Active Automotive Suspension

The last topic to be addressed in this chapter is that of semi-active frictional suspension systems for automotive applications. As mentioned above, the idea makes use of semi-active friction forces to develop an improved vehicle suspension system. The concept is similar to that proposed by Ferri and Heck for flexible structure control; feedback is used to augment the passive damping of a suspension system by varying the normal force across a frictional interface. To test this concept, a quarter car experimental apparatus was constructed. Figure 28 shows a schematic of the experimental configuration. The experimental system approximates one wheel/axle plus the associated carbody mass supported by that wheel. The primary suspension spring stiffness is k_s while the effective stiffness of the tire is denoted k_t. The masses and stiffness of the experimental system were chosen to be approximately 1/100th of their values in a full-size automobile. Thus, the scale system had the same dominant frequencies as that of a full-size system: carbody heave mode natural frequency of 1.25 Hz and a wheel hop mode natural frequency of 8.5 Hz. The semi-active friction damping force is generated through an electromechanical clutch (EMC) with computer-controlled coil voltage. The resistive moment generated by the EMC is converted into a damping force through means of crank-follower system. The supplied resistive torque is measured by means of a bending beam affixed to the output shaft of the EMC on which four strain gages were attached.

The control structure for the semi-active suspension is shown in Figure 29. It is seen that the control system is made up of an inner loop that controls the EMC, and an

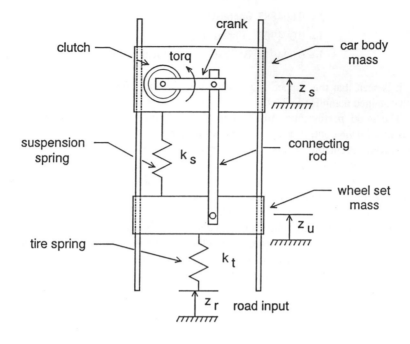

Figure 28: Quarter-car model with semi-active frictional suspension. From Ref. 97.

outer loop that supplies the suspension force command. The inner loop is made possible by the real-time measurement of the resistive torque using the bending beam. The significance of this innovation cannot be overemphasized. It removes the necessity of predicting the frictional moment based on an indirect measurement of coil current or normal force. Furthermore, it greatly lessens the system's sensitivity to variations in friction coefficient and to variations of friction coefficient with interface and environmental factors. The quarter-car experiment served as an effective proof-of-concept for the semi-active frictional suspension. The reader is referred to Lane et al.[97] and Huynh[98] for a more detailed discussion.

It should be mentioned that there is a fundamental difference between friction-based and viscous-based semi-active suspension elements. In many cases, damping forces are needed even though the stroke velocity (the relative velocity between the carbody mass and the wheelset mass) is zero or nearly so. This happens, for instance when a "skyhook damping" control algorithm is employed, which commands the damper force to be proportional to the *absolute carbody velocity* rather than the relative stroke velocity[79]. A related problem occurs when the control algorithm demands a zero force. This is

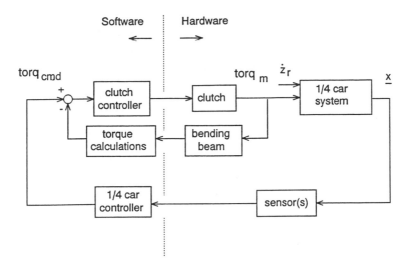

Figure 29: Control diagram for semi-active frictional suspension system. From Ref. 97.

sometimes called the zero-signal-zero-force problem and occurs whenever the relative velocity across the suspension is opposite of that necessary to supply the commanded damping force. In the case of clipped LQ control, for example, when the LQ controller calls for energy to be input into system, the command is clipped or set to zero as discussed in Section 3.3. Obviously, the force can be made zero in a friction-based element just by setting the normal force equal to zero. The damping force in a viscous-based element cannot be made zero, identically, due to the fact that a force will exist as so long as the fluid has some viscosity and the orifice has a finite opening. Both of these limitations degrade the performance of the viscous-based elements.

4. Conclusions

The behavior of dry friction damped systems with motion-dependent normal forces has been discussed. The variation in normal force occurs to some extent in all friction-damped systems. When the dependence of the normal force on the system motion becomes significant, however, it is seen that the systems can develop dynamic behavior that is substantially different from that typically associated with dry friction damped systems. The chapter discusses how the behavior of these systems can be analyzed and how normal force variation can be put to good use in vibration control and shock isolation.

In passive dry friction damped systems, the normal force can be made to vary with slip displacement in a relatively straightforward manner as discussed above. However, many unresolved issues remain. Among them are: What is the optimal manner in which the normal force should vary with slip displacement? This chapter considers only linear dependence, but are other functional forms better? It is certainly possible to envision a system for which the normal force, though passively controlled, depends on *slip velocity and slip acceleration* in addition to slip displacement. This could be accomplished, for example, by including a viscous dashpot parallel to the normal spring K in Figure 1 and by accounting for the "roller mass" in the model. What effect would this have on the system behavior? Finally, though one means of varying a friction force is through changing the normal force, it is also possible to do so by changing the friction coefficient along the frictional interface. For example, the surface roughness or material properties can vary along the contact plane. How might this be exploited to improve damping and vibration isolation performance?

Research on semi-active friction damping is just in its infancy. There are many topics that are yet to be studied. For instance, what control laws are appropriate for semi-active friction dampers? Although semi-active friction dampers are mathematically similar to semi-active dampers based on other energy dissipation mechanisms, the physics of friction dampers is different. It should be possible to exploit this difference in the design of control laws. Finally, while EMC's provide one method by which a semi-active friction damper can be constructed, other force actuation systems such as those based on piezoelectric and magnetorestrictive elements should be explored.

5. Acknowledgment

This work was partially supported by the Office of Naval Research under contract number N00014-94-1-0517. Dr. Geoffrey L. Main is the contract monitor. The author would also like to thank his many co-authors on the work cited in this chapter.

6. References

1. Beards, C.F., 1992, "Damping in Structural Joints," *The Shock and Vibration Digest*, Vol. 24, No. 7, pp. 3-7.

2. Ungar, E.E., 1973, "The Status of Engineering Knowledge Concerning the Damping of Built-Up Structures," *Journal of Sound and Vibration*, Vol. 26, No. 1, pp. 141-154.

3. Guran, A., Feeny, B., Hinrichs, N., Popp, K., 1996, "Dynamics of Friction: Modeling, Analysis, and Experiment," Chapter 12, Dynamics With Friction: Modeling Analysis, and Control, Vol. 7, Series on Stability, Vibration, and Control of Structures, World Scientific Publishing.

4. Ibrahim, R.A., 1994, "Friction-Induced Vibration, Chatter, Squeal, and Chaos- Part I: Mechanics of Contact and Friction; - Part II: Dynamics and Modeling," *Applied Mechanics Reviews*, Vol. 47, No. 7, pp. 209-253.

5. Armstrong-Hélouvry, B., Dupont, P., and Canudas de Wit, C., 1994, "A Survey of Models, Analysis Tools and Compensation Methods for the Control of Machines with Friction," *Automatica*, Vol. 30, No. 7, pp. 1083-1138.

6. Ferri, A.A., 1995, "Friction Damping and Isolation Systems," *ASME Journal of Vibration and Acoustics*, Vol. 117B, June, pp. 196-206.

7. Jones, D.I.G., 1990, "Damping of Dynamic Systems," *The Shock and Vibration Digest*, Vol. 22, No. 4, pp. 3-10.

8. Bindemann, A.C., and Ferri, A.A., 1995, "The Influence of Alternative Friction Models on the Passive Damping and Dynamic Response of a Flexible Structure," Proceedings of the 36th Structures, Structural Dynamics, and Materials Conference, New Orleans, LA, April 10-12, pp. 180-189.

9. Bielawa, R.L., 1978, "An Analytical Study of the Energy Dissipation of Turbomachinery Bladed-Disk Assemblies due to Inter-Shroud Segment Rubbing," *ASME Journal of Mechanical Design*, Vol. 100, April, pp. 222-228.

10. Jacobsen, L.S., 1930, "Steady Forced Vibration as Influenced by Damping," *Transactions of the ASME*, Vol. APM-52-15, pp. 169-181.

11. Den Hartog, J.P., 1956, Mechanical Vibrations, McGraw-Hill, NY.

12. Earles, S.W.E., and Williams, E.J., 1972, "A Linearized Analysis for Frictionally Damped Systems," *Journal of Sound and Vibration*, Vol. 24, No. 4, pp. 445-458.

13. Menq, C.-H., Griffin, J.H., and Bielak, J., 1986, "The Forced Response of Shrouded Fan Stages," *ASME Journal of Vibration, Acoustics, Stress, and Reliability in Design*, Vol. 108, No. 1, January, pp. 50-55.

14. Menq, C.-H., Griffin, J.H., and Bielak, J., 1986, "The Influence of a Variable Normal Load on the Forced Vibration of a Frictionally Damped Structure," *ASME Journal of Engineering for Gas Turbines and Power*, Vol. 108, No. 2, April, pp. 300-305.

15. Den Hartog, J.P., 1931, "Forced Vibrations with Combined Coulomb and Viscous Friction," *Transactions of the ASME*, APM-53-9, pp. 107-115.

16. Sinha, A., and Griffin, J.H., 1983, "Friction Damping of Flutter in Gas Turbine Engine Airfoils," *AIAA Journal of Aircraft*, Vol. 20, No. 4, pp. 372-376.

17. Sinha, A, and Griffin, J.H., 1985, "Stability of Limit Cycles in Frictionally Damped and Aerodynamically Unstable Rotor Stages," *Journal of Sound and Vibration*, Vol. 103, No. 3, pp. 341-356.

18. Sinha, A., and Griffin, J.H., 1985, "Effects of Friction Dampers on Aerodynamically Unstable Rotor Stages," *AIAA Journal*, Vol. 23, No. 2, pp. 262-270.

19. Sinha, A., Griffin, J.H., and Kielb, R.E., 1986, "Influence of Friction Dampers on Torsional Blade Flutter," *ASME Journal of Engineering for Gas Turbines and Power*, Vol. 108, April, pp. 313-318.

20. Ferri, A.A. and Dowell, E.H., 1985, "The Behavior of a Linear, Damped Modal System with a Nonlinear Spring-Mass-Dry Friction Damper System Attached, Part II," *Journal of Sound and Vibration*, Vol. 101, No. 1, pp. 55-74.

21. Hertz, T.J., and Crawley, E.F., 1983, "The Effects of Scale on the Dynamics of Flexible Space Structures," Report # SSL 18-83, Space Systems Laboratory, Department of Aeronautics and Astronautics, Massachusetts Institute of Technology, Cambridge, MA, September.

22. Hertz, T.J., and Crawley, E.F., 1984, "Damping in Space Structure Joints," AIAA Dynamics Specialists Conference, Palm Springs, CA, May 17-18, AIAA Paper No. AIAA-84-1039-CP.

23. Hertz, T.J., and Crawley, E.F., 1985, "Displacement Dependent Friction in Space Structural Joints," *AIAA Journal*, Vol. 23, No. 12, pp. 1998-2000.

24. Folkman, S.L., and Redd, F.J., 1990, "Gravity Effects on Damping of a Space Structure with Pinned Joints," *AIAA Journal of Guidance, Control, and Dynamics*, Vol. 13, No. 2, March-April, pp. 228-233.

25. Beucke, K.E., and Kelly, J.M., 1985, "Equivalent Linearizations for Practical Hysteretic Systems," *International Journal of Non-Linear Mechanics*, Vol. 23, No. 4, pp. 211-238.

26. Ferri, A.A., 1988, "Modeling and Analysis of Nonlinear Sleeve Joints at Large Space Structures," *AIAA Journal of Spacecraft and Rockets*, Vol. 25, No. 5, September-October, pp. 354-360.

27. Anderson, J.R., and Ferri, A.A., 1990, "Behavior of a Single-Degree-of-Freedom System with a Generalized Friction Law," *Journal of Sound and Vibration*, Vol. 140, No. 2, pp. 287-304.

28. Makris, N., and Constantinou, M.C., 1991, "Analysis of Motion Resisted by Friction. I. Constant Coulomb and Linear/Coulomb Friction," *Mechanics of Structures and Machines*, Vol. 19, No. 4, pp. 477-500.

29. Ferri, A.A., and Heck, B.S., 1992, "Analytical Investigation of Damping Enhancement Using Active and Passive Structural Joints," *AIAA Journal of Guidance, Control, and Dynamics*, Vol. 15, No. 5, September-October, pp. 1258-1264.

30. Whiteman, W.E., and Ferri, A.A., 1995, "Analysis of Beam-Like Structures with Displacement-Dependent Friction Forces, Part I: Single-Degree-Of-Freedom Model; and Part II: Multi-Degree-Of-Freedom Model," Proceedings of the ASME 15th Biennial Conference on Vibration and Noise, Boston, MA, September 17-20, DE-Vol. 84-1, Vol. 3, Part A, pp. 1093-1108.

31. Inaudi, J.A., Nims, D.K., and Kelly, J.M., 1994, "Linear-Friction Dissipators for Truss Structures," Proceedings of SPIE- International Society of Optical Engineering, Orlando, FL, February 14-16, Vol. 2193, pp. 213-224.

32. Beards, C.F., 1983, "The Damping of Structural Vibration by Controlled Interfacial Slip in Joints," *ASME Journal of Vibrations, Acoustics, Stress and Reliability in Design*, Vol. 105, July, pp. 369-373.

33. Jézéquel, L., 1983, "Structural Damping by Slip in Joints," *ASME Journal of Vibration, Acoustics, Stress, and Reliability in Design*, Vol. 105, pp. 497-504.

34. Dowell, E.H., 1986, "Damping in Beams and Plates due to Slipping at the Support Boundaries," *Journal of Sound and Vibration*, Vol. 105, No. 2, pp. 243-253.

35. Tang, D.M., and Dowell, E.H., 1986, "Damping in Beams and Plates due to Slipping at the Support Boundaries, Part 2: Numerical and Experimental Study" *Journal of Sound and Vibration*, Vol. 108, No. 3, pp. 509-522.

36. Tang, D.M., and Dowell, E.H., 1986, "Random Response of Beams and Plates With Slipping at the Support Boundaries" *AIAA Journal*, Vol. 24, No. 8, pp. 1354-1361.

37. Tang, D.M., Dowell, E.H., and Albright, J.E., 1987, "Damping in Beams with Multiple Dry Friction Supports," *AIAA Journal of Aircraft*, Vol. 24, No. 11, pp. 820-832.

38. Feeny, B.F., and Moon, F.C., 1989, "Autocorrelation on Symbol Dynamics for a Chaotic Dry-Friction Oscillator," *Physics Letters A*, Vol. 141, No. 8-9, November 20, pp. 397-400.

39. Feeny, B., 1992, "A Nonsmooth Coulomb Friction Oscillator," *Physica D*, Vol. 59, pp. 25-38.

40. Feeny, B., and Moon, F.C., 1994, "Chaos in a Forced Dry-Friction Oscillator: Experiments and Numerical Modelling," *Journal of Sound and Vibration*, Vol. 170, No. 3, pp. 303-323.

41. Dupont, P.E., and Bapna, D., 1994, "Stability of Sliding Frictional Surfaces with Varying Normal Force," *ASME Journal of Vibration and Acoustics*, Vol. 116, April, pp. 237-242.

42. Dupont, P.E., and Bapna, D., 1995, "Perturbation Stability of Frictional Sliding with Varying Normal Force," submitted to the *ASME Journal of Vibration and Acoustics*.

43. Hess, D.P., and Soom, A., 1991, "Normal Vibrations and Friction Under Harmonic Loads: Part I- Hertzian Contacts," *ASME Journal of Tribology*, Vol. 113, January, pp. 80-86.

44. Hess, D.P., and Soom, A., 1991, "Normal Vibrations and Friction Under Harmonic Loads: Part II- Rough Planar Contacts," *ASME Journal of Tribology*, Vol. 113, January, pp. 87-92.

45. Hess, D.P., Soom, A., and Kim, C.H., 1992, "Normal Vibrations and Friction at a Hertzian Contact Under Random Excitation: Theory and Experiments," *Journal of Sound and Vibration*, Vol. 153, No. 3, pp. 491-508.

46. Caughey, T.K., and Vijayaraghavan, A., 1970, "Free and Forced Oscillations of a Dynamic System With 'Linear Hysteretic Damping' (Non-Linear Theory)", *International Journal of Non-Linear Mechanics*, Vol. 5, No. 3, pp. 533-555.

47. Pierre, C., Ferri, A.A., and Dowell, E.H., 1985, "Multi-Harmonic Analysis of Dry Friction Damped Systems Using an Incremental Harmonic Balance Method," *ASME Journal of Applied Mechanics*, Vol. 51, No. 4, pp. 958-964.

48. Ferri, A.A., and Dowell, E.H., 1988, "Frequency Domain Solutions to Multi-Degree-of-Freedom, Dry Friction Damped Systems," *Journal of Sound and Vibration*, Vol. 124, No. 2, pp. 207-224.

49. Gelb, A., and Vander Velde, W.E., 1968, <u>Multiple-Input Describing Functions and Nonlinear System Design</u>, McGraw-Hill Book Co., New York.

50. Timoshenko, S., Young, D.H., and Weaver, W., Jr., 1974, <u>Vibration Problems in Engineering, Fourth Edition</u>, John Wiley & Sons, New York.

51. Meirovitch, L., 1967, <u>Analytical Methods in Vibrations</u>, Macmillan Publishing Co., NY, NY.

52. Shaw, S.W., 1986, "On the Dynamic Response of a System with Dry Friction," *Journal of Sound and Vibration*, Vol. 108, No. 2, pp. 305-325.

53. Marui, E., and Kato, S., 1984, "Forced Vibration of a Base-Excited Single-Degree-of-Freedom System with Coulomb Friction," *ASME Journal of Dynamic Systems, Measurement, and Control*, Vol. 106, December, pp. 280-285.

54. Whiteman, W.E., and Ferri, A.A., 1996, "Suppression of Bending-Torsion Flutter Through Displacement-Dependent Dry Friction Damping," Proceedings of the 37th Structures, Structural Dynamics, and Materials Conference, Salt Lake City, UT, April 15-17, pp. 1578-1584.

55. Whiteman, W.E., and Ferri, A.A., 1996, "Stability and Forced Response of Beam-Like Structures Having Negative Viscous Damping and Displacement-Dependent Dry Friction Damping," Proceedings of the 1996 ASME International Congress and Exposition, Atlanta, GA, November 17-22.

56. Kelly, J.M., 1986, "Aseismic Base Isolation: Review and Bibliography," *Soil Dynamics and Earthquake Engineering*, Vol. 5, No. 3, pp. 202-216.

57. Levitan, E.S., 1960, "Forced Oscillation of a Spring-Mass System Having Combined Coulomb and Viscous Damping," *Journal of the Acoustical Society of America*, Vol. 32, No. 10, pp. 1265-1269.

58. Yeh, G.C.K., 1966, "Forced Vibrations of a Two-Degree-of-Freedom System with Combined Coulomb and Viscous Damping," *Journal of the Acoustical Society of America*, Vol. 39, pp. 14-24.

59. Hundal, M.S., 1979, "Response of a Base Excited System With Coulomb and Viscous Friction," *Journal of Sound and Vibration*, Vol. 64, No. 3, pp. 371-378.

60. Schlesinger, A., 1979, "Vibration Isolation in the Presence of Coulomb Friction," *Journal of Sound and Vibration*, Vol. 63, No. 2, pp. 213-224.

61. Westermo, B., and Udwadia, F., 1983, "Periodic Response of a Sliding Oscillator System to Harmonic Excitation," *Earthquake Engineering and Structural Dynamics*, Vol. 11, pp. 135-146.

62. Parnes, R., 1984, "Response of an Oscillator to a Ground Motion With Coulomb Friction Slippage," *Journal of Sound and Vibration*, Vol. 94, No. 4, pp. 469-482.

63. Schwarz, B., Weinstock, H., Greif, R., and Briere, R., 1988, "An Analytical Study of the Bounce Motion of a Freight Car Model in Response to Profile Irregularities," Proceedings of the ASME Winter Annual Meeting, Chicago, IL, Nov. 27-Dec. 2; AMD-Vol. 96 (RTD-Vol. 2) pp. 191-198.

64. Ravindra, B., and Mallik, A.K., 1993, "Hard Duffing-Type Vibration Isolator with Combined Coulomb and Viscous Damping," *International Journal of Non-Linear Mechanics*, Vol. 28, No. 4, pp. 427-440.

65. Caughey, T.K., 1960, "Random Excitation of a System with Bilinear Hysteresis," *ASME Journal of Applied Mechanics*, Vol. 27, Dec., pp. 649-652.

66. Crandall, S.H., Lee, S.S., and Williams, J.H., 1974, "Accumulated Slip of a Friction-Controlled Mass Excited by Earthquake Motions," *ASME Journal of Applied Mechanics*, Vol. 41, No. 4, Dec., pp. 1094-1098.

67. Ahmadi, G., 1983, "Stochastic Earthquake Response of Structures on Sliding Foundation," *International Journal of Engineering Science*, Vol. 21, No. 2, pp. 93-102.

68. Karnopp, D.C., and Trikha, A.K., 1969, "Comparative Study of Optimization Techniques for Shock and Vibration Isolation," ASME *Journal of Engineering for Industry*, Vol. 91, No. 4, pp. 1128-1132.

69. Mercer, C.A., and Rees, P.L., 1971, "An Optimum Shock Isolator," *Journal of Sound and Vibration*, Vol. 18, No. 4, pp. 511-520.

70. Tadjbakhsh, I.G., and Lin, B.C., 1987, "Displacement-Proportional Friction (DPF) in Base Isolation," *Earthquake Engineering and Structural Dynamics*, Vol. 15, pp. 799-813.

71. Ferri, A.A., and Bindemann, A.C., 1992, "Damping and Vibration of Beams with Various Types of Frictional Support Conditions," *ASME Journal of Vibration and Acoustics*, Vol. 114, No. 3, pp. 289-296.

72. Bindemann, A.C., 1993, Dry Friction Damping of Built-Up Structures, Ph.D. Thesis, School of Mechanical Engineering, Georgia Institute of Technology.

73. Bindemann, A.C. and Ferri, A.A., 1995, "Large Amplitude Vibration of a Beam Restrained by a Nonlinear Sleeve Joint," *Journal of Sound and Vibration,* Vol. 184, No. 1, pp. 19-34.

74. Karnopp, D., Crosby, M.J., and Harwood, R.A., 1974, "Vibration Control Using Semi-Active Force Generators," ASME *Journal of Engineering for Industry*, Vol. 96, No. 2, pp. 619-626.

75. Karnopp, D.C., 1987, "Force Generation in Semi-Active Suspensions Using Modulated Dissipative Elements," *Vehicle System Dynamics*, Vol. 16, pp. 333-343.

76. Hrovat, D., Margolis, D.L., and Hubbard, M., 1988, "An Approach Toward the Optimal Semi-Active Suspension," *ASME Journal of Dynamic Systems, Measurement, and Control*, Vol. 110, Sept., pp. 288-296.

77. Butsuen, T., Hedrick, J.K., 1989, "Optimal Semi-Active Suspensions For Automotive Vehicles: The 1/4 Car Model," Advanced Automotive Technologies- 1989, A.M. Karmel, E.H. Law, and S.R. Velinski, eds., pp. 305-319, ASME WAM, San Francisco, CA, Dec. 10-15.

78. Karnopp, D.C., 1990, "Design Principles for Vibration Control Systems Using Semi-Active Dampers," *ASME Journal of Dynamic Systems, Measurement, and Control*, Vol. 112, Sept., pp. 448-455.

79. Karnopp, D.C., 1995, "Active and Semi-Active Vibration Isolation," *ASME Journal of Vibration and Acoustics*, Vol. 117(B), June, pp. 177-185.

80. Davis, P., Cunningham, D., Bicos, A., and Enright, M., 1994, "Adaptable Passive Viscous Damper (An Adaptable D-Strut^{TM})," Proceedings of SPIE- International Society of Optical Engineering, Orlando, FL, February 14-16, Vol. 2193, pp. 47-58.

81. Pinkos, A., Shtarkman, E., and Fitzgerald, T., 1993, "Actively Damped Passenger Car Suspension System With Low Voltage Electro-Rheological Magnetic Fluid," Vehicle Suspension and Steering Systems, SAE Special Publication No. 92, pp. 87-93.

82. Coulter, J.P., Don, D.L., Yalcintas, M., and Biermann, P.J., 1993, "Experimental Investigation of Electrorheological Material Based Adaptive Plates," Adaptive Structures and Material Systems, ASME AD Vol. 35, pp. 287-296.

83. McClamroch, N.H., Gavin, H.P., Oritz, D.S., and Hanson, R.D., 1994, "Electrorheological Dampers and Semi-Active Structural Control," Proceedings of the 33rd Conference on Decision and Control, Lake Buena Vista, FL, Dec. 14-16, pp. 97-102.

84. Wang, K.W., Kim, Y.S., and Shea, D.B., 1994, "Structural Vibration Control Via Electrorheological-Fluid-Based Actuators with Adaptive Viscous and Frictional Damping," *Journal of Sound and Vibration*, Vol. 177, No. 2, pp. 227-237.

85. Masri, S.F., Miller, R.K., Dehghanyar, T.J., and Caughey, T.K., 1989, "Active Parameter Control of Nonlinear Vibrating Structures," *ASME Journal of Applied Mechanics*, Vol. 56, No. 3, pp. 658-666.

86. Cao, S., and Semercigil, S.E., 1994, "A Semi-Active Controller for Excessive Transient Vibrations of Light Structures," *Journal of Sound and Vibration*, Vol. 178, No. 2, pp. 145-161.

87. Shen, I.Y., 1993, "Intelligent Constrained Layer: An Innovative Approach," Intelligent Structures, Materials, and Vibration, ASME DE-Vol. 58, pp. 75-82.

88. Wang, K.W., Lai, J.S., and Yu, W.K., 1994, "Structural Vibration Control via Piezoelectric Materials with Real-Time Semi-Active Electrical Networks," Adaptive Structures and Composite Materials: Analysis and Application, ASME AD-45, pp. 219-226.

89. Ferri, A.A., and Heck, B.S., 1990, "Active and Passive Joints for Damping Augmentation of Large Space Structures ," Proceedings of the American Control Conference, San Diego, CA, May 23-25, pp. 2978-2983.

90. Lane, J.S. and Ferri, A.A., 1992, "Optimal Control of a Semi-Active, Frictionally Damped Joint," Proceedings of the American Control Conference, Chicago, IL, June 24-26, pp. 2754-2759.

91. Lane, J.S., Ferri, A.A., and Heck, B.S., 1992, "Vibration Control Using Semi-Active Friction Damping," Proceedings of the ASME Winter Annual Meeting, Anaheim, CA, November 8-13, Friction-Induced Vibration, Chatter, Squeal, and Chaos, Ibrahim, R.A., and Soom, A., eds., ASME Press, DE-Vol. 49, pp. 165-171.

92. Lane, J.L., 1993, Control of Dynamic Systems Using Semi-Active Friction Damping, Ph.D. Thesis, School of Mechanical Engineering, Georgia Institute of Technology.

93. Lane, J.S., and Ferri, A.A., 1995, "Control of a Flexible Structure Using Combined Active and Semi-Active Elements," Proceedings of the 1995 Structures, Structural Dynamics and Materials Conference, New Orleans, LA, April 10-12, pp. 719-728.

94. Onoda, J., and Minesugi, K., 1994, "Semiactive Vibration Suppression of Truss Structures by Coulomb Friction," *AIAA Journal of Spacecraft and Rockets*, Vol. 31, No. 1, Jan.-Feb., pp. 67-74.

95. Dupont, P.E., and Stoker, A., 1995, "Semi-Active Control of Friction Dampers," Proceedings of the 34th Conference on Decision and Control, New Orleans, LA, Dec. 13-15.

96. Ferri, A.A., and Heck, B.S., 1992, "Semi-Active Suspension Using Dry-Friction Energy Dissipation," Proceedings of the American Control Conference, Chicago, IL, June 24-26, 1992, pp. 31-35.

97. Lane, J.S., Ferri, A.A., and Heck, B.S., 1993, "Analytical and Experimental Investigation of a Semi-Active Dry Friction Shock Absorber," Proceedings of the International Federation of Automatic Control World Congress, Sydney, Australia, July 18-23, Vol. 2, pp. 245-248.

98. Huynh, D.Q., 1995, Optimization of Coulombic Semi-Active Automotive Suspension Systems, M.S. Thesis, School of Mechanical Engineering, Georgia Institute of Technology.

99. Ryba, D., 1993, "Semi-Active Damping with an Electromagnetic Force Generator," *Vehicle System Dynamics*, Vol. 22, pp. 79-95.

100. Rezeka, S.F., and Saafan, A.A., 1994, "Adaptive Semi-Active Suspension with Combined Viscous and Dry Energy Dissipation," Proceedings of the 1994 ASME International Mechanical Engineering Congress and Exposition, Chicago, IL, November 6-11. DE-Vol. 75, pp. 519-525.

101.Utkin, V.I., 1977, "Variable Structure Systems with Sliding Modes," *IEEE Transactions on Automatic Control*, Vol. AC-22, April, pp. 212-222.

102.Junkins, J.L., and Kim, Y., 1993, Introduction to Dynamics and Control of Flexible Structures, AIAA Education Series, Washington, DC.

103.Oberle, H.J., and Grimm, W., 1985, "BNDSCO- A Program for the Numerical Solution of Optimal Control Problems," English Translation of DFVLR-Mitt. 85-05

Subject Index

Author Index

311

ABOUT THE EDITORS

ARDÉSHIR GURAN is the Director of the Institute for Structronics, Canada. He received his M.Eng. from McGill University, Canada; M.S. in Mathematics and Ph.D. in Mechanical Engineering from the University of Toronto. He is the Editor-in-Chief of *Stability, Vibration and Control of Systems* and served as Associate Editor of *International Journal of Modeling and Simulations*. Dr. Guran authored or co-authored over 200 publications in the areas of nonlinear dynamics, structural stability, structronics, acoustics, wave propagations, gyroscopic systems, structural control, impact and friction, locomotion, and history of science and technology. Dr. Guran chaired several international scientific meetings including: International Symposium on Impact and Friction of Solids, Structures and Intelligent Machines (Ottawa, Canada, 1998); International Congress on Dynamics and Control of Systems (Chateau Laurier, Canada, 1999); International Conference on Acoustics, Noise and Vibrations (McGill University, Canada, 2000).

FRIEDRICH PFEIFFER was born in Wiesbaden (Germany). Dipl.-Ing. (in Mechanical Engineering) 1961, Dr.-Ing. (in Mechanical Engineering) 1965, both from Technical University of Darmstadt. Messerschmidt-Boelkow-Blohm 1966. Head of theoretical and technical mechanics department (MBB space division), 1968–75. Project Manager Anti-Ship Guided Missile (MBB guided missile division), 1975. Technical secretary to Dr. Boelkow 1976–77. General Manager Bayern-Chemie GmbH, 1978–80. Vice President R&D (MBB guided missile division), 1980–82. Since 1982, Professor of Mechanics and head of the Institute of Mechanics at Technical University of Munich. Serves on the advisory board of *Ingenieur Archiv, Nonlinear Dynamics, Machine Vibration, Chaos, European Journal of Solid Mechanics, Autonomous Robotics, Stability, Vibration and Control of Systems*. Research interests: nonlinear systems, dynamics and control of mechanical systems, optimization, transmission and robotic system.

KARL POPP was born in Regensburg (Germany). Dipl.-Ing. (in Mechanical Engineering) 1969, Dr.-Ing. (in Mechanical Engineering) 1972, both Technical University of Munich. Assistant to Professor Kurt Magnus (Technical University of Munich), 1969–76. Habilitation in Mechanics, 1978. Visiting Professor at University of California Berkeley and at Universidade Estadual de Campinas (Brasil). Since 1981 Professor of Mechanics and since 1985 head of the Institute of Mechanics at University of Hannover. Research interests: nonlinear vibrations, dynamics, systems and control theory with applications in vehicle dynamics, machine dynamics and mechatronics.